Driving Organizational Transformation Through Human–Robotic Collaboration

Muhammad Anshari
Universiti Brunei Darussalam, Brunei

Patricia Ordóñez de Pablos
University of Oviedo, Spain

Mohammad Nabil Almunawar
https://orcid.org/0000-0001-5296-2576
Universiti Brunei Darussalam, Brunei

IGI Global
Scientific Publishing
Publishing Tomorrow's Research Today

Vice President of Editorial — Melissa Wagner
Director of Acquisitions — Mikaela Felty
Director of Book Development — Jocelynn Hessler
Production Manager — Mike Brehm
Cover Design — Jose Rosado

Published in the United States of America by
IGI Global Scientific Publishing
701 East Chocolate Avenue
Hershey, PA, 17033, USA
Tel: 717-533-8845 | Fax: 717-533-7115
Website: https://www.igi-global.com E-mail: cust@igi-global.com

Library of Congress Cataloging-in-Publication Data

Names: Anshari, Muhammad, 1977- editor | Ordóñez de Pablos, | Almunawar, Mohammad Nabil, 1960- editor
Title: Driving organizational transformation through human-robotic collaboration / [edited by] Muhammad Anshari, Patricia Ordóñez de Pablos, Mohammad Nabil Almunawar.
Description: Hershey, PA : IGI Global Scientific Publishing, [2026] | Includes bibliographical references and index. | Summary: "This book aims to Investigate the opportunities and challenges posed by HRC in organizational contexts"-- Provided by publisher.
Identifiers: LCCN 2025013509 (print) | LCCN 2025013510 (ebook) | ISBN 9798337334042 hardcover | ISBN 9798337334059 paperback | ISBN 9798337334066 ebook
Subjects: LCSH: Organizational change http://id.loc.gov/authorities/subjects/sh85095525 | Human-robot interaction http://id.loc.gov/authorities/subjects/sh2011002448
Classification: LCC HD58.8 .D75 2026 (print) | LCC HD58.8 (ebook) | DDC 658.4/06--dc23/eng/20250623
LC record available at https://lccn.loc.gov/2025013509
LC ebook record available at https://lccn.loc.gov/2025013510

British Cataloguing in Publication Data
A Cataloguing in Publication record for this book is available from the British Library.

Editorial Advisory Board

Table of Contents

Detailed Table of Contents

for creating AI-augmented workplaces. By leveraging AI-driven workforce transformation, companies can position themselves for long-term competitiveness while fostering a sustainable and inclusive work environment.

This review explores the transformative role of Human-Robot Collaboration in driving organizational transformation, particularly within the framework of Industry 4.0. The rise of advanced technologies has revolutionized organizational dynamics by fostering enhanced productivity, innovation, and adaptability. Human-Robot Collaboration emerges as a critical enabler of adaptive and resilient work environments, empowering human workers while optimizing operations across various industries. Recognizing the urgency of understanding Human-Robot Collaboration impact, this study employs a systematic bibliometric analysis of peer-reviewed articles from the Scopus database, focusing on publications from 2020–2025. Through rigorous screening and eligibility criteria, 233 articles were selected to provide a comprehensive foundation for examining the implications of Human-Robot Collaboration. By synthesizing insights from existing literature, this review underscores the pivotal role of Human-Robot Collaboration in organizational transformation and its potential to shape the future of work.

This study investigates the influence of service quality dimensions on customer satisfaction regarding the use of self-service kiosks (SSKs) in Brunei. SSKs are a type of self-service technology that allows customers to perform tasks without the assistance of service employees. The research adopts the SSTQUAL scale, which consists of seven dimensions: functionality, enjoyment, security, design, assurance, convenience, and customization. The study also addresses the challenges and concerns associated with SSKs, such as theft and identity theft. Additional security measures to deter theft could make honest customers feel uncomfortable or distrusted, which could also lead to longer waiting times, negating the benefits of SSKs. The data has been collected from 163 respondents who may or may not have used SSKs. The result of this study has revealed that only 2 constructs have a significant impact towards customer satisfaction towards using SSKs in which are considered the most important. Functionality has been found to be the most statistically significant while customers perceive security to be the most important. These findings offer guidance to Brunei's service sector on where to invest in recent technology to improve customer satisfaction and experience.Along with adding to the body of knowledge on self-service technologies and customer satisfaction, the study is relevant to Brunei, a nation undergoing digital transformation and working toward realizing its 2035 vision as it improves understanding of customer behavior in Brunei, which can be used in comparative studies and to better understand cultural influences on technological adoption. From a business perspective, the study helps firms understand the elements impacting customer happiness, enabling them to improve their SSKs and so promote customer satisfaction and loyalty.

ABSTRACT This chapter examines the ethical boundaries in automation, concentrating on the incorporation of robots inside organizational structures and the ethical challenges arising from this transition. In an age where automation is transforming company operations, it is essential to assess the equilibrium between technological progress and the welfare of the human labor. The chapter explores human-robot collaboration, highlighting the necessity for ethical frameworks to govern the deployment of automated technology. The chapter provides insights into best practices and solutions for promoting ethical automation by examining conceptual and theoretical models, ethical difficulties, and organizational consequences. This document will examine recommendations for addressing new trends, improving accessibility, and fostering collaboration between humans and robots, thereby offering a framework for ethical automation in corporate contexts.

In the age of AI and robotics, human-machine trust plays a pivotal role in the successful integration of technology into organizational workflows. This chapter explores the

Preface

In the era of Industry 5.0, the integration of human and robotic intelligence represents a profound paradigm shift in how organisations evolve, adapt, and sustain competitiveness. This edited volume, Driving Organisational Transformation Through Human–Robotic Collaboration, explores the interplay between humans, artificial intelligence (AI), and robotics in shaping the future of work, management, and innovation. The book situates itself at the intersection of technology, organisational studies, and human–machine interaction, addressing both the opportunities and challenges of achieving meaningful synergy between human cognition and machine intelligence.

Human–Robot Collaboration (HRC) is no longer a futuristic ideal but a contemporary imperative that redefines productivity, efficiency, and creativity. In a world increasingly characterised by automation and digitalisation, organisations must rethink their operational and strategic models to remain agile, ethical, and human-centric. This volume aims to contribute to the growing discourse on how AI and robotics can be harnessed to complement, rather than replace, human capability, thereby fostering collaborative intelligence for sustainable organisational growth.

The intended audience for this book includes scholars, policymakers, and practitioners in business information systems, management, and engineering, as well as postgraduate students exploring emerging intersections between technology and organisational transformation. It offers both theoretical insights and empirical evidence relevant to academia and industry.

Each chapter in this collection makes a unique contribution to the overarching theme of human–robot synergy. The opening chapters (1–4) introduce the concept of collaborative intelligence, emphasising the rise of human–machine synergy in Industry 5.0 and the transformative potential of AI and robotics in shaping future-ready enterprises. Chapter 5 presents a bibliometric analysis that situates current research trends within the broader context of organisational transformation. Chapter 6 extends this discussion by exploring AI integration in public policy development, highlighting inclusivity and responsiveness as key outcomes.

Chapters 7 and 8 explore employee collaboration and trust mechanisms, offering empirical and conceptual insights into how human–AI cooperation can improve performance and decision-making. Chapter 9 shifts focus to Islamic governance and AI ethics, demonstrating the adaptability of AI governance frameworks within diverse sociocultural contexts.

Collectively, these chapters offer an interdisciplinary perspective that bridges technology, management, and social science. The book underscores that the future of organisational transformation lies not merely in technological adoption but in cultivating trust, ethics, and shared intelligence between humans and machines. By synthesising diverse research contributions, this volume advances understanding of how collaborative technologies can drive sustainable, inclusive, and adaptive organisational ecosystems.

We extend our appreciation to all contributing authors for their scholarly commitment and to reviewers whose feedback has significantly enhanced the quality of the chapters in this book. To our readers, we hope this book provides valuable knowledge and insights, and your engagement with it will inspire you to further shape the evolving dialogue on human–robot collaboration and its transformative impact on organisations worldwide.

Muhammad Anshari
Patricia Ordóñez de Pablos
Mohammad Nabil Almunawar
Editors

Acknowledgment

The editors gratefully acknowledge all contributors for their valuable research and commitment to this volume, Driving Organisational Transformation Through Human–Robotic Collaboration. We extend sincere appreciation to the reviewers for their constructive feedback, and to the IGI Global editorial and production teams for their professional guidance throughout the publication process.

Special thanks are also due to the UBD School of Business and Economics and the Institute of Policy Studies, UBD, for their support and encouragement. This book is the result of collective effort and shared dedication to advancing knowledge in human–robot collaboration and organisational transformation.

Muhammad Anshari

Patricia Ordóñez de Pablos

Mohammad Nabil Almunawar

Editors

Introduction

FROM AUTOMATION TO COLLABORATION: THE FUTURE OF HUMAN–ROBOT COLLABORATION

The integration of humans and intelligent machines marks one of the most transformative developments of the twenty-first century. Human–Robot Collaboration (HRC) has evolved beyond industrial automation to become a cornerstone of organisational transformation, redefining how people and technology interact to create value. As artificial intelligence (AI) and robotics continue to mature, their convergence with human expertise is reshaping processes, strategies, and the very structure of work. This edited volume, Driving Organisational Transformation Through HRC, explores this dynamic phenomenon through multidisciplinary perspectives, reflecting how collaboration between humans and machines is becoming integral to organisational innovation, resilience, and sustainability.

The rise of HRC is closely associated with the broader vision of Industry 5.0, which emphasises human-centric, sustainable, and intelligent collaboration between people and technology. The chapters in this book collectively demonstrate that the future of organisational success depends not only on automation and efficiency but also on cultivating synergy between human creativity and machine intelligence. This relationship fosters what may be named collaborative intelligence, a fusion of human empathy, judgment, and adaptability with robotic precision, consistency, and analytical capacity.

Figure 1 shows HRC Process Framework outlines a structured approach to integrating humans and robots effectively. It begins with preparation, including capability assessment, safety planning, and workspace design. During execution, tasks are allocated, interactions are managed, and safety programming is implemented. The monitoring phase involves real-time data collection, performance tracking, and feedback loops to ensure smooth collaboration. Finally, evaluation focuses on efficiency metrics, anomaly detection, safety reviews, and process optimisation. Together, these stages promote safe, efficient, and adaptive collaboration between humans and robots.

Figure 1. Human-Robot Collaboration Process Framework

The early chapters introduce the theoretical and conceptual foundations of human–machine synergy, situating it within the context of Industry 5.0 and knowledge management. They explore how AI and robotics can enhance organisational learning, responsiveness, and growth. Subsequent chapters examine empirical and bibliometric studies, illustrating the tangible effects of human–robot collaboration on performance, inclusivity, and policy development. These contributions provide a strong foundation for understanding how AI-driven collaboration influences both strategic and operational dimensions of organisations.

Several chapters address critical human factors, including trust, ethics, and decision-making. They discuss how employee–AI collaborations can improve productivity and knowledge sharing, while also emphasising the importance of transparency and reliability in fostering trust. Other contributions extend this discourse into governance, sustainability, and sector-specific contexts such as banking, agriculture, and logistics, showing that HRC's applications are broad and rapidly expanding. The inclusion of chapters on AI-driven personalisation, blockchain-enabled transformation, and augmented organisations further highlights the cross-sectoral potential of human–robot collaboration in achieving adaptive, data-driven, and resilient enterprises.

As automation advances, the question is no longer whether humans and robots will collaborate, but how this collaboration can be optimised to ensure mutual augmentation rather than substitution. The insights presented in this volume underscore that the evolution of HRC will continue to rise and expand, driven by technological progress, organisational learning, and societal adaptation. Human–Robot synergy will increasingly define how businesses innovate, how policies are formulated, and how societies navigate the complexities of digital transformation.

In bringing together theoretical models, empirical analyses, and sectoral applications, this book contributes to the growing body of knowledge on HRC. It offers scholars, practitioners, and policymakers an integrated understanding of how collaborative technologies can be leveraged to build more intelligent, ethical, and human-centred organisations.

Muhammad Anshari
Patricia Ordóñez de Pablos
Mohammad Nabil Almunawar
Editors

Chapter 1
Collaborative Intelligence:
The Rise of Human-Machine Synergy in Industry 5.0

Mohammad Nabil Almunawar
https://orcid.org/0000-0001-5296-2576
Universiti Brunei Darussalam, Brunei

Heru Kurnianto Tjahjono
Universitas Muhammadiyah of Yogyakarta, Indonesia

ABSTRACT

Industry 4.0, or the Fourth Industrial Revolution, emphasizes full automation but often sidelines human involvement, leading to various challenges. In contrast, Industry 5.0 is not a mere continuation but a corrective shift that places humans back at the center of value creation. The rapid rise of AI is a double-edged sword, empowering yet threatening. While AI replaces routine jobs and may cause unemployment if mishandled, it also enables humans to work faster and more efficiently. However, AI outputs must be supervised to ensure alignment with intended outcomes and to avoid bias. Human intelligence remains essential, especially in areas involving ethics, emotions, and complex judgments. This chapter provides an overview of Industry 5.0, clarifying how it differs from Industry 4.0 and emphasizing human-machine collaboration. It concludes with real-world case studies demonstrating the integration of human-centric principles across various industries.

DOI: 10.4018/979-8-3373-3404-2.ch001

INTRODUCTION

The term *Industrial Revolution (IR) 4.0*, which is often used interchangeably with *Industry 4.0*, has become a buzzword in recent years. Actually, they are not the same, as Industry 4.0 is a smaller scope of IR 4.0, which focuses on the manufacturing and industrial sector. IR 4.0 was introduced as we have been shifting to a new revolution, as compared to the previous one, IR 3.0. It is characterized by computer automation, interconnected systems through the Internet, and the use of Artificial Intelligence (AI).

Although IR 3.0 marked the shift from analog and mechanical systems to digital ones, the use of digital electronics and information and communication technology (ICT) in business processes was largely focused on digitization or computerization to support operations, with minimal intelligence and limited connectivity between human and non-human entities. The emphasis was on reaching a broader audience, but with limited personalization.

In contrast, supported by rapid technological advancements, Industry 4.0 has significantly improved business processes through extensive connectivity enabled by the Internet. Almost everything can now be connected via the Internet of Things (IoT), while human connectivity has been enhanced through mobile broadband and the widespread use of Web 2.0, such as social media. This era is also marked by a strong trend toward personalization and decentralization, supported by technologies such as blockchain. Industry 4.0 emphasizes full automation enhanced with AI, making it smart automation; hence, this may sideline human involvement, shifting the human role from central to peripheral.

The impacts and disruptions caused, as well as the limitations and challenges faced by approaches in Industry 4.0, have attracted widespread criticism. Among these are job displacement, the digital divide, overemphasis on technology, dehumanization of activities, cybersecurity and data privacy concerns, and sustainability issues (Dieste et al., 2023).

Smart automation in Industry 4.0 displaces workers, particularly those performing routine and repetitive tasks. This may significantly disadvantage developing economies and widen the digital divide between technologically advanced and less developed countries, thereby deepening socio-economic inequality. An overemphasis on technology and automation can also lead to the diminishing of the human or social dimension and neglect of ethical considerations, especially in AI-based systems. In addition, automation may increase the risk of catastrophic failure during technology malfunctions, as there is no one who can guarantee that human-made systems are free from failures.

Moreover, excessive reliance on smart machines for decision-making may erode human capabilities in judgment and problem-solving, potentially resulting in a loss of creativity, engagement, and the human touch in daily life. The growing use of IoT and dependence on automation make Industry 4.0 systems increasingly vulnerable to cyberattacks. At the same time, there are rising concerns about mass surveillance and the misuse of automatically gathered data (Sima et al., 2020).

The widespread adoption of Industry 4.0 may also lead to high energy consumption, raising sustainability issues. Additionally, the disposal of obsolete digital devices contributes to environmental challenges that are often overlooked in planning and implementation.

The concerns surrounding Industry 4.0 have led to the emergence of a new concept: Industry 5.0. Unlike the clear technological shift that occurred between Industry 3.0 and Industry 4.0, Industry 5.0 is not characterized by a shift but rather by a correction, with the emphasis putting humans back at the center of value creation, or human-centric value creation and innovation (Alves et al., 2023; Eriksson et al., 2024; Fachrunnisa et al., 2019).

The advancement of smart automation in various forms, such as intelligent robots, should go hand in hand with the development of human knowledge and skills to effectively manage these technologies. This aims to foster a collaborative environment where humans and machines work together to achieve optimal and sustainable solutions that enhance human well-being.

The objective of this chapter is to present the concept of collaborative intelligence, the synergy between human and machine intelligence working hand in hand to create value and solve problems in the context of rapid technological advancement, particularly in the rapid adoption of AI and machine learning.

The chapter covers six sections and is organized as follows. The next Section (Section 2) discusses the pillars of Industry 5.0 while highlighting Industry 4.0, digital transformation, and integration of Internet of Things (IoT), AI, and collaborative robots for value creation. Section 3 discusses the major human-machine collaboration technologies, such as chatbots, collaborative robots (cobots), AI, machine learning, and augmented reality (AR) & virtual reality (VR). Section 4 presents applications with case studies. Section 5 presents challenges and ethical issues and Section 6 is the conclusion.

THE PILLARS OF INDUSTRY 5.0

As previously mentioned, Industry 5.0 is not a complete shift from Industry 4.0. It is rather a correction, especially to bring the human back to the center of value creation while addressing other important aspects that are overlooked in Industry 4.0, such as sustainability and resiliency, while recognizing the advancement of technology and its utilization through cooperation between humans and machines in achieving goals (Ghobakhloo et al., 2024). It is important to note that technology should serve humans in advancing their civilization and improving their well-being without compromising the future of life on the planet. Through fostering collaboration between humans and machines, Industry 5.0 emphasizes incorporating technology in enhancing human well-being and ensuring a sustainable future for the planet.

We will briefly highlight the core focus or pillars of Industry 4.0 before moving to discuss those of Industry 5.0 to highlight the key differences. Industry 4.0 places a strong emphasis on automation of production and value creation in light of the fast development of smart technologies, especially those based on AI. The main goal of Industry 4.0 is to automate production and value creation as much as possible, boosting efficiency by shifting many tasks from humans to machines.

Industry 4.0 relies on Cyber-Physical Systems (CPS). These systems integrate physical processes that contain connected machines, devices, and sensors with a computer system that controls and monitors them to enable real-time automation, including machine-based decision-making (Elnadi & Abdallah, 2024; Folgado et al., 2024). For example, a CPS in a smart factory comprises an assembly line equipped with smart robot arms that can adjust their movements based on the feedback from sensors. By automatically acquiring data from sensors, the system optimizes its performance and reduces defects through predictive data analytics.

The advancement in communication technology has greatly enhanced bandwidth capabilities across the Internet. Fast communication methods like fiber optics, as well as 4G and the more recent 5G wireless technologies, have transformed the Internet into a leading and economical communication platform. The emergence of IPv6, featuring a 128-bit address space, can nearly support an infinite number of IP addresses, allowing virtually anything to receive an IP address and therefore connect to the Internet or the IoT (Ashraf et al., 2019; Emad-ul-Haq et al., 2015; Kane, 2025). It is no surprise that a vast array of machines, devices, and sensors are now linked to the Internet.

Additionally, the widespread and high-speed connectivity provided by the Internet fosters a novel business model centered around computing resources such as cloud computing. Cloud computing removes the need for significant investments and ongoing maintenance of computing resources, resulting in only manageable and foreseeable operating expenses (Almunawar & Almunawar, 2018; Almunawar & Anshari, 2018; Armbrust et al., 2010; Marston et al., 2011). As a result, similar to its predecessor, the Internet, cloud computing has seen global adoption. The implementation of IPv6 and broadband connections allows cloud services to accommodate the continually expanding digital ecosystem of connected devices. This model removes the necessity for large capital expenditures and continuing upkeep of local computing infrastructure, substituting them with predictable and manageable operational expenses. Comparable to a utility service, cloud computing provides on-demand scalability across various resources, including processing power, storage, platforms, and applications, without the complications of establishing local infrastructure, which can be excessively costly. Consequently, devices and sensors are able to regularly send real-time data to cloud storage, where AI-driven analytics can conduct diagnostics and generate predictions. Hence, cloud computing is an important infrastructure in Industry 4.0.

Cloud computing plays an important role in Industry 4.0 by providing centralized data storage and powerful processing capabilities. However, sending data from remote or distributed sources to cloud data centers can introduce latency, which may be unacceptable in time-critical applications. Furthermore, the transmission of sensitive data over long distances may pose security and privacy risks. Edge computing addresses these challenges by enabling data to be processed locally or near the source, such as sensors or IoT devices. In essence, edge computing complements cloud computing by shifting processing closer to the data source, thereby reducing latency and enhancing data security.

However, there are a few concerns if Industry 4.0 is fully adopted and implemented. First, extensive use of automation will result in minimal or mitigated human labor, resulting in mass unemployment, especially in developing countries. Some labor-intensive industries that absorb many laborers will disappear. This may end up solving a problem, but it will create a bigger problem with no easy way out. Industry 4.0 may create win-lose solutions for humanity. We want technology to help us improve our wellbeing, not to replace us!

Automation with human or very minimal human intervention, as envisioned by Industry 4.0, is good in some areas but dangerous in other areas, as there is no guarantee that the automation can perform perfectly without failure all the time. Serious consequences will result from failures of overreliance on automation (Brocal et al., 2019; Min & Borch, 2022). For example, a robotic assembly line with very minimal human intervention and a fully automated trading system.

In an advanced manufacturing system, to maximize speed and precision, the assembly line utilizes robots for full automation. This may improve productivity. However, there is no guarantee that the robots perform flawlessly. A sensor malfunction or undetected software bugs will create a big problem for the assembly line, which may take hours or days for the expert to identify and fix the problem, causing lost revenue because of the downtime.

A fully automated trading system in a financial market can help execute thousands of trades per second. It seems very efficient. However, there is no perfect system; unintended bugs may cause the system to experience a flash crash, or unnoticed unintended behavior resulting in destabilizing the market, which can cause massive financial loss and loss of trust in minutes.

Industry 5.0 is built upon Industry 4.0, an improvement. As such, many components of Industry 4.0 are still relevant; however, the approach or principle in solving human problems with technology is different as Industry 5.0 realigns the utilization of technology toward human well-being, promoting sustainability. We can summarize that there are three pillars of Industry 5.0: a human-centric approach, sustainability, and resilience (Leng et al., 2022). Note that these pillars are closely related. Figure 1 shows the pillars of Industry 5.0.

Figure 1. The Pillars of Industry 5.0 (author's own work)

The human-centric pillar ensures that technology and its advancements serve to enhance the well-being, creativity, and safety of human beings, rather than sidelining them. This aligns with the sustainability pillar, which emphasizes the long-term or future well-being of humans, avoiding the side effects of industries that can harm society and the environment while creating wealth. The resilience pillar is highly connected to human and sustainability, as it can absorb or endure unexpected events such as outbreaks, pandemics, natural disasters, or geopolitical tensions, making production and supply chains survive in abnormal situations. Hence, these three pillars establish a cohesive framework of Industry 5.0

Human-centric approach

A human-centric approach is a main principle in Industry 5.0, which essentially means that any innovation, advanced technology and their applications should not merely be aimed at improving productivity and efficiency. Automation must be directed at a higher objective: to serve human needs and improve their wellbeing. This approach harnesses technology to empower people, improve quality of life, foster inclusion, and preserve human dignity. This is in contrast to Industry 4.0, which heavily emphasizes digitalization, automation, and machine-to-machine communication, with a strong focus on improving operational efficiency and productivity.

It is important to note that Industry 5.0 does not reject automation but rather prevents the dehumanization of value creation through excessive automation in pursuit of efficiency and productivity (Nahavandi, 2019). Automation remains valuable for improving efficiency and is essential in executing processes in a dangerous environment or the case of a shortage of human labor. In essence, Industry 5.0 adopts automation to support rather than to replace humans in an industry.

Adoption of technology, especially advanced technology capable of automation, needs people's trust, as without trust, people will hesitate to use such technologies, limiting their benefits to society (Hoff & Bashir, 2015). Therefore, it is very important to foster trust in emerging technologies like AI and machine learning as these technologies take over human intelligence in performing tasks and making decisions. Although advanced machine learning like deep learning can learn and improve through feedback, bias and mistakes cannot be eliminated. Hence, close human supervision and proper training must be in place to avoid unintended problems and misuse and to establish public trust (Langer et al., 2024; Purnama et al., 2021).

How to foster trust in Industry 5.0, especially for emerging technologies like AI and machine learning? First, we need to incorporate the human aspect in designing systems, or human-in-the-loop design. Application of technologies must be designed to work with humans, not be independent of them. As such, AI or machine learning-assisted decision-making must be designed to be supervised by humans and allow humans to override machine decisions. A notable example is the application of AI and machine learning in healthcare. People will not trust a fully AI or machine learning decision in healthcare. AI is handy in helping diagnose diseases and suggest treatment. But the final decision must be made by a doctor. So we will see it is the doctor who is responsible for the decision.

Second, transparency and explainability must be in place. Incorporating AI systems in Industry 5.0 must be transparent, and any decision involved can be explained to the users, assuring them that the system is not arbitrary or biased. For example, in hiring or finance, an AI-based system must be able to explain why a candidate is rejected or a loan is denied.

Third, ethical and responsible innovation. AI systems do not understand ethics and responsibility, but people do. Therefore, ethical guidance and human rights considerations must be in place to avoid bias, data privacy breaches, and unconsented acts. The safety standards must be in place, too. For example, a facial recognition system must meet ethical standards to avoid discrimination or surveillance abuse.

Fourth, safety and reliability must be enforced. As machines will work hand-in-hand with humans, all systems, especially AI systems and robots, must be designed ergonomically with built-in safety protocols to prevent harm. Hence, they must be programmed with safety in mind to work with people, such as an automatic shutdown in the case of unexpected contact. For example, collaborative robots (cobots) must detect human proximity and adapt their movements accordingly.

Fifth, the importance of digital literacy. Users and workers must be educated or trained to master working hand-in-hand with the technology. So, there is a need to fully understand how this technology functions and how to control it. This is important in trust building, so humans need to confidently know that they are in control of machines and are happy that machines are designed to empower them. For example, AI-predictive maintenance may train operators to interpret AI alerts rather than fear job displacement.

Sixth, regulatory frameworks and standards must be in place. It is important that the government or regulatory institutions play a major role in establishing and enforcing standards for the ethical use of technology, especially AI and robots. Accountability and compliance will boost public trust. For example, the EU's AI Act is a step toward regulating AI use for transparency and risk management.

Besides trust, another element of a human-centric approach is to ensure inclusion and reduce the risk of digital inequality. Application of technology in Industry 5.0 must consider the diverse needs of people, such as people with special needs, marginalized, and underserved communities. This is to ensure that everyone is accessible to and benefits from technologies. Digital transformation in Industry 5.0 should address the digital divide, providing equal opportunities for people to participate in and benefit from the advancement of technology regardless of their demography, socioeconomic status, and physical abilities. By actively addressing barriers to access, affordability, and digital literacy, policymakers and organizations involved can foster a more equitable and human-centered digital future.

In addition, the human-centric approach needs to be linked with the UN Sustainable Development Goals (SDGs), aligning industrial growth with human and environmental needs, especially with SDG 4 (quality education), SDG 8 (decent work and economic growth), SDG 10 (reduced inequalities), and SDG 13 (climate action). All these mentioned SDGs are highly related to human dignity, inclusiveness, and sustainability, which need to be incorporated into Industry 5.0 to ensure that industrial progress enhances quality of life for all, without compromising the needs of future generations. We will shortly discuss sustainability, another pillar of Industry 5.0.

Sustainability

Sustainability is a recent global issue that needs to be taken care of. As the second pillar of Industry 5.0, it addresses a crucial issue in shaping a responsible industry that not only aims for profit but also equally cares for the people and the planet. For this reason, Industry 5.0 integrates environmental consciousness and social responsibility in solving human problems with technology. Sustainability in Industry 5.0 gears transformation toward a regenerative, responsible, and inclusive industry (Adhiatma et al., 2021; Ghobakhloo et al., 2024; Shabur et al., 2025).

Unlike Industry 4.0, which merely facilitates companies to boost efficiency and productivity for increasing competitiveness and profits, the sustainability pillar in Industry 5.0 challenges companies to move beyond efficiency in making profits to actively design processes and systems that are beneficial not only to shareholders and people but also to the environment to secure the well-being of future generations while supporting economic growth. This pillar aligns closely with the UN SDGs, offering structural and systemic change on how industries create, deliver, and capture value. Key elements of this pillar are the *circular economy model*, *carbon-neutral and renewable energy*, *eco-friendly manufacturing*, and *sustainable supply chains*.

Circular economy is an economic model that aims to create value with minimal waste by emphasizing the principles of reduce, reuse, and recycle in creating value and conserving natural resources. This is in contrast with the traditional economy characterized by the "take, make, dispose" model, which does not support sustainability, as value creation generates waste and negatively influences the environment. Consequently, the traditional linear economy, particularly in activities involving natural resources, is under increased scrutiny (Gubeladze & Pavliashvili, 2020; Kanda et al., 2021; Sariatli, 2017). The linear economic model prioritizes immediate gains, primarily benefiting shareholders through substantial profits. However, it generates waste during production and further waste after production when products are discarded because they are no longer functional, ultimately harming stakeholders beyond just shareholders and negatively affecting society and the environment, which undermines sustainability (Kennedy & Linnenluecke, 2022; Rejeb et al., 2025).

The first key element of sustainability is circular economy. The concept of the circular economy was examined by numerous researchers in the 1970s. It embodies a combination of various ideas in Industrial Ecology (Allenby, 2006; Kapur & Graedel, 2004). Performance Economy, Cradle to Cradle (McDonough & Braungart, 2002), Blue Economy (Pauli, 2010), and Biomimicry (Benyus, 2002). However, Pearce & Turner (1989) were the pioneers in formally introducing the term *circular economy* along with its theoretical framework. The fundamental principle of circular economy is conservation, which aims to prevent waste and harmful effects on society and the environment by responsibly and optimally utilizing all resources. Conservation is achieved by minimizing the inputs taken from the biosphere, eliminating waste through the transformation of materials into products, or by designing high-quality items that are easy to maintain, reuse, and recycle (Kanda et al., 2021). The business models present in the circular economy differ from those in a linear economy. For instance, value is generated more through recurring services instead of repeated purchases. Therefore, businesses operating within a circular economy must create high-quality, durable products while generating revenue through leasing and maintenance services, as well as reclaiming outdated products or reintegrating them into production once they are no longer functional.

The second element of sustainability is carbon-neutral and renewable energy. This element is closely related to the circular economy, focusing more on carbon emissions and renewable energy. Besides practicing a circular economy model, industries are encouraged to reduce greenhouse gas emissions, particularly carbon dioxide (CO_2), as this emission contributes to the increase in global temperature, negatively influencing the global environment and potentially causing climate change (Murakami et al., 2022).

Major traditional energy sources such as oil, gas, and coal are non-renewable and contribute significantly to greenhouse gas emissions. Innovation in generating alternative energy, particularly from renewable sources such as solar, hydro, and wind, must be actively encouraged and recognized in Industry 5.0. Smart AI-based technology for saving energy, smart grids, and energy monitoring systems for practicing optimal use of energy are highly welcome (Agrawal & Bhat, 2025; Alqahtani & Afy-Shararah, 2025).

The third element of sustainability is eco-friendly manufacturing, which is closely linked to the circular economy and the carbon-neutral, renewable economy. Eco-friendly manufacturing aims to minimize waste, reduce water consumption, and utilize clean energy while avoiding the release of harmful byproducts. It emphasizes resource efficiency and environmental protection throughout the production process. Advanced technologies such as 3D printing and AI can help optimize material use and reduce waste generation. Ultimately, eco-friendly manufacturing promotes green practices that conserve resources and generate value without compromising environmental integrity (Agrawal & Bhat, 2025; Alqahtani & Afy-Shararah, 2025).

The fourth element of sustainability is sustainable supply chains. Circular economy, sustainable carbon-neutral and renewable energy, and eco-friendly manufacturing cannot function effectively in isolation without being supported by sustainable supply chains. These supply chains serve as the vital link that connects producers and manufacturers who practice sustainability with other like-minded companies or directly with consumers (Theeraworawit et al., 2022).

Sustainable supply chains should preserve the three important dimensions, which are people and community through social responsibility, the planet through environmental stewardship, and profit through economic viability throughout the entire value chain, from raw materials sourcing to the final product's delivery. Some key elements of a sustainable supply chain are traceability, empowering local sources, and adhering to ethical production.

Traceability promotes transparency, accountability, and hence trust among stakeholders, as every component or product can be traced to its origin. Recent digital technologies like IoT and blockchain are handy to monitor the traceability of products in real time. Next, the use of local sources for production will help the local economy grow and save transportation costs, hence reducing the use of energy and carbon emissions. Ethical production practices or standards will adhere to human rights, fair wages, safe working conditions, and the prohibition of child and forced labor. Consequently, supply chains are not only monitored for their environmental impacts and social responsibility but also for their ethical production.

Resilience

The COVID-19 pandemic imparted a hard lesson to industries, as they were not prepared to adapt and quickly recover from disruptions. The pandemic is only one type of disruption; there are many other types, including natural disasters such as earthquakes, fires, and floods, as well as man-made disruptions such as cyberattacks, supply chain breakdowns, and political instability. Industry 5.0 addresses resilience by designing robust and flexible systems that can continue operating in difficult and unforeseen situations. A few concepts, such as digital twins, decentralization, and agility, along with supporting technologies, can be incorporated to improve resiliency (Barata & Kayser, 2024; Castillo et al., 2025; Ivanov et al., 2021).

Digital twins. Digital twins are virtual replicas of physical systems. They help to perform simulations of production environments, real-time monitoring, and regular diagnostics. By incorporating IoT and AI, they can be used to perform predictive maintenance and more accurate early warning systems to anticipate possible equipment or system failures and fix the possible problems before they actually occur, enhancing the reliability and fostering agility.

Decentralization. Decentralization includes localization, aims to reduce the dependency on global supply chains, hence avoiding risk disruption beyond reach, such as experienced during COVID-19. Localization utilizes local resources to produce based on what local people need, enhances economic growth, and empowers and absorbs local workforces. Blockchain technology strongly supports decentralization. In addition, 3D printing technology, modular and small-scale, adaptive, and customized production systems enable this model.

Agility. Agility is a concept of resiliency in facing turbulence. For example, an agile supply chain system is the ability of the system to quickly respond to change and disruption and adapt flexibly in real-time. Agility can be supported with the current and future advanced technology, like data analytics, AI, and integrated platforms that can find an alternative on the spot in case of a problem or change.

HUMAN-MACHINE COLLABORATIONS

Unlike Industry 4.0, where machines work independently in the automation of various tasks in industry, the Industry 5.0 approach empowers humans with digital machines, such as robots or smart machines that can learn (through machine learning), to form integrated intelligence. This enables human-machine collaborations to execute tasks under various conditions in industry. Relaying solely on machines for decision-making is avoided, as making decisions is considered a domain best suited to humans, who possess multiple forms of intelligence (Demir et al., 2019; Nahavandi, 2019). AI-powered machines can definitely assist humans in improving both the efficiency of business processes and decision-making.

A typical example of human-machine collaboration is managing crops in a large farming area. Applying a large number of workers to diagnose possible problems is costly and has limited capacity, as workers are unable to work 24 hours a day, 7 days a week. This problem can be solved by incorporating IoT devices or drones, coupled with AI-powered diagnostic tools, into the data center. IoT devices or drones can capture changes in the physical appearance of crops in a typical location and automatically send the data to the data center, which is analyzed by AI-powered diagnostic tools. If an anomaly is detected, a message—along with an analysis of the potential issue—is sent to a human collaborator. The human then makes a timely decision to resolve the problem before it spreads to a broader area.

Collaborative Framework

Industry 5.0 encourages applying a collaborative framework for humans and machines to work collaboratively in solving problems or identifying opportunities. Figure 2 is a typical collaborative framework in Industry 5.0 (Floridi & Cowls, 2019; Maddikunta et al., 2022; Nahavandi, 2019).

The collaborative framework shown in Figure 2 has seven components to enable humans to work seamlessly with intelligent machines in the form of robots, AI-powered systems and digital interfaces. Note that, besides technologies, this framework encompasses organizational culture, workflow design, user experience, ethics, and governance. This framework is aimed at maximizing synergy between humans and machines, enabling them to perform much better than if they worked independently.

Figure 2. Collaborative Framework in Industry 5.0 (authors' own work)

Human-centered design. This is the heart of the framework, putting humans in control and incorporating humans' needs, abilities, and limitations. All systems designed based on this framework focus on people enhanced with technology. For example, they are ergonomically designed interfaces; intuitive interaction methods such as through gesture, touch, voice, and support accessibility needs (Fachrunnisa et al., 2018, 2019; Na'imah et al., 2023).

Shared control decision-making. Humans and machines collaborate in making decisions. In this collaboration, machines automate repetitive tasks, collecting data and performing data analytics, and may recommend decisions. Humans make decisions or can override, adjust and guide machine behaviors based on knowledge and experiences. AI-powered machines should offer explanations in making decisions and allow humans' input to adjust the recommendations.

Role allocation and task division. Role allocation and task division are designed based on their respective strength, and they are complementary. Humans are good at empathy, critical thinking, creativity, and contextual while machines are good at performing repetitive tasks, speed, data analytics, and endurance. However, these can change depending on the tasks and scenarios.

Communication and feedback loops. There must be two-way or bidirectional communication between humans and machines for effective collaboration. The feedback loops enable machines to learn from human input, humans to digest or understand machines' reasoning, tendencies, behavior, and error corrections for continuous improvement.

Trust and transparency mechanisms. Any collaborative environment must be built on trust and trust needs transparency. Therefore, any AI decision process must be explainable and understood by humans. Machines' performances must be consistent. Any anomaly can be traced; safety protocols and fallback mechanisms must be in place.

Safety and risk mitigation protocols. There is no perfection in human-made systems. Hence, human-machine collaborative environment (both physical and digital environments) must be designed with safety in mind. Sensors can be used to detect human presence and the emergency stop function must be activated automatically in the case of an emergency. To prevent data breaches and malfunctions, suitable cybersecurity protocols must be implemented. It is important to comply with existing safety regulations, such as ISO 10218 for collaborative robotics.

Ethical and inclusive governance. The framework must include ethical considerations with strong respect for human dignity and autonomy in decision-making. Since data or information is the lifeblood of the human-machine collaboration, data privacy and security must be safeguarded at all times. Any intelligence system that involves AI must be guided from bias. Finally, diverse groups such as gender, age, and disability must be given fair access and opportunity to participate.

Key Technologies Enabling Human-Machine Collaboration

There are several advanced technologies that make human-machine collaboration viable in the context of Industry 5.0, including collaborative robots (cobots), AI-powered systems, extended reality (XR), and natural language processing (NLP). These technologies enhance human-machine collaboration by combining human cognitive and creative capabilities with speed, precision, scalability, and learning capabilities of advanced machines, to create a more adaptive, efficient, and human-centric industrial environment (Javaid et al., 2020; Keshvarparast et al., 2024; Nahavandi, 2019; Rinaldi et al., 2025; Sharma et al., 2025; Wang et al., 2024).

Cobots. Cobots are robots that are specially designed machines to enable them to work collaboratively with humans in performing their tasks. As such, cobots behave differently compared to traditional industrial robots that work in isolation for safety. Unlike conventional robots, to interact with humans, cobots are equipped with safety protocols, advanced sensors, and compliant actuators. Conventional robots are machines designed to perform repetitive jobs with speed and precision. Cobots are designed differently to assist humans by executing physically demanding, hazardous, repetitive jobs while leaving human operators to focus on creative, high-value, supervisory and decision-making roles to enhance productivity. With the advancement of AI, many cobots are powered by AI. As such, they have the ability to learn, adapt, and interact intelligently within dynamic industrial environments.

AI-powered systems. In Industry 5.0, AI-powered systems are handy in providing feedback for intelligent decision support to come up with more effective decision-making. AI-powered systems work interactively with humans by taking human feedback, adapting to preferences. It can perform data analytics to provide predictions. Hence, AI-powered systems can help detect potential problems earlier, acting as an intelligent early warning system. This is important in predictive maintenance, quality assurance, and risk mitigation.

XR. Extended reality or XR is a general term that covers several real and virtual integrated technologies, such as augmented reality (AR), virtual reality (VR), and mixed reality (MR), that can be utilized to support human-computer interactions. XR creates immersive interfaces between humans and machines to collaborate, leading to better real-time communication and interactions. XR can be used in training workers in virtual or mixed environments safely and efficiently. It supports remote maintenance or assistance through AR. It also improves product design through virtual prototyping and testing.

NLP. NLP allows intuitive communication between machines and humans, as through NLP machines communicate with humans through human language. Hence, voice interface can be practiced with multiple language support. By incorporating large language models (LLM) in machine learning, NLP not only helps in enhancing communication between machines and humans but also helps solve problems through knowledge discovery in a large dataset available in machine learning.

Important Factors for Implementation

There are a few important factors that need to be considered before adopting human-machine collaborations, which are organizational readiness, training and upskilling, and interdisciplinary integration.

Organizational readiness. Successful machine-human collaboration requires alignment across organizational dimensions, including culture, leadership commitment, change management, and worker engagement. It is important to make sure that an organization pursuing human-machine collaboration in its business process assesses its readiness in both technological and non-technological aspects, especially the potential change in organizational culture, before implementing it.

Training and upskilling. Workers, especially those involved in human-machine collaboration, need to understand their new roles and learn to master in performing collaborative work with intelligent machines effectively. Training and upskilling are necessary for digital fluency, collaborative problem-solving, and tech-human interface skills.

Interdisciplinary integration. Collaboration should not only be conducted between humans and machines, but also among workers, especially those with diverse disciplines, including engineers, psychologists, designers, and human resource specialists. They need to work collaboratively to solve problems and create value, ultimately achieving the goal of supporting human wellbeing, sustainability, and resiliency.

Finally, for successful human-machine collaboration in the context of Industry 5.0, it is essential to adhere to a well-constructed collaborative framework, as this helps ensure that technology is deployed to support and empower humans, enhancing productivity, creativity, and dignity while they work to create value, rather than replace them. As such, effective human-machine collaboration can become a source of innovation, societal progress, and competitive advantage.

APPLICATIONS AND CASE STUDIES

This Section discusses applications of human-machine collaboration, along with a mini case study for each application. The main objective is to demonstrate recent practices in human-machine collaboration and highlight their benefits.

Human-robot collaboration in healthcare

Robot-assisted surgery is a specially designed robot system to assist a surgeon in performing invasive procedures with enhanced precision, flexibility, and control. One of the most well-known systems is the da Vinci Surgical System that assists surgeons in performing complex surgical procedures (Probst, 2023). More recent development is an AI-powered robot for laser eye surgery (Rana et al., 2025; Sachdev, 2025).

Several studies indicate encouraging results of performing operations using the robot-like da Vinci Surgical System (dVSP). Li et al. (2022) reported that operating with a surgical robot could produce effective therapeutic results in individuals with early-stage cervical cancer while maintaining a low rate of complications. Another study (Kim et al., 2023) showed the surgical and oncological safety and practicality of dVSP, which could serve as a new surgical platform for colorectal procedures.

Robotic technology is advancing fast; new robots are introduced every year and existing robots are being applied to different procedures. Robotic advancements enhance the skills of surgeons across various fields, including urology, gynecology, neurology, spine surgery, orthopedic reconstruction (knee and shoulder), hair restoration, oral surgery, thoracic surgery, laparoscopic procedures, and endoscopic techniques (Cepolina & Razzoli, 2024). Furthermore, incorporating AI and machine learning in the robot system will improve decision-making capability. According to Chatterjee et al. (2024), AI and machine learning will lead to faster recovery and fewer complications in patients. However, the cost of investment, operation, maintenance, and training for robot surgeon systems is substantial, leading to high costs for the patients, which may pose a major challenge in incorporating such systems for healthcare.

Collaborative robots on the assembly line

Tasks in an assembly line, such as tightening screw, applying adhesive, and inserting components, are repetitive and physically demanding. These tasks are time-consuming, ergonomically strenuous (leading to long-term musculoskeletal issues), hence prone to human error due to fatigue. Conventional industrial robots are placed in the assembly line to replace humans. However, these robots are not safe for humans; hence, they need to be handled with safety or put in safety cages, separated from human workers, thus limiting the flexibility (Cherubini et al., 2016; Michalos et al., 2018; Schmiedchen & Beutnagel, 2024).

Unlike conventional industrial robots, collaborative robots (cobots) are designed to work closely and collaboratively with humans. For this reason, many automobile manufacturers are interested in adopting cobots. For example, BMW, a world-leading automobile manufacturer, has deployed cobots, such as cobots produced by Universal Robot (UR10), at several production plants in Germany. The cobots meet ISO 10218 safety standards. They can work collaboratively with human operators without safety fences, like conventional industrial robots mentioned above. Some recent humanoid cobots have been developed to serve the same purpose with better capabilities such XPeng Iron (Isaac, 2025).

Some interesting features the cobots are human-centric design, lightweight and easy to program, allowing workers to teach them new tasks through hand-guiding; flexible, cobots can be easily re-deployed to different stations based on production needs; safe, as they are equipped with sensors to detect human presence and stop instantly when needed; and focus on repetitive and monotonous tasks to be automated, allowing human workers to focus on quality control and complex tasks, hence, improve productivity.

As a result, cobots have reduced physical strain on workers, thus improving ergonomics by automating awkward and repetitive motions. They reduced cycle times and enhanced throughput, hence increasing productivity. They also enhance quality by consistently performing repetitive tasks and reducing defects. More importantly, operators were satisfied, appreciated cobots and they did not see cobots as a threat, but as a helpful “third hand.”

UR10 cobots or similar assembly line cobots have helped the company achieve a smarter, safer, and more adaptive assembly line. This is a good example of how humans and machines work in harmony in Industry 5.0.

Human-chatbot collaboration in customer service

Customer service is an important function in a company in retaining the existing customers and attracting new ones. One of the main factions of customer service is customer support, which helps answer customers' queries about products or services offered, resolves customer complaints and helps customers solve problems with the products or services that they purchased.

Traditionally, companies use toll-free numbers to communicate with customers regarding their queries, problems with products or services, and complaints. As toll-free facilities are very limited, a web-based customer support, which later developed into customer relationship management (CRM), was deployed to handle customers' queries, complaints or solve their problems.

Further, with the advancement of web-based technology to Web 2.0, CRM developed into interactive CRM or CRM 2.0 (Anshari et al., 2012, 2019), facilitating real-time communication between customers with customer support staff. Although this advancement is very good, it has limitations as human customer support is unable to handle many cases at the same time, around the clock.

The solution is to employ a chatbot. A chatbot is a software application designed to simulate human conversation, allowing interaction with humans via text or voice at any time. Recent chatbots incorporate natural language processing (NLP) and AI to understand user inputs to provide relevant and reasonable responses.

However, even advanced chatbots are not able to solve customers' complaints, queries or problems; they need to be supervised. Some difficult problems need to be solved by human experts or trained workers. However, many problems can be handled by chatbots most of the time; hence, cooperation between humans and chatbots is necessary, not only to solve customers' problems but also to maintain or increase customers' trust.

For example, AirAsia, a leading low-cost airline, utilizes a chatbot called AskBo (Lacsamana, 2023) to address numerous customer queries and issues. AskBo can handle most customers' queries; however, it needs to pass to the customer support staff to handle more complex problems, such as handling requests for cancellation with a refund, changing a flight because of the airline cancellation or substantial postponement.

Chatbots are not a replacement for customer support staff, but they complement the employees who work hand-in-hand to answer customers' queries or solve their problems. Therefore, chatbots are basically a human-machine collaboration to support customers.

AI for fraud detection and decision support

Frauds are a major problem in many financial institutions like banks, fintechs, and credit card firms. Manually detecting fraud is difficult and tedious. More importantly, most of today's financial transactions are done digitally. Hence effective digital solution is badly needed. Advances in AI and machine learning bring good news as they can be used to help with fraud detection in the digital space.

Mastercard, a world-leading payment network, processes billions of payments yearly across the globe. The 2023 annual report stated that Mastercard processed over 90 billion payments across more than 210 countries. Given this huge scale of payments to be processed, fraud in payments with increasing sophistication needs to be addressed accordingly to avoid financial loss and damage to reputation.

The traditional solution for fraud detection used rule-based systems. Rule-based systems lack adaptability and often produce high false positives, frustrating customers and merchants. As such solution to real-time and more accurate decision-making across billions of transactions is required.

Mastercard decided to employ an AI-powered and machine learning-based solution called Decision Intelligence (DI) (Mastercard, 2023, 2024). DI integrates neural networks in machine learning, behavioral analytics and anomaly detection algorithms to detect fraud and generate risk for scoring transactions in making decisions to approve, flag for review, or block transactions.

Using DI helps MasterCard perform much better. For example, fraud detection improves significantly, from roughly 70% to more than 90% with false positives reduced to 50%. In addition, transaction speed is much higher.

In conclusion, the use of DI, an AI-powered decision support for detecting fraud by MasterCard, has significantly improved the performance of payment processes, as it enables smarter, faster, and more accurate decision-making.

CHALLENGES AND ETHICAL ISSUES

This Section examines several challenges in adopting Industry 5.0, such as the risk of displacement, trust and transparency in AI-human collaboration, concerns about data privacy and security, as well as ethical implications and potential bias related to AI-based recommendations.

Job displacement vs. job augmentation debates

As previously discussed, Industry 5.0 is a correction to Industry 4.0. The main one is the correction of human involvement in value creation, from emphasizing automation in Industry 4.0, replacing humans with machines, ending up with the major issue of job displacement, to human-machine collaboration, bringing humans back to the center of value creation, changing the issue from job displacement to job augmentation.

Job displacement refers to job loss due to the automation of tasks with machines or AI systems, replacing human labor due to efficiency or productivity issues. However, job displacement will lead to a major problem: an increase in unemployment, especially if Industry 4.0-led automation spreads to many sectors, not only manufacturing, but also finance, logistics, and retail. Workers who are displaced because of automation may lack of digital or technical skills required to find new jobs. This can create a larger problem compared to the benefit of efficiency gained from automation. For example, a report published by the McKinsey Global Institute in 2023 (Creative Dock, 2024) projected that as much as 25% of existing work tasks might be automated by the year 2030, with lower-income and less-educated employees likely to be more severely impacted.

In contrast, job augmentation focuses on empowering human workers with technology, enabling them to work hand-in-hand, which is in line with Industry 5.0's human-centric approach. For example, cobots, AI-powered systems and intelligent decision support are designed to work collaboratively with humans to improve productivity, effectiveness, and efficiency in decision-making. There must be upskilling opportunities for the existing workers. With an effective training program, new roles that are needed by the new systems can be crafted, promoting the existing workers to better roles instead of displacing them.

However, to realize the job augmentations in Industry 5.0, a proper transformation strategy guided by the collaborative framework needs to be in place. A carefully planned transition with proper change management needs to ensure that incorporating advanced technology is in harmony with human values, reflecting the principles of Industry 5.0.

Trust and transparency in AI-human collaboration

As AI-powered systems are increasingly adopted in industry, ranging from performing repetitive tasks intelligently to supporting decision-making, a number of issues related to trust and transparency have emerged. Many human workers perceive these systems as "black boxes," meaning that they do not understand how the AI reaches conclusions. For example, if an AI system is unable to explain the reason behind a conclusion proposed and how it reached the conclusion, the human collaborators will be in hesitation, hence a trust problem may emerge when collaborating with AI systems. In the case of MasterCard's use of AI to recommend risk scores for fraud detection, some users may question the basis of these recommendations, especially if the rationale behind the scoring is not transparent.

Why are trust and transparency important for any AI-powered system to be operated effectively? To make an AI-powered system accepted, trusted, and perform seamlessly with users, first and foremost, trust in an AI-powered system must be established. This means its users or collaborators believe that it can perform reliably, ethically, and in line with human values to reach the expected objectives. Next, trust implies user acceptance. In other words, users work confidently with the trusted AI-powered system. This will help the AI-powered system or its collaboration with humans lead to the expected performance. For an AI-powered system that works physically with humans, such as autonomous vehicles, cobots in manufacturing, and robot surgery, trust means being confident that the AI-powered system is safe or harmless and behaves as expected in an emergency. In the case of an AI-powered system with decision-making capability, trust means users' or collaborators' confidence that the decision or recommendation proposed can be relied on (Hancock et al., 2011; Lee & See, 2004).

Transparency is highly related to trust. In AI-powered systems, transparency means explainability, understandability, and accountability, especially related to the outputs of the systems. Explainability means any AI-powered output such as a decision or recommendation can be explained on how the system reach to such a decision; understandability means the output can be understood by humans, while accountability means any output or conclusion is traceable, including who programmed the model and the data used, and more importantly, who is responsible for the outcome.

Without transparency, AI-powered systems are fragile. For example, in healthcare diagnostics, clinicians require AI-powered tools to come up with reasons or confidence levels on the predictions or recommendations so that they can confidently incorporate the AI insights with their medical judgment.

Although trust and transparency are very important, achieving high trust and transparency in AI-powered systems is not easy due to several problems. First, as mentioned previously, most AI-powered systems, due to their complexity, are black-box systems, meaning that the process that leads to the conclusion is opaque; hence, it is difficult to explain the conclusion. Second, a lack of transparency in training data and algorithms utilized will lead to bias. This will reduce trust, especially in sensitive domains like hiring, lending, or law enforcement. Third, dynamic learning of advanced AI-powered systems means there is a possible change of behavior unpredictably, which may raise the consistency problem. Fourth, mismatch or incompatibility between human and machine reasoning leads to difficulty for humans to trust AI systems.

Several strategies can help to overcome the above challenges. In the case of explainability and understandability problems, an AI system needs to be equipped with an explainer that can be understood by humans. In the case of accountability, the backward chain mechanism and transparency on training data can be used to trace the conclusions. It is important to keep humans in the decision process (human-in-the-loop) to increase confidence in the conclusion produced. Next, an ethical framework, such as the EU's AI Act, needs to be followed. In addition, the capabilities and limitations of an AI system used need to be transparently communicated to users. Last but not least, the AI system needs to be regularly tested and audited for performance, fairness, and unintended behaviors (Glikson & Woolley, 2020; Siau & Wang, 2018; Tjahjono, 2011; Tjahjono et al., 2019).

Data privacy and security concerns

Data is the lifeblood of human-machine collaboration in Industry 5.0. Most systems in Industry 5.0 involve a continuous and vast amount of data transmission from sensors, cobots, and industrial equipment, often including personal and sensitive information for further processing. These data and information are prone to cyber threats; thus, proper safeguards need to be in place. As AI-powered systems are utilized, they introduce substantial risk as they need extensive data regarding behavior, preferences, biometrics, or contextual information required for them to function effectively. In addition, data transmitted through internet-connected devices, like IoT, is vulnerable not only to breaches of confidentiality but also to threats to data integrity (Boisrond et al., 2024; Santos et al., 2024; Tawalbeh et al., 2020).

Privacy and security concerns of Industry 5.0 can be examined systematically through the triad of information security: confidentiality, integrity, and availability (CIA). Confidentiality means ensuring data or information, especially sensitive ones such as personal information, operational parameters, or proprietary designs, can only be accessed by authorized users and systems. Therefore, data or information protection from industrial espionage and property rights theft is critical. Further challenges are that real-time data sharing across connected devices and partners increases the risk of leaks and the use of AI and cloud platforms exposes more attack surfaces.

Integrity means that all data and information must be kept accurate, trustworthy, and unaltered. Integrity is vital in AI systems that make decisions based on input data. If the data is tampered with, it can lead to catastrophic outcomes, as decisions are wrong. Integrity is also vitally important for digital twins, cobots, diagnostic systems, and predictive maintenance. Some notable challenges are that malicious modifications to sensor data, AI training models, or control instructions can occur. These modifications could lead to safety hazards, production errors, or sabotage.

The last one is availability, which means all systems and data are ready and available when needed. It is important to note that the Industry 5.0 environment depends on real-time collaboration between digital systems, including AI systems and human operators. Unavailability because of downtime or disruption can stop production and may cause safety issues. Some notable challenges are ransomware, DDoS attacks, and infrastructure failure, which can disrupt availability.

To address these challenges, it is important for organizations adopting Industry 5.0 to ensure privacy-by-design and security-by-design principles in all systems, particularly AI-powered systems. Failing to do this will result in dire consequences later. Therefore, ensuring encryption, access controls, anonymization techniques, and transparent consent mechanisms is essential. It is also important to comply with security and privacy regulations and standards, as well as to practice good data governance.

Ethical AI and bias mitigation

Advances in AI technology, especially machine learning, have made AI a powerful technology. As such, it has now been adopted in various fields across the world. In Industry 5.0, AI plays a major role, especially in enhancing human-machine collaboration, ranging from smart automation and cobots to support in decision-making. Like other digital technologies, applications of AI, including applications in industry, present ethical challenges such as ethical issues in bias, data privacy, and accountability.

One of the widely mentioned ethical issues in AI systems is algorithmic bias. This bias emerges from incorrect model assumptions, unbalanced training data, or the personal biases of developers. If not addressed, the output of AI-powered systems used for collaboration with humans will skew toward something not in line with the principles of Industry 5.0. For instance, an AI-powered recruitment tool trained on past hiring data might unintentionally prefer male candidates if previous hiring practices were influenced by gender bias. Even more, it may lead to a trust problem, which is difficult to resolve, such as in the case of an improper model use in an early warning system that creates many false positive warnings.

Bias mitigation needs to be incorporated carefully before and during the development phase of AI-powered systems. First, the developers of any AI-powered system need to understand and adhere to ethical principles to personal tendencies. Second, to avoid bias during development, the algorithms used need to be carefully scrutinized or tested for bias. Third, the datasets for training need to be properly chosen to reflect the populations the AI system will be applied to, and proper data preprocessing techniques must be employed to reduce skewness. Additionally, the system must be carefully tested to detect and correct bias before deployment.

As previously highlighted when discussing trust, explainability and transparency need to be incorporated in any ethical AI-powered system. This is to allow humans to understand how a decision was made and the reasons behind it. It is also important to inform users or collaborators about the capabilities and limitations of the system. In addition, potential harms need to be identified and the appropriate safeguards must be in place before an AI system is developed.

Finally, it is important to adhere to ethical regulations or an existing regulatory framework. This is to make sure that the deployment and utilization of AI systems comply with ethical principles.

CONCLUSION

Industry 5.0 is not a complete shift from Industry 4.0. It is rather a correction or improvement, especially to bring the human back to the center of value creation while addressing other important aspects that are overlooked in Industry 4.0, such as sustainability and resiliency, while recognizing the advancement of technology and its utilization through cooperation between humans and machines in achieving goals. Industry 5.0 realigns the utilization of technology toward human wellbeing, promoting sustainability. The three pillars of Industry 5.0 are a human-centric approach, sustainability, and resilience.

It is important to note that Industry 5.0 does not reject automation but rather prevents the dehumanization of value creation through excessive automation in pursuit of efficiency and productivity. Automation remains valuable for improving efficiency and is essential in executing processes in a dangerous environment, improving ergonomics, or in the case of a shortage of human labor. In essence, Industry 5.0 adopts automation to support rather than to replace humans in an industry. Unlike Industry 4.0, where machines independently work in the automation of various tasks in industry, the Industry 5.0 approach is to empower humans with digital machines like robots or smart machines that can learn (machine learning) to form integrated intelligence that enables human-machine collaborations to execute tasks under various conditions in industry. Relying solely on machines in making decisions is avoided, as making decisions is considered a domain best suited to humans, who possess multiple forms of intelligence.

Machine-human collaboration needs to be thought out carefully. An attempt to do this systematically is to have a collaborative framework. A typical collaborative framework has seven components to enable humans to work seamlessly with intelligent machines in the form of robots, AI-powered systems and digital interfaces. The seven components are: Human-centered design, shared control decision-making, role allocation and task division, communication and feedback loops, trust and transparency mechanisms, safety and risk mitigation protocols, and ethical and inclusive governance. Note that, besides technologies, this framework encompasses organizational culture, workflow design, user experience, ethics, and governance. This framework is aimed at maximizing synergy between humans and machines, enabling them to perform much better than if they worked independently.

Some challenges of human-machine collaboration in the context of Industry 5.0 include the risk of displacement, trust and transparency in AI-human collaboration, concerns about data privacy and security, as well as ethical implications and potential bias related to AI-based recommendations. These challenges need to be addressed accordingly before deploying AI systems.

ACKNOWLEDGMENT

This research was supported by the Universitas Muhammadiyah Yogyakarta (UMY) Research Fellowship Program.

REFERENCES

Adhiatma, A., Fachrunnisa, O., & Tjahjono, H. K. (2021). A Value Creation Process for Sustainability of Knowledge Based-Society. In Barolli, L., Poniszewska-Maranda, A., & Enokido, T. (Eds.), *Complex, Intelligent and Software Intensive Systems* (pp. 307–314). Springer International Publishing., DOI: 10.1007/978-3-030-50454-0_29

Agrawal, K., & Bhat, A. R. (2025). Advances in 3D printing with eco-friendly materials: A sustainable approach to manufacturing. *RSC Sustainability*, *3*(6), 2582–2604. DOI: 10.1039/D4SU00718B

Allenby, B. (2006). The ontologies of industrial ecology? *Progress in Industrial Ecology*, *3*(1–2), 28–40. DOI: 10.1504/PIE.2006.010039

Almunawar, M. N., & Almunawar, H. J. (2018). From Information Systems Outsourcing to Cloud Computing. In *Encyclopedia of Information Science and Technology* (4th ed., Vol. 1–10, pp. 1101–1115). IGI Global.

Almunawar, M. N., & Anshari, M. (2018). Moving to the Cloud. *Journal of Business and Economic Analysis*, *01*(01), 65–80. DOI: 10.1142/S2737566821800049

Alqahtani, M. F., & Afy-Shararah, M. (2025). Integrating Circular Economy Practices into Renewable Energy in the Manufacturing Sector: A Systematic Review of the Literature. *Sustainability (Basel)*, *17*(16), 7301. DOI: 10.3390/su17167301

Alves, J., Lima, T. M., & Gaspar, P. D. (2023). Is Industry 5.0 a Human-Centred Approach? A Systematic Review. *Processes (Basel, Switzerland)*, *11*(1), 1. Advance online publication. DOI: 10.3390/pr11010193

Anshari, M., Almunawar, M. N., Lim, S. A., & Al-Mudimigh, A. (2019). Customer relationship management and big data enabled: Personalization & customization of services. *Applied Computing and Informatics*, *15*(2), 2. Advance online publication. DOI: 10.1016/j.aci.2018.05.004

Anshari, M., Almunawar, M. N., & Low, P. K. C. (2012). CRM 2.0 Within E-Health Systems: Towards Achieving Health Literacy & Customer Satisfaction. *International Journal on Informatics for Development*, *1*(2), 90476. DOI: 10.14421/ijid.2012.01201

Armbrust, M., Stoica, I., Zaharia, M., Fox, A., Griffith, R., Joseph, A. D., Katz, R., Konwinski, A., Lee, G., Patterson, D., & Rabkin, A. (2010). A view of cloud computing. *Communications of the ACM*, *53*(4), 4. Advance online publication. DOI: 10.1145/1721654.1721672

Ashraf, Z., Sohail, A., & Yousaf, M. (2019). Performance Analysis of Network Applications on IPv6 Cloud Connected Virtual Machine. *International Journal of Computer Network and Information Security*, *11*(12), 1–9. DOI: 10.5815/ijcnis.2019.12.01

Barata, J., & Kayser, I. (2024). How will the digital twin shape the future of industry 5.0? *Technovation*, *134*, 103025. DOI: 10.1016/j.technovation.2024.103025

Benyus, J. M. (2002). *Biomimicry: Innovation Inspired by Nature*. Harper Perennial.

Boisrond, K., Tardif, P. M., & Jaafar, F. (2024). Ensuring the Integrity, Confidentiality, and Availability of IoT Data in Industry 5.0: A Systematic Mapping Study. *IEEE Access : Practical Innovations, Open Solutions*, *12*, 107017–107045. DOI: 10.1109/ACCESS.2024.3434618

Brocal, F., González, C., Komljenovic, D., Katina, P. F., & Sebastián, M. A. (2019). Emerging Risk Management in Industry 4.0: An Approach to Improve Organizational and Human Performance in the Complex Systems. *Complexity*, *2019*(1), 2089763. DOI: 10.1155/2019/2089763

Castillo, C., Otero-Romero, T., & Alvarez-Palau, E. J. (2025). Navigating the transition to industry 5.0: Advancing sustainability, resilience, and human-centricity in Spanish supply chain management. *Discover Sustainability*, *6*(1), 331. DOI: 10.1007/s43621-025-01190-0

Cepolina, F., & Razzoli, R. (2024). Review of robotic surgery platforms and end effectors. *Journal of Robotic Surgery*, *18*(1), 74. DOI: 10.1007/s11701-023-01781-x PMID: 38349595

Chatterjee, S., Das, S., Ganguly, K., & Mandal, D. (2024). Advancements in robotic surgery: Innovations, challenges and future prospects. *Journal of Robotic Surgery*, *18*(1), 28. DOI: 10.1007/s11701-023-01801-w PMID: 38231455

Cherubini, A., Passama, R., Crosnier, A., Lasnier, A., & Fraisse, P. (2016). Collaborative manufacturing with physical human–robot interaction. *Robotics and Computer-integrated Manufacturing*, *40*, 1–13. DOI: 10.1016/j.rcim.2015.12.007

Creative Dock. (2024). *The future of work aka is your position on the list of jobs most at risk from AI?* https://www.creativedock.com//blog/the-future-of-work

Demir, K. A., Döven, G., & Sezen, B. (2019). Industry 5.0 and Human-Robot Co-working. *Procedia Computer Science*, *158*, 688–695. DOI: 10.1016/j.procs.2019.09.104

Dieste, M., Orzes, G., Culot, G., Sartor, M., & Nassimbeni, G. (2023). The "dark side" of Industry 4.0: How can technology be made more sustainable? *International Journal of Operations &amp. International Journal of Operations & Production Management*, *44*(5), 900–933. DOI: 10.1108/IJOPM-11-2022-0754

Elnadi, M., & Abdallah, Y. O. (2024). Industry 4.0: Critical investigations and synthesis of key findings. *Management Review Quarterly*, *74*(2), 711–744. DOI: 10.1007/s11301-022-00314-4

Emad-ul-Haq, Q., Aboalsamh, H., Belghith, A., Hussain, M., Abdul, W., Dahshan, M. H., & Ghouzali, S. (2015). Challenges and solutions for Internet of Things Driven by IPv6. [TIIS]. *KSII Transactions on Internet and Information Systems*, *9*(12), 4739–4758. DOI: 10.3837/tiis.2015.12.001

Eriksson, K. M., Olsson, A. K., & Carlsson, L. (2024). Beyond lean production practices and Industry 4.0 technologies toward the human-centric Industry 5.0. *Technological Sustainability*, *3*(3), 286–308. DOI: 10.1108/TECHS-11-2023-0049

Fachrunnisa, O., Adhiatma, A., & Tjahjono, H. K. (2019). Collective Engagement and Spiritual Wellbeing in Knowledge Based Community: A Conceptual Model. In Barolli, L., Hussain, F. K., & Ikeda, M. (Eds.), *Complex, Intelligent, and Software Intensive Systems* (pp. 899–906). Springer International Publishing., DOI: 10.1007/978-3-030-22354-0_83

Fachrunnisa, O., Tjahjono, H. K., & Palupi, M. (2018). Cognitive collective engagement in virtual collaborative team. *2018 7th International Conference on Industrial Technology and Management (ICITM)*, 390–393. DOI: 10.1109/ICITM.2018.8333981

Floridi, L., & Cowls, J. (2019). A Unified Framework of Five Principles for AI in Society. *Harvard Data Science Review*, *1*(1). Advance online publication. DOI: 10.1162/99608f92.8cd550d1

Folgado, F. J., Calderón, D., González, I., & Calderón, A. J. (2024). Review of Industry 4.0 from the Perspective of Automation and Supervision Systems: Definitions, Architectures and Recent Trends. *Electronics (Basel)*, *13*(4), 4. Advance online publication. DOI: 10.3390/electronics13040782

Ghobakhloo, M., Iranmanesh, M., Fathi, M., Rejeb, A., Foroughi, B., & Nikbin, D. (2024). Beyond Industry 4.0: A systematic review of Industry 5.0 technologies and implications for social, environmental and economic sustainability. *Asia-Pacific Journal of Business Administration*. Advance online publication. DOI: 10.1108/APJBA-08-2023-0384

Glikson, E., & Woolley, A. W. (2020). Human Trust in Artificial Intelligence: Review of Empirical Research. *The Academy of Management Annals*, *14*(2), 627–660. DOI: 10.5465/annals.2018.0057

Gubeladze, D., & Pavliashvili, S. (2020). LINEAR ECONOMY AND CIRCULAR ECONOMY - CURRENT STATE ASSESSMENT AND FUTURE VISION. *International Journal of Innovative Technologies in Economy, 5(32). Article*, *5*(32). Advance online publication. DOI: 10.31435/rsglobal_ijite/30122020/7286

Hancock, P. A., Billings, D. R., Schaefer, K. E., Chen, J. Y. C., de Visser, E. J., & Parasuraman, R. (2011). A Meta-Analysis of Factors Affecting Trust in Human-Robot Interaction. *Human Factors*, *53*(5), 517–527. DOI: 10.1177/0018720811417254 PMID: 22046724

Hoff, K. A., & Bashir, M. (2015). Trust in Automation: Integrating Empirical Evidence on Factors That Influence Trust. *Human Factors*, *57*(3), 407–434. DOI: 10.1177/0018720814547570 PMID: 25875432

Isaac, S. (2025, April 28). *XPENG Iron: Advanced Humanoid Robot Unveiled - Qviro Blog*. https://qviro.com/blog/xpeng-launches-iron-advanced-humanoid-robot/

Ivanov, D., Blackhurst, J., & Das, A. (2021). Supply chain resilience and its interplay with digital technologies: Making innovations work in emergency situations. *International Journal of Physical Distribution & Logistics Management*, *51*(2), 97–103. DOI: 10.1108/IJPDLM-03-2021-409

Javaid, M., Haleem, A., Singh, R. P., Haq, M. I. U., Raina, A., & Suman, R. (2020). Industry 5.0: Potential Applications in COVID-19. *Journal of Industrial Integration and Management*, *05*(04), 507–530. DOI: 10.1142/S2424862220500220

Kanda, W., Geissdoerfer, M., & Hjelm, O. (2021). From circular business models to circular business ecosystems. *Business Strategy and the Environment*, *30*(6), 2814–2829. DOI: 10.1002/bse.2895

Kane, A. (2025). Navigating the Transition: Challenges and Benefits of Shifting from IPv4 to IPv6 in a Rapidly Evolving Internet Landscape. *International Journal of Internet and Distributed Systems*, *7*(2), 2. Advance online publication. DOI: 10.4236/ijids.2025.72002

Kapur, A., & Graedel, T. E. (2004). Industrial Ecology. In Cleveland, C. J. (Ed.), *Encyclopedia of Energy* (pp. 373–382). Elsevier., DOI: 10.1016/B0-12-176480-X/00533-7

Kennedy, S., & Linnenluecke, M. K. (2022). Circular economy and resilience: A research agenda. *Business Strategy and the Environment*, *31*(6), 2754–2765. DOI: 10.1002/bse.3004

Keshvarparast, A., Battini, D., Battaia, O., & Pirayesh, A. (2024). Collaborative robots in manufacturing and assembly systems: Literature review and future research agenda. *Journal of Intelligent Manufacturing*, *35*(5), 2065–2118. DOI: 10.1007/s10845-023-02137-w

Kim, H. S., Oh, B.-Y., Cheong, C., Park, M. H., Chung, S. S., Lee, R.-A., Kim, K. H., & Noh, G. T. (2023). Single-incision robotic colorectal surgery with the da Vinci SP® surgical system: Initial results of 50 cases. *Techniques in Coloproctology*, *27*(7), 589–599. DOI: 10.1007/s10151-023-02791-5 PMID: 36971849

Langer, M., Baum, K., & Schlicker, N. (2024). Effective Human Oversight of AI-Based Systems: A Signal Detection Perspective on the Detection of Inaccurate and Unfair Outputs. *Minds and Machines*, *35*(1), 1. DOI: 10.1007/s11023-024-09701-0

Lee, J. D., & See, K. A. (2004). Trust in Automation: Designing for Appropriate Reliance. *Human Factors*. https://journals.sagepub.com/doi/abs/10.1518/hfes.46.1.50_30392

Leng, J., Sha, W., Wang, B., Zheng, P., Zhuang, C., Liu, Q., Wuest, T., Mourtzis, D., & Wang, L. (2022). Industry 5.0: Prospect and retrospect. *Journal of Manufacturing Systems*, *65*, 279–295. DOI: 10.1016/j.jmsy.2022.09.017

Li, J., Gong, X., Li, P., Xiao, L., Chang, X., Ouyang, X., & Tang, J. (2022). Application of Da Vinci robotic surgery system in cervical cancer: A single institution experience of 557 cases. *Asian Journal of Surgery*, *45*(2), 707–711. DOI: 10.1016/j.asjsur.2021.07.052 PMID: 34373166

Maddikunta, P. K. R., Pham, Q.-V., B, P., Deepa, N., Dev, K., Gadekallu, T. R., Ruby, R., & Liyanage, M. (2022). Industry 5.0: A survey on enabling technologies and potential applications. *Journal of Industrial Information Integration*, *26*, 100257. DOI: 10.1016/j.jii.2021.100257

Marston, S., Li, Z., Bandyopadhyay, S., Zhang, J., & Ghalsasi, A. (2011). Cloud computing—The business perspective. *Decision Support Systems*, *51*(1), 1. Advance online publication. DOI: 10.1016/j.dss.2010.12.006

Mastercard. (2023). *Mastercard leverages its AI capabilities to fight real-time payment scams.* https://newsroom.mastercard.com/news/press/2023/july/mastercard-leverages-its-ai-capabilities-to-fight-real-time-payment-scams/

Mastercard. (2024). *Mastercard supercharges consumer protection with gen AI.* https://newsroom.mastercard.com/news/press/2024/february/mastercard-supercharges-consumer-protection-with-gen-ai/

McDonough, W., & Braungart, M. (2002). *Cradle to Cradle: Remaking the Way We Make Things* (1st ed.). North Point Press.

Michalos, G., Kousi, N., Karagiannis, P., Gkournelos, C., Dimoulas, K., Koukas, S., Mparis, K., Papavasileiou, A., & Makris, S. (2018). Seamless human robot collaborative assembly – An automotive case study. *Mechatronics*, *55*, 194–211. DOI: 10.1016/j.mechatronics.2018.08.006

Min, B. H., & Borch, C. (2022). Systemic failures and organizational risk management in algorithmic trading: Normal accidents and high reliability in financial markets. *Social Studies of Science*, *52*(2), 277–302. DOI: 10.1177/03063127211048515 PMID: 34612758

Murakami, S., Shimizu, K., Tokoro, C., & Nakamura, T. (2022). Role of Resource Circularity in Carbon Neutrality. *Sustainability (Basel)*, *14*(24), 16408. DOI: 10.3390/su142416408

Na'imah, T., Tjahjono, H. K., & Madjid, A. (2023). Workplace Well-Being: The Roles of Perceived Organizational Support, Organizational Justice and Workplace Spirituality. *Quality - Access to Success*, *24*(193). Advance online publication. DOI: 10.47750/QAS/24.193.29

Nahavandi, S. (2019). Industry 5.0—A Human-Centric Solution. *Sustainability (Basel)*, *11*(16), 16. Advance online publication. DOI: 10.3390/su11164371

Pauli, G. (2010). *Blue Economy-10 Years, 100 Innovations, 100 Million Jobs.* Paradigm Pubns.

Pearce, D. W., & Turner, R. K. (1989). *Economics of Natural Resources and the Environment.* JHU Press. DOI: 10.56021/9780801839863

Probst, P. (2023). A Review of the Role of Robotics in Surgery: To DaVinci and Beyond! *Missouri Medicine*, *120*(5), 389–396. PMID: 37841561

Purnama, Y. H., Tjahjono, H. K., Elqadri, Z. M., & Prajogo, W. (2021). Innovative Work Behavior: The Role of Self-efficacy and Organizational Climates. In Barolli, L., Poniszewska-Maranda, A., & Enokido, T. (Eds.), *Complex, Intelligent and Software Intensive Systems* (pp. 477–484). Springer International Publishing., DOI: 10.1007/978-3-030-50454-0_49

Rana, M., Akter, J., & Hanif, M. A. (2025). Next-gen vision: A systematic review on robotics transforming ophthalmic surgery. *Journal of Robotic Surgery*, *19*(1), 452. DOI: 10.1007/s11701-025-02623-8 PMID: 40762731

Rejeb, A., Rejeb, K., Keogh, J. G., & Süle, E. (2025). When Industry 5.0 Meets the Circular Economy: A Systematic Literature Review. *Circular Economy and Sustainability*, *5*(4), 2621–2652. Advance online publication. DOI: 10.1007/s43615-025-00570-y

Rinaldi, M., Di Pasquale, V., Farina, P., Iannone, R., Macchiaroli, R., & Grosse, E. H. (2025). Human–robot interaction in industry: A tertiary study. *Procedia Computer Science*, *253*, 1691–1701. DOI: 10.1016/j.procs.2025.01.231

Sachdev, R. (2025, July 2). How AI & Robotics Are Transforming Laser Eye Surgery. *Planet Lasik - Best Eye Care Center | Lasik & Refractive Surgeries*. https://planetlasik.com/blog/ai-robotics-in-laser-eye-surgery

Santos, B., Costa, R. L. C., & Santos, L. (2024). Cybersecurity in Industry 5.0: Open Challenges and Future Directions. *2024 21st Annual International Conference on Privacy, Security and Trust (PST)*, 1–6. DOI: 10.1109/PST62714.2024.10788065

Sariatli, F. (2017). Linear Economy Versus Circular Economy: A Comparative and Analyzer Study for Optimization of Economy for Sustainability. *Visegrad Journal on Bioeconomy and Sustainable Development*, *6*(1), 31–34. DOI: 10.1515/vjbsd-2017-0005

Schmiedchen, R., & Beutnagel, W. (2024, September 24). *How cobots can support automotive production*. https://www.automotivemanufacturingsolutions.com/digitalisation/how-cobots-can-support-automotive-production/533015

Shabur, M., Shahriar, A., & Ara, M. A. (2025). From automation to collaboration: Exploring the impact of industry 5.0 on sustainable manufacturing. *Discover Sustainability*, *6*(1), 341. DOI: 10.1007/s43621-025-01201-0

Sharma, S., Sharma, C., Batra, I., Malik, A., Kaswan, M. S., Du, D., & Kumar, V. (2025). Navigating the Frontiers of Industry 5.0: Predictive Analysis Using Natural Language Processing. *Operations Research Forum, 6*(3), 93. DOI: 10.1007/s43069-025-00501-5

Siau, K., & Wang, W. (2018). Building trust in artificial intelligence, machine learning, and robotics. *Cutter Business Technology Journal*, *31*(2), 47–53.

Sima, V., Gheorghe, I. G., Subić, J., & Nancu, D. (2020). Influences of the Industry 4.0 Revolution on the Human Capital Development and Consumer Behavior: A Systematic Review. *Sustainability (Basel)*, *12*(10), 10. Advance online publication. DOI: 10.3390/su12104035

Tawalbeh, L., Muheidat, F., Tawalbeh, M., & Quwaider, M. (2020). IoT Privacy and Security: Challenges and Solutions. *Applied Sciences (Basel, Switzerland)*, *10*(12), 12. Advance online publication. DOI: 10.3390/app10124102

Theeraworawit, M., Suriyankietkaew, S., & Hallinger, P. (2022). Sustainable Supply Chain Management in a Circular Economy: A Bibliometric Review. *Sustainability (Basel)*, *14*(15), 9304. DOI: 10.3390/su14159304

Tjahjono, H. K. (2011). The Configuration among Social Capital, Distributive and Procedural Justice and Its Consequences to Individual Satisfaction. *International Journal of Information and Management Sciences*, 87–103. DOI: 10.6186/IJIMS.2011.22.1.6

Tjahjono, H. K., Fachrunnisa, O., & Palupi, M. (2019). Configuration of organisational justice and social capital: Their impact on satisfaction and commitment. *International Journal of Business Excellence*, *17*(3), 336–360. DOI: 10.1504/IJBEX.2019.097957

Wang, B., Zheng, L., Wang, Y., Fang, W., & Wang, L. (2024). Towards the industry 5.0 frontier: Review and prospect of XR in product assembly. *Journal of Manufacturing Systems*, *74*, 777–811. DOI: 10.1016/j.jmsy.2024.05.002

Chapter 2

Using Artificial Intelligence for Human–AI Collaboration in Knowledge Management Systems to Improve Organizational Efficiency and Learning

Hamza Yusuf
https://orcid.org/0009-0007-8029-5660
RMIT University, Australia

Annisa Fitri
https://orcid.org/0009-0005-0942-7443
University of Melbourne, Australia

Abdur Razzaq
https://orcid.org/0000-0002-8267-5700
Universitas Islam Negeri Raden Fatah Palembang, Indonesia

ABSTRACT

In today's knowledge driven economy, effective information management is key to organizational success. Knowledge Management Systems (KMS) have evolved from static repositories into dynamic ecosystems enabling knowledge creation, sharing, and application. AI integration enhances KMS by boosting collaboration, informa-

DOI: 10.4018/979-8-3373-3404-2.ch002

tion access, and decision making. This study explores how AI driven KMS improve efficiency, learning, and adaptability via a qualitative design using secondary data from academic and professional sources. It highlights AI's role in transforming knowledge processes, with human-AI collaboration driving continuous learning. Microsoft Viva exemplifies AI streamlining knowledge discovery in hybrid work settings. AI based KMS enhance performance and offer strategic value by turning information into actionable insights. AI-human interaction is vital for adaptive, knowledge centered organizations to sustain competitiveness.

1. INTRODUCTION

In the digital age, knowledge has become a central element of organizational success (Alavi & Leidner, 2001). The ability to manage information effectively determines how well a company can respond to market changes, make informed decisions, and maintain long term competitiveness (Pinho et al., 2012). Knowledge Management Systems (KMS) play an essential role in supporting these outcomes by helping organizations collect, organize, share, and apply knowledge in structured ways (Alavi & Leidner, 2001). With most organizations now operating in technology driven environments, KMS provide an important foundation for collaboration, problem solving, and efficient knowledge sharing across dispersed teams (Cerchione et al., 2024).

AI has added new dimensions to how knowledge is created, stored, and accessed (Dwivedi et al., 2021). Within the knowledge economy, where decision making depends on accurate and timely information, the combination of KMS and AI has become increasingly important (Dogru & Keskin, 2020). AI supports KMS by identifying patterns, processing large datasets, and helping users locate relevant information more efficiently (Wang et al., 2022). This connection between AI and KMS allows organizations to convert existing data into practical insights, improving how knowledge is applied to daily operations and strategic planning (Banasiewicz, 2021). As a result, KMS are now viewed as part of an integrated digital ecosystem that links human understanding with machine assisted analysis (Mahmood, 2025).

Earlier generations of KMS focused mainly on collecting and storing information for future use (Alavi & Leidner, 2001). These traditional systems were often static, with limited capacity to interpret or connect knowledge across different contexts (Pinho et al., 2012). Current developments, supported by AI technologies such as natural language processing and semantic analysis, have made it possible for KMS to handle more dynamic forms of knowledge (Wang et al., 2022). Modern systems can suggest relevant materials, connect related information sources, and support users through context based recommendations (Dwivedi et al., 2021). This shift represents a move from storage oriented systems toward adaptive knowledge environments that respond to user needs and organizational goals (Cerchione et al., 2024).

The purpose of this paper is to examine how AI driven Knowledge Management Systems contribute to efficiency, learning, and collaboration in organizations. It aims to understand how these systems function as part of wider digital strategies that support knowledge sharing and continuous improvement (Vining et al., 2022). The study also considers how human interaction with AI shapes knowledge development, and what implications this has for workforce skills and management practices (Kim, 2022).

2. LITERATURE REVIEW

In recent years, organizations have increasingly recognized that the management of knowledge is not just an administrative task but a strategic capability that drives adaptability and innovation (Pinho et al., 2012). As workplaces become more distributed and data intensive, Knowledge Management Systems have evolved to meet new expectations for speed, accessibility, and contextual awareness (Cerchione et al., 2024). This section examines the evolution of KMS, the integration of AI in knowledge processes, the interaction between humans and intelligent systems, and the emerging trends shaping modern knowledge ecosystems.

2.1 Knowledge Management Systems

Knowledge Management Systems (KMS) have evolved significantly since their early introduction in organizational practice (Alavi & Leidner, 2001). The initial purpose of KMS was to collect and store knowledge in centralized databases, allowing information to be retrieved when needed (Alavi & Leidner, 2001). These systems primarily focused on explicit knowledge, which could be documented and transferred easily (Pinho et al., 2012). Over time, however, scholars and practitioners recognized that knowledge also exists in tacit forms, embedded in employees' experiences, skills, and social interactions (Cerchione et al., 2024). This realization broadened the function of KMS from mere information storage to mechanisms that facilitate communication, collaboration, and organizational learning (Banasiewicz, 2021).

As organizations expanded globally and became more dependent on digital processes, the need for systems capable of integrating knowledge across geographic and functional boundaries grew (Pinho et al., 2012). The development of intranets, collaborative software, and enterprise portals during the 2000s represented a major step toward connected knowledge environments (Alavi & Leidner, 2001). These technologies enabled employees to share best practices, access lessons learned and contribute new ideas in real time (Cerchione et al., 2024). KMS became central to building organizational memory and preserving knowledge continuity in fast changing industries (Banasiewicz, 2021).

Modern KMS now operate as multi layered environments that incorporate communication tools, learning platforms, and analytics (Mahmood, 2025). They are designed to support both individual and collective learning through shared access to knowledge assets (Pinho et al., 2012). Instead of serving as isolated repositories, they function as ecosystems that connect data, processes, and human expertise (Cerchione et al., 2024). This transformation reflects a shift from managing documents to managing relationships between people and knowledge (Alavi & Leidner, 2001). The focus has therefore moved toward enabling collaboration, contextual understanding, and problem solving, particularly within hybrid and distributed work settings (John et al., 2022).

2.2 Integration of AI in Knowledge Management

AI has become a critical factor in extending the capacity of KMS (Dwivedi et al., 2021). Traditional systems required manual input for organizing and retrieving information, which was time consuming and limited in scope (Alavi & Leidner, 2001). AI technologies such as natural language processing, semantic analysis, and machine learning now automate many of these tasks (Wang et al., 2022). They enable systems to process large volumes of unstructured data, recognize underlying themes, and suggest relevant resources to users (Pai et al., 2022). This automation allows knowledge to be delivered more efficiently and reduces the cognitive burden associated with information search and retrieval (Dogru & Keskin, 2020).

One of the most important contributions of AI to KMS is the ability to understand context (Wang et al., 2022). AI powered systems can interpret how knowledge is used and adjust recommendations accordingly (Dwivedi et al., 2021). For example, they can identify frequently discussed topics, link them to associated projects, and provide employees with related documents or expert contacts (Pai et al., 2022). This capability allows knowledge to be surfaced within the user's workflow rather than requiring them to navigate multiple systems (John et al., 2022). In doing so, AI changes the nature of knowledge management from a reactive activity to a proactive one (Mahmood, 2025).

AI also supports decision making by synthesizing insights from different data sources (Dogru & Keskin, 2020). Predictive analytics can identify future information needs based on emerging trends or organizational changes (Alshadoodee et al., 2022). Such capabilities are especially useful in fast paced industries where timely access to information can influence performance and innovation (Dwivedi et al., 2021). However, effective use of AI within KMS requires proper data governance, as the quality of outcomes depends on the accuracy and consistency of underlying information (Vining et al., 2022). When well implemented, AI based KMS can significantly enhance organizational responsiveness and strategic planning (Banasiewicz, 2021).

2.3 Human-AI Collaboration in Knowledge Processes

Human-AI collaboration is increasingly viewed as a key feature of effective knowledge management. AI systems excel in processing large scale data and identifying hidden relationships, but human expertise remains essential for interpretation and decision making (Kim, 2022). The value of knowledge depends not only on access to information but also on the ability to understand its relevance to specific situations (Pinho et al., 2012). In this sense, the interaction between human reasoning and machine intelligence creates a complementary relationship that supports deeper understanding (Vining et al., 2022).

In practice, this collaboration manifests through feedback loops. Humans train AI models through their actions and corrections, while AI supports human tasks by identifying relevant information and automating routine work (Wang et al., 2022). This process allows both the system and its users to improve over time (Kim, 2022). It also encourages reflective learning, where employees gain new insights from system generated patterns (Banasiewicz, 2021). For organizations, this interaction enhances the adaptability of knowledge systems and ensures that learning becomes an ongoing process integrated into everyday work (Cerchione et al., 2024).

However, for collaboration to succeed, organizations must develop trust in AI systems (Vining et al., 2022). Transparency in how AI functions and clarity about its decision making processes are necessary for users to rely on its outputs (Dwivedi et al., 2021). Training and digital literacy are equally important, as employees must understand how to evaluate AI recommendations critically (Kim, 2022). When these conditions are met, human-AI collaboration transforms KMS into environments where knowledge creation, sharing, and refinement occur continuously across both human and technological components.

2.4 Rising Trends and Research Perspectives

Current research highlights several emerging directions in knowledge management (Cerchione et al., 2024). One significant trend is the movement toward intelligent ecosystems that combine data integration, AI analytics, and collaboration tools (Mahmood, 2025). These systems not only capture and distribute information but also learn from user interactions to improve their recommendations (Wang et al., 2022). They enable organizations to manage knowledge dynamically, ensuring that relevant insights remain accessible as conditions change (Cerchione et al., 2024). This approach is particularly beneficial for hybrid and remote work models, where knowledge must flow smoothly across virtual environments (John et al., 2022).

Another trend involves the growing use of AI driven personalization (Pai et al., 2022). Systems now have the ability to adapt to individual learning preferences and work patterns, offering customized access to knowledge resources (Wang et al., 2022). This personalization improves engagement and reduces information overload by focusing on materials most relevant to each user (Dwivedi et al., 2021). Additionally, organizations are exploring cross platform integration, where KMS connect directly with communication tools, project management systems, and data visualization software (John et al., 2022). This ensures that knowledge is embedded within operational workflows rather than existing as a separate system (Mahmood, 2025).

At the same time, ethical and governance challenges are attracting attention in the literature (Vining et al., 2022). The use of AI in KMS raises questions about data privacy, algorithmic transparency, and accountability (Dwivedi et al., 2021). Scholars argue that responsible implementation is essential to maintain trust and fairness (Vining et al., 2022). Future research is expected to focus on balancing efficiency with ethical considerations, ensuring that intelligent KMS not only improve performance but also uphold values such as inclusivity, security, and user empowerment (Kim, 2022). These trends indicate that KMS will continue to evolve as adaptive, human centered systems designed to support learning and knowledge sharing in increasingly digital workplaces (Cerchione et al., 2024).

3. METHODOLOGY

This section discusses the research design, data sources, and analytical framework used to examine the role of AI in developing KMS. Since the topic involves understanding conceptual developments, organizational practices, and technological integration, a qualitative approach was selected (Selvi, 2019). The study aims to explore how AI based KMS improve collaboration, learning, and decision making, and how these systems contribute to strategic knowledge use in modern organizations (Dodla & Jones, 2023). The Microsoft Viva platform is used as a case study to illustrate these principles in action within the hybrid work environment (John et al., 2022).

3.1 Research Design

The study uses qualitative research design grounded in secondary data analysis (Selvi, 2019). This approach is suitable for examining conceptual and contextual relationships rather than numerical trends (Selvi, 2019). By analyzing academic studies, industry reports, and professional documentation, the research identifies how AI integration transforms knowledge management processes and what implications this holds for organizational efficiency and learning (Dodla & Jones, 2023). A qualitative approach allows for interpretation of meanings and underlying mechanisms that cannot be easily captured through quantitative metrics (Selvi, 2019).

This design is particularly relevant to the study's exploratory nature (Selvi, 2019). The goal is not to measure specific outcomes but to understand how AI driven KMS function, evolve, and influence organizational practices (Dodla & Jones, 2023). The research builds on conceptual synthesis and interpretation, offering a comprehensive understanding of the link between knowledge systems, AI technologies, and organizational performance (Banasiewicz, 2021). This interpretive orientation aligns with prior studies in knowledge management that emphasize understanding systems within their social and technological contexts (Alavi & Leidner, 2001).

3.2 Data Collection and Selection Criteria

The study relies on secondary data from academic journals, books, and professional reports (Selvi, 2019). Sources include scholarly research on knowledge management, digital transformation, and AI applications, as well as industry publications from Microsoft, Gartner, and McKinsey (Dodla & Jones, 2023). Selection was based on relevance, credibility, and recency (Selvi, 2019). Only materials published between 2000s and 2025 were considered to ensure that the discussion reflects current developments in AI enabled KMS and hybrid work environments (John et al., 2022).

3.3 Analytical Approach

A qualitative content analysis method was employed to synthesize information across the selected literature (Selvi, 2019). This method involves identifying key concepts, comparing perspectives, and grouping findings into thematic categories (Selvi, 2019). The analysis focused on four major themes: (1) the evolution of KMS, (2) the integration of AI technologies, (3) human-AI collaboration, and (4) organizational outcomes such as learning and strategic value creation (Dodla & Jones, 2023). Through iterative reading and coding, patterns were identified regarding how

AI enhances knowledge accessibility, workflow efficiency, and decision making quality (Banasiewicz, 2021).

The Microsoft Viva case is incorporated as an illustrative example rather than the central object of study (John et al., 2022). It is used to demonstrate how theoretical concepts are applied in real organizational settings (John et al., 2022). The case helps visualize how AI based tools contribute to knowledge discovery, reduce information fragmentation, and enable knowledge flow in hybrid workplaces (John et al., 2022). This interpretive approach allows for linking abstract ideas with observable organizational practices, reinforcing the theoretical arguments made in the literature review (Selvi, 2019).

4. ANALYSIS AND DISCUSSION

This section focuses on how AI driven KMS operate in real organizational settings, using Microsoft Viva as a representative example (John et al., 2022). It explores how Viva integrates AI to enhance knowledge accessibility, streamline collaboration, and strengthen human-machine interaction (John et al., 2022). The discussion extends to the broader implications of AI integration for knowledge flow, workforce adaptability, and strategic value creation within digital organizations (Mahmood, 2025).

4.1 AI integration in Microsoft Viva

Microsoft Viva serves as a contemporary example of how organizations apply AI within KMS to manage knowledge in hybrid work environments (John et al., 2022). Designed to integrate with Microsoft Teams and other Microsoft 365 applications, Viva provides a unified workspace that combines communication, content management, and learning (John et al., 2022). Its AI powered feature, Viva Topics, scans corporate data sources to identify recurring subjects, such as project names or product lines, and automatically generates topic cards (John et al., 2022). These cards summarize information, list related files, and connect users to internal experts who possess relevant experience (John et al., 2022). This process reduces the fragmentation that often occurs in digital collaboration tools and ensures that critical knowledge is easily accessible to all employees (John et al., 2022).

Viva also includes Viva Answers, which centralizes the exchange of questions and expertise (John et al., 2022). When an employee submits a query, the system uses AI to identify suitable experts and suggest existing answers from previous interactions (John et al., 2022). This not only accelerates problem solving but also transforms informal conversations into formalized knowledge assets that others can reuse (John et al., 2022). The integration across SharePoint, OneDrive, and Teams creates a seamless knowledge layer that spans multiple data sources (John et al., 2022). As a result, employees interact with knowledge directly in the course of their daily tasks rather than searching through separate systems (John et al., 2022).

4.2 Knowledge Flow, Collaboration, and Productivity

Viva's design demonstrates how AI can enhance the flow of knowledge across large, distributed organizations (John et al., 2022). By linking diverse information sources, it helps overcome the challenge of knowledge silos that commonly emerge in hybrid workplaces (John et al., 2022). Employees benefit from a more consistent and contextual access to knowledge, which supports both individual efficiency and team collaboration (John et al., 2022). In addition, the platform aids new employees in quickly understanding organizational processes and resources, reducing the time required for onboarding and adaptation (John et al., 2022).

The automation of knowledge discovery allows workers to focus more on critical thinking, creativity, and decision making (Pai et al., 2022). Viva's approach illustrates that AI driven KMS can act as facilitators of collaboration rather than replacements for human expertise (John et al., 2022). By connecting employees to the right information at the right time, the system enhances productivity without increasing cognitive workload (Dogru & Keskin, 2020). Over time, this process contributes to a more informed and agile workforce, capable of responding effectively to change and complexity in the digital business environment (Mahmood, 2025).

4.3 Continuous Learning and Human-AI Interaction

The interaction between humans and AI within Microsoft Viva represents a continuous learning process (John et al., 2022). Employees contribute knowledge through documents, discussions, and feedback, which the system analyzes to refine its recommendations (John et al., 2022). AI, in turn, exposes users to related topics and resources that expand their understanding (John et al., 2022). This reciprocal exchange supports both organizational learning and personal development (Banasiewicz, 2021). It also aligns with the concept of knowledge ecosystems, where learning evolves through repeated cycles of human contribution and machine supported synthesis (Cerchione et al., 2024).

Another important aspect of this interaction is the reinforcement of collective intelligence. By surfacing expert knowledge and connecting people with similar interests, Viva encourages cross departmental collaboration (John et al., 2022). Employees not only consume knowledge but also contribute to its development, creating a self sustaining system that grows in relevance and accuracy over time (John et al.,

2022). This collaboration fosters a culture of openness and learning that strengthens both human and technological components of the organization (Kim, 2022).

4.4 Strategic Value and Organizational Transformation

Beyond operational improvements, AI driven KMS deliver long term strategic value by reshaping how organizations use knowledge as a competitive asset (Mahmood, 2025). Systems such as Viva transform information from a static resource into a continuously evolving source of strategic insight (John et al., 2022). By analyzing patterns in communication and content usage, AI helps identify emerging trends, expertise gaps, and potential areas for innovation (Dogru & Keskin, 2020). These insights inform leadership decisions, allowing organizations to anticipate challenges and align their strategies with evolving market needs (Alshadoodee et al., 2022).

The integration of AI within KMS also supports organizational transformation by embedding learning and adaptability into everyday workflows (Banasiewicz, 2021). As employees interact with AI enhanced systems, they develop digital skills and analytical thinking that strengthen the organization's overall capability (Kim, 2022). This contributes to resilience in rapidly changing environments where continuous learning is essential (Cerchione et al., 2024). From a management perspective, such systems enable leaders to base decisions on real time knowledge rather than historical data, making planning more responsive and evidence based (Alshadoodee et al., 2022).

Furthermore, AI driven KMS enhance the scalability of organizational knowledge (Mahmood, 2025). They allow expertise developed in one part of the business to be quickly disseminated across others, improving consistency and reducing duplication of effort (John et al., 2022). In doing so, they not only support productivity but also reinforce a culture where knowledge is viewed as a shared organizational resource (Pinho et al., 2012). Over time, this creates a more connected, intelligent, and adaptive enterprise that can sustain competitive advantage in the evolving digital economy (Cerchione et al., 2024).

5. CONCLUSION

The integration of AI into Knowledge Management Systems (KMS) marks a turning point in how organizations capture, share, and apply knowledge (Dwivedi et al., 2021). The findings of this study indicate that AI driven KMS transform knowledge management from a static process into a dynamic, continuous, and adaptive system that operates across human and technological boundaries (Cerchione et al., 2024). These systems no longer act only as repositories for organizational memory but as intelligent partners that facilitate communication, learning, and collaboration in real time (Mahmood, 2025).

Through the case of Microsoft Viva, it becomes clear that AI plays a central role in overcoming traditional barriers to knowledge sharing, particularly in hybrid and distributed work environments (John et al., 2022). Viva's ability to automatically identify key topics, connect employees to expertise, and embed knowledge within daily workflows demonstrates how AI enables seamless information access (John et al., 2022). This automation improves efficiency while ensuring that knowledge remains contextual and relevant (Pai et al., 2022). Importantly, AI does not replace human decision making but strengthens it by providing structured insights that help employees act with greater confidence and accuracy.

Beyond operational benefits, the study highlights the strategic impact of AI powered KMS (Mahmood, 2025). By identifying patterns and linking knowledge across departments, these systems enable leaders to make informed decisions and anticipate future needs (Alshadoodee et al., 2022). The result is a shift toward data informed management and evidence based strategy formulation (Dogru & Keskin, 2020). Moreover, as employees engage with AI systems, they develop digital literacy and analytical competencies that contribute to long term organizational adaptability (Kim, 2022). This aligns with the growing need for continuous learning in a digital economy where innovation and responsiveness define success (Cerchione et al., 2024).

The interaction between humans and AI also establishes a feedback loop that enhances both system intelligence and user capability. As employees provide input and feedback, AI systems learn to refine their recommendations, while users gain access to new perspectives and resources that expand their knowledge base (Wang et al., 2022). This reciprocal relationship creates a self improving knowledge ecosystem that supports both individual development and collective intelligence (Banasiewicz, 2021). In such an environment, knowledge becomes an evolving organizational asset rather than a static collection of information (Cerchione et al., 2024).

In conclusion, AI driven KMS represent more than a technological advancement; they signify a strategic transformation in how organizations operate and compete (Mahmood, 2025). Their ability to integrate knowledge, automate analysis, and support collaboration positions them as foundational tools for the modern enterprise (Dwivedi et al., 2021). However, successful implementation depends on balancing automation with human insight and maintaining transparency in how AI is used (Vining et al., 2022). Organizations that invest in these systems with a clear focus on trust, governance, and learning culture will be best positioned to leverage knowledge as a sustainable source of value in the evolving digital economy (Kim, 2022).

REFERENCES

Alavi, M., & Leidner, D. E. (2001). Review: Knowledge Management and Knowledge Management Systems: Conceptual Foundations and Research Issues. *Management Information Systems Quarterly*, *25*(1), 107–136. DOI: 10.2307/3250961

Alshadoodee, H. A. A., Mansoor, M. S. G., Kuba, H. K., & Gheni, H. M. (2022). The role of artificial intelligence in enhancing administrative decision support systems by depend on knowledge management. *Bulletin of Electrical Engineering and Informatics*, *11*(6), 3577–3589. DOI: 10.11591/eei.v11i6.4243

Banasiewicz, A. (2021). Organizational Learning in the Age of Data. In *EAI/Springer Innovations in Communication and Computing*. Springer International Publishing., DOI: 10.1007/978-3-030-74866-1

Cerchione, R., Centobelli, P., Borin, E., Usai, A., & Oropallo, E. (2024, November 19). Piera Centobelli, Borin, E., Usai, A., & Oropallo, E. (2024). The WISED knowledge-creating company: Rethinking SECI model in light of the digital transition. *Journal of Knowledge Management*, *28*(10), 2997–3022. Advance online publication. DOI: 10.1108/JKM-02-2024-0133

Dodla, T. R., & Jones, L. A. (2023). Identifying knowledge management strategies for knowledge management systems. *Access Journal - Access to Science, Business. Innovation in the Digital Economy*, *4*(2), 261–277. DOI: 10.46656/access.2023.4.2(8)

Dogru, A. K., & Keskin, B. B. (2020). AI in operations management: Applications, challenges and opportunities. *Journal of Data, Information and Management*, *2*(2), 67–74. DOI: 10.1007/s42488-020-00023-1

Dwivedi, Y. K., Hughes, L., Ismagilova, E., Aarts, G., Coombs, C., Crick, T., Duan, Y., Dwivedi, R., Edwards, J., Eirug, A., Galanos, V., Ilavarasan, P. V., Janssen, M., Jones, P., Kar, A. K., Kizgin, H., Kronemann, B., Lal, B., Lucini, B., & Williams, M. D. (2021). Artificial Intelligence (AI): Multidisciplinary Perspectives on Emerging challenges, opportunities, and Agenda for research, Practice and Policy. *International Journal of Information Management*, *57*(101994), 101994. DOI: 10.1016/j.ijinfomgt.2019.08.002

John, B., Alsamarra'l, Z., & Panteli, N. (2022). Enhancing Employee Experience in the Era of Hybrid Work: The Case of Microsoft Viva. *IEEE Software*, *40*(2), 1–9. DOI: 10.1109/MS.2022.3229956

Kim, S. (2022). Working With Robots: Human Resource Development Considerations in Human–Robot Interaction. *Human Resource Development Review*, *21*(1), 153448432110688. DOI: 10.1177/15344843211068810

Mahmood. (2025). AI-Driven Knowledge Ecosystems: Transforming Continuous Innovation in Cloud Data Engineering. *Journal of Computer Science and Technology Studies, 7*(8), 401–407. https://doi.org/DOI: 10.32996/jcsts.2025.7.8.44

Pai, R. Y., Shetty, A., Shetty, A. D., Bhandary, R., Shetty, J., Nayak, S., Dinesh, T. K., & D'souza, K. J. (2022). Integrating artificial intelligence for knowledge management systems – synergy among people and technology: A systematic review of the evidence. *Ekonomska Istrazivanja, 35*(1), 1–23. DOI: 10.1080/1331677X.2022.2058976

Pinho, I., Rego, A., & Pina e Cunha, M. (2012). Improving knowledge management processes: A hybrid positive approach. *Journal of Knowledge Management, 16*(2), 215–242. DOI: 10.1108/13673271211218834

Selvi, A. F. (2019). Qualitative content analysis. *The Routledge Handbook of Research Methods in Applied Linguistics*, 440–452. https://doi.org/DOI: 10.4324/9780367824471-37

Vining, R., McDonald, N., McKenna, L., Ward, M., Doyle, B., Liang, J., Hernandez, J., & Guilfoyle, J. Arwa Shuhaiber, Geary, U., Fogarty, M., & Brennan, R. (2022). Developing a Framework for Trustworthy AI-Supported Knowledge Management in the Governance of Risk and Change. *Lecture Notes in Computer Science*, 318–333. https://doi.org/DOI: 10.1007/978-3-031-17615-9_22

Wang, X., Zhang, X., Cheng, Y., Tian, F., Chen, K., & Ordóñez, P. (2022). Artificial Intelligence-Enabled Knowledge Management. *Routledge EBooks*, 153–168. https://doi.org/DOI: 10.4324/9781003112150-13

Chapter 3
Synergizing Human Intelligence With AI and Robotics for Organizational Growth and Success

L. B. Muralidhar
https://orcid.org/0000-0003-3453-613X
Jain University, India

R. Shilpa
https://orcid.org/0009-0002-9805-1481
RJS Institute of Management Studies, India

N. Sathyanarayana
https://orcid.org/0000-0002-4185-7751
Jain University, India

N. Rakesh
https://orcid.org/0000-0002-6400-4954
Jain University, India

R. Arun
https://orcid.org/0009-0001-1571-3201
Jain University, India

DOI: 10.4018/979-8-3373-3404-2.ch003

ABSTRACT

This chapter explores the integration of human intelligence with artificial intelligence (AI) and robotics to drive organizational growth and success. It examines the evolution from traditional organizational methods to the adoption of advanced technologies, emphasizing how AI and robotics complement human capabilities. The synergy between human creativity, empathy, and decision-making, and the precision, scalability, and automation offered by AI and robotics, enhances operational efficiency, innovation, and decision-making processes. Practical strategies for effective integration, leadership roles, and ethical considerations are discussed. The chapter outlines how organizations can strategically blend these forces to foster sustainable growth, innovation, and competitive advantage in a rapidly evolving business landscape.

1. INTRODUCTION

The incorporation of AI and robotics into organizational structures signals a fundamental shift away from mechanistic models to a more fluid, networked approach to work, strategy, and value generation—a reconception of what work entails on all levels. As the digital economy becomes more present, the focus on human cognition, emotional intelligence, and ethical reasoning is enhanced, not replaced, by the computational power of machines. This man-and-machine hybrid system inverts traditional roles, where human judgment aligns with machine efficiency. For example, in the healthcare domain, the machine precision of robotic-assisted surgeries is combined with human expertise to yield better results than either alone (Yang et al., 2017), and in finance, algorithmic trading is controlled under human supervision to maintain ethical values. The potential importance of agency relations between humans and machines have been highlighted by theories such as socio-technical systems theory (Trist & Emery, 1973) and Actor-Network Theory (Latour, 2005), which stress the mutual shaping of human and machine roles with roles, agency or responsibility being a matter of joint enactment, not being accommodated to either one or the other.

Nevertheless, the AI advance raises fears of job loss, especially in professions that can be automated, and some people believe these losses will be catastrophic (Frey & Osborne, 2017). However, work such as Bessen (2019) also suggests that, in knowledge-intensive industries, AI is as much a complement to humans as a substitute for them, thereby tempering the breathless narrative of its potential. Policies, such as the European Commission's "Coordinated Plan on Artificial Intelligence" and projects in Singapore and India, prioritize striking a balance between innovation and human-centered values, including upskilling and inclusivity (European Commission, 2021; Smart Nation and Digital Government Office, 2019).

Cultural values also influence the use of AI, such as carebots in Japan, due to the country's aging society and the pursuit of human-machine harmony (Robertson, 2018). In the West, on the other hand, AI is viewed with distrust, and the focus is more on job security and the morality of automation. These contrasting views highlight the importance of culturally sensitive AI strategies that align with local customs and values. Despite the benefits of AI's increased performance, however, ethical concerns remain, especially with algorithmic bias. The opaqueness concerns are that AI systems are "black-boxes," and operate in ways that are not open to verification and can therefore lead to discrimination due to lack of transparency and accountability (Buolamwini & Gebru, 2018).

Challenges like these require robust human supervision and regulation to ensure that AI is used to benefit society and is fair. With AI and robotics embedded in daily work and contributing directly to your bottom line, leadership is key to steering the company through this maze. The adaptive leadership required to deal with technological change is the leadership that enables the management of uncertainty and the mobilization of collective intelligence (Heifetz & Linsky, 2002).

Also important in the integration of AI is to ensure organisational coherence (O'Reilly & Tushman, 2013), which can be achieved through ambidextrous organisations that balance innovation and operational efficiency—the role of training organizations and professional organizations. Training organizations and professional organizations play a crucial role in shaping a workforce that is both technically competent and ethically principled, with an emphasis on critical thinking and ethical decision-making. In this rapidly evolving landscape, the future of productivity does not involve machines replacing humans, but instead developing ecosystems where one complements the other, cultivating resilience, inclusivity, and intelligent, sustainable ecosystems.

2. STATEMENT OF THE PROBLEM

The convergence of human intelligence, AI, and robotics in enterprises presents an unresolved, complex, and multifaceted problem with numerous gaps and challenges. The sweet spot would be a convergence of human cognition, emotional intelligence, and technological precision that 'transforms' the (operational) climate, decision-making ability, and ethical foresight in the decision-management process without a glitch. However, the converse is also true with systemic bias in AI decision-making, a lack of human-machine interaction, and ethical challenges related to job loss, control, and data privacy. These challenges suggest that substantial gaps remain in the theory of AI, particularly in reconciling AI autonomy with human oversight and respecting cultural and economic diversity during its deployment.

There are still gaps in knowledge regarding the formative processes of governance frameworks that adequately balance innovation with ethical considerations, as well as the long-term consequences of using AI and robotics on employee welfare and organizational identity. There are also contextual gaps to consider, as the policies that should support AI use are, on the whole, only partially formed in most settings, often in ways that exacerbate any uneven access to technology. Finally, real-world challenges of integrating technology development with human resource strategies, training, and leadership models remain considerable.

This book chapter aims to bridge this gap by providing a comprehensive understanding of how organizations can address the ethical, theoretical, and practical challenges associated with AI-human collaboration. This document aims to inform the efforts of future researchers and policymakers interested in developing human-centric artificial intelligence strategies, governance models, and ethical principles. By combining case studies, empirical knowledge, and theoretical approaches, it is intended that organizations create systems that integrate human creativity and emotional intelligence with AI-generated automation, promoting operational excellence while fostering a future of work that is both equitable and sustainable.

3. RESEARCH OBJECTIVES

- To explore the synergies and tensions between human cognitive and emotional intelligence and AI/robotic systems.
- To examine the ethical, cultural, and organizational challenges of AI-human integration.
- To identify and analyse the knowledge, theoretical, and practical gaps in current AI-human collaboration models.

4. METHODOLOGY

First, a comprehensive literature review was conducted to establish a solid theoretical foundation, drawing on socio-technical systems theory (Trist & Emery, 1973), human-computer interaction (Norman, 2013), actor-network theory (Latour, 2005), and hybrid intelligence models (Daugherty & Wilson, 2018). These models offered some insight into the application of AI and robotics to augment human intelligence in organizational settings. They then examined case studies across healthcare, finance, and manufacturing to offer real-world examples of human-AI collaboration, including AI-fueled diagnostics in healthcare or robotic process automation in banking.

Primary and secondary data from industry reports and academic studies provided quantitative insight into the performance and impact of AI implementation. A conceptual framework emerged, focusing on the relationship between human-machine complementarity, receptive leadership, board oversight, ethical leadership, adaptive leadership, and organizational culture. This approach was refined through numerous rounds of feedback from peers and experts, resulting in a chapter that combines theoretical robustness with practical insight, informing organizational decision-making on how to successfully adopt AI and robotics and scale for growth and success.

4.1 Origin and Evolution of the Concepts

The integration of HI and AI, rather than the comparison of HI versus AI, has progressed through different stages, from early automation to a hybrid system that enables human cognition and machine precision. This convergence was first witnessed with the Industrial Revolution, which saw, for the first time, the meaning of work transform with the shift towards mechanization, labour markets, and organizational structures. It was only relatively recently, in the mid-20th century, that machines were perceived as a practical complement to human intelligence in organizations.

The concept of automation first emerged during the Industrial Revolution, when machines began to replace human labor in tasks that were monotonous and repetitive. The intention behind the first mechanization process was to increase productivity and reduce the physical effort required for work (Frey & Osborne, 2017). However, it was still purely mechanized and had no involvement in cognitive or decision-making processes. The invention of the computer in the late 1940s and 1950s, combined with early theories of artificial intelligence, set the stage for the application of computers to higher-level cognitive processing.

The formal academic discipline of AI was established in the 1950s, with the groundbreaking work of scholars such as Alan Turing, who introduced the concept of machine intelligence and the well-known Turing Test. This represented the first step toward imagining machines that could perform tasks usually associated with human intelligence, such as solving problems, learning new things, and adapting to new situations. Over the next few decades, AI research focused on developing algorithms that could emulate specific cognitive functions, including natural language processing (NLP), expert systems, and machine learning (Russell & Norvig, 2016).

Meanwhile, robotics, which was initially mechanical, integrated higher-order functions such as perception and control. Early robots, such as the Unimate 18, created in the 1960s for industrial use, were repetitive and straightforward (Hodges, 2017) – they were designed to assemble products on factory lines. However, these robots lacked intelligence; they could not interact with the environment as effectively as a human worker. The true synergy between human intelligence and AI/robotics emerged in the late 20th and early 21st centuries, driven by advances in computational power, machine learning, and robotics.

With the shift in AI from knowledge-based systems to machine learning (ML) and deep learning (DL) models, its scope was dramatically extended from automated pattern recognition and predictive analytics to decision-making in complex environments. Robotics continued to advance, with cobots able to work and collaborate with human employees on a shared floor and to adjust to human movements in real-time. This change began to close the cognitive gap between human workers and machines, moving automation from a tool to a partnership between human and machine intelligence.

Among the most groundbreaking models for human-AI collaboration is to consider them as part of a 'fusion skills' model, as suggested by Daugherty and Wilson (2018). This model underscores the need for the next-generation workforce to excel in 'human and machine teaming' -human intelligence enhancing AI's precision and scalability. This kind of cooperation, rather than replacement, came to occupy the centre of the discussion on the AI invasion of workplaces, rebutting previous worries about extensive job loss. This "symbiotic intelligence", leveraging the strengths of humans and machines for mutual advantage, was a guiding principle when we discussed how organisations can grow (Haselberger, 2020).

5. THEORETICAL AND CONCEPTUAL FRAMEWORK

Human Intelligence Enrichment Artificial Intelligence (AI) and Robotics in an Organizational Context. Several conceptual theories underpin the integration of human intelligence with AI and robotics within an organizational context, which, when combined, provide an overview of best practices for ensuring an optimal partnership between humans and machines. Developed by Trist and Emery (1973), socio-technical systems theory suggests that the success of an organization does not come from one subsystem—human or machine—taking precedence, but rather from a harmonious relationship between social and technical systems. In contrast, AI and robotics automate rote functions while leaving people in control of vital decisions, particularly in ethically ambiguous or context-dependent settings (Daugherty & Wilson, 2018).

Likewise, HCI theory emphasizes user-friendly, natural interfaces in AI systems, allowing humans who operate them to interact with, comprehend, and intervene in AI-driven routines without being overwhelmed (Norman, 2013). For example, the precision of robotic-assisted surgery is enhanced in healthcare; however, additional human interpretability of data and consideration of ethical issues remain (Yang et al., 2017). Actor-Network Theory (ANT), as proposed by Latour (2005), also challenges the dichotomy of human and machine roles, but this time it emphasizes both as actants in organizational networks. This approach advocates for a shift in perspective, departing from the traditional view of machines as mere tools and acknowledging their active role in the decision-making process, particularly in ethical decisions where responsibility is shared among machines, developers, and the organizations that deploy them (Pasquale, 2015).

Extending these approaches, hybrid intelligence, as introduced by Dellermann et al. (2021), highlights the synergies between human and machine abilities. The latter enhances the data processing and efficient performance of the former, supporting human creativity, emotional intelligence, and ethical reasoning. The fusion skills model (Daugherty & Wilson, 2018), for example, argues for developing problem-solving skills that enable workers to collaborate with AI to achieve an optimized, ethical, and beneficial work experience.

The theory of extended cognition (Clark & Chalmers, 1998) extends this understanding, stating that human cognitive processes extend into artificial tools, thereby allowing AI to enhance human decision-making. A related computationalist position, referred to as computational rationality, posits that AI's data-handling capabilities help overcome cognitive limitations, such as bounded rationality, and support more well-informed and efficient decision-making processes (Simon, 1957).

Cumulatively, these theories provide a strong foundation for integrating human intelligence with AI and robotics, suggesting that organizations can benefit from the combined advantages of human and machine capabilities. This ethical integration can generate greater productivity, and, more importantly, perhaps a future, movable, and inclusive environment for human-AI partnership, in which both sides contribute to synergizing value in a reflexive and adaptive space.

6. UNDERSTANDING HUMAN INTELLIGENCE AND ITS ROLE IN ORGANIZATIONS

It is no secret that the key to an organisation's success has always been related to human intelligence, encompassing not just cognitive abilities such as reasoning and memory, but also emotional intelligence (EI), which has become increasingly crucial in leadership and fostering an integrated company culture. These theories, such as Sternberg's triarchic model (1985), expand our understanding of intelligence by recognizing creativity and practical intelligence as crucial to leadership in complex, culturally diverse organizations like Unilever and SAP. Highlighted by Goleman (1995), emotional intelligence (EI) is becoming an accepted construct for understanding the determinants of leadership effectiveness. This is also demonstrated in Project Oxygen at Google, where EI proved a better predictor of management success than technical skills (Garvin et al., 2013).

However, human intelligence is also" imperfect, to say the least. Cognitive illusions, such as confirmation bias and anchoring, as discussed by Kahneman and Tversky (1979), are frequently observed, leading to critical errors in judgment. This is evident in cases like the corporate collapses of Enron and Lehman Brothers, where flawed risk perceptions substantially influenced the outcomes. Bounded rationality (Simon, 1957) introduces an additional layer of complexity to decision-making, as individuals often make satisfactory decisions when cognitive resources are limited, a scenario that frequently occurs in fast-paced environments where rapid and accurate decision-making is crucial.

These limits were revealed during the COVID-19 pandemic, with leaders such as Jacinda Ardern demonstrating how it is possible to balance cognitive and emotional challenges through empathetic leadership. In contrast, others failed due to either overconfidence or poor communication (Wilson & Wilson, 2020). Given these deficiencies, AI and predictive analytics are being used to supplement human intellect. Advocates of AI-supported decision-making, such as the centaur model of chess, claim that human instinct and machine accuracy are complementary.

However, the crisis surrounding the Boeing 737 Max has highlighted the risks of relying too heavily on AI without sufficient human oversight, particularly when human judgment fails and there is no backup system in place. The crisis highlights the crucial role of human decision-makers in safety, ethics, and accountability, and is followed by the development of regulatory frameworks that promote human-centric AI (European Commission, 2020). With the advent of AI, organizations are taking it a step further and examining how to distribute the cognitive load while maintaining human intelligence at the strategic level. Neurodiverse hiring at Accenture and Amazon's "bar raisers" hiring technique symbolize an understanding that cognitive diversity is an asset to strategy, ushering in better, more inclusive decision-making. Simultaneously, learning systems are adapting to provide leaders with metacognitive skills, training them in reflection that is anchored in awareness of cognitive biases, which have increasingly been incorporated into business education (OECD, 2020).

However, tensions remain, particularly with emotional labour, which is often neglected despite its significant impact on the well-being of organisations. Working from home has blurred emotional lines, leading to burnout and disengagement. Also, the disingenuous use of EI in unhealthy cultures, or "dark" EI, highlights the importance of leading ethically (Schneider et al., 2015). Ultimately, human intelligence serves as the foundation of organizational resilience and ethics; however, its limitations can be mitigated by mechanisms that underpin and augment human decision-making, fostering a culture of trust, ethical deliberation, and learning.

This enables organizations to develop environments where human intelligence and machine intelligence reinforce one another, resulting in better, more rational, ethical, and sustainable results.

7. AI AND ROBOTICS: TRANSFORMING ORGANIZATIONAL DYNAMICS

The combination of AI and robotics is fundamentally changing organizational structures, decision-making processes, business models, and supply chains. The evolution of AI from early heuristic-based systems to sophisticated cognitive architectures that incorporate ML, DL, and NLP enables enterprises to analyse massive datasets in real-time, facilitating predictive decision-making without waiting and reacting. In industries from retail (for example, Amazon's recommendations team) to banking (JPMorgan Chase's COIN) recommendation engine already make real-time predictions about consumers, perform predictions of behaviour and scrub inventory, automate loan agreement analysis, and so on—in the process, subsuming and destroying jobs that were wasteful of human labour (McKinsey, 2020).

This said, as AI and robotics continue to grow, important ethical and epistemological questions emerge, particularly related to the opacity of computer algorithms and the biases rooted in the training data, which erode trust and accountability, especially in domains such as legal decision-making (Pasquale, 2015). Regulations, such as the AI Act introduced by the European Commission, require transparency and human supervision to mitigate and alleviate these problems, promoting the ethical use of AI (European Commission, 2021).

Robotics, combined with AI, enhances operational efficiencies, particularly in processes such as automotive manufacturing, where collaborative robots (cobots) work alongside humans to achieve faster throughput and fewer on-the-job accidents (IFR, 2022). Within the healthcare field, da Vinci Surgical Systems utilizes AI-enhanced machines that enhance accuracy and mitigate postoperative risks, resulting in shorter recovery times (Yang et al., 2021). However, there are also concerns about the erosion of that "human touch" in the caregiving field, where empathy and personal interactions are crucial.

Furthermore, the impact of automation on the job market is a source of concern, with some forecasts predicting a mass loss of jobs due to AI and robotics (Frey & Osborne, 2017). However, it is suggested that AI will most likely supplement human activities, creating an increased demand for creativity, judgment, and personal interaction (Autor et al., 2020); therefore, systemic re-skilling is needed. Some countries, such as Singapore and Germany, have implemented policies to promote lifelong learning in AI, data analytics, and robotics, to address potential inequities (SkillsFuture, 2021).

From an organizational perspective, AI challenges traditional decision-making models, such as bounded rationality (Simon, 1957), by enabling better and more optimized decisions. As AI and robotics progress in their journey toward autonomous cognition and decision-making, they raise a set of new ethical issues around accountability, particularly when these systems unintentionally cause harm. Multilateral efforts are also working to foster global cooperation, leveraging the expertise of AI developers from diverse nationalities and avoiding a technology controlled by a few nations (GPAI, 2021).

The success of organizations in this emerging era depends not on replacing human workers with machines, but on creating new models of human-machine collaboration that ethically extend human capability. As both Microsoft and Hitachi illustrate, AI can enhance customer engagement and security; however, the human element remains crucial in ensuring informed decision-making (Microsoft, 2019). Ultimately, the strategic alignment of this emergent space requires a delicate balance between technological affordance, ethical leadership, and human-centered leadership in a new work domain where AI and robotics support rather than replace humans.

8. THE SYNERGY: MERGING HUMAN INTELLIGENCE WITH AI AND ROBOTICS

The incorporation of AI and robotics into organizational structures represents more than simply technological progress - it is a visceral transformation of decision-making, innovation, and productivity. Here, human intelligence and AI are no longer independent of each other; instead, they coexist in a mutually supportive relationship, combining human cognitive capabilities with artificial intelligence's computational accuracy and ability to recognize patterns. This pairing, represented by the "human-in-the-loop" (HITL) framework, ensures that AI augments human judgment, especially in high-stakes sectors such as healthcare, finance, and defense, where decision-making is mission-critical (Dellermann et al., 2019).

Existentially, this partnership may be inspired by theoretical orientations such as extended cognition (Clark & Chalmers, 1998) and sociomateriality (Orlikowski, 2007), which assert that humans and non-human actors contribute equally to organizational success. For example, in healthcare, AI systems enhance diagnostic accuracy; yet, clinicians incorporate moral judgment and patient stories, such that a combined approach of AI as a supplement to human-based care is evident (Yang et al., 2017).

This is no less true in the legal and creative worlds, where AI can assist in research or content creation. However, human judgment and socio-political understanding are required to contextualise and enhance those outputs. The synergy is demonstrated in practice where AI is used to comprehend large-scale medical data, as in cases like IBM's Watson for Oncology at the Mayo Clinic. On the other hand, oncologists consider the patient's preferences and psychosocial issues when making treatment choices (Somashekhar et al., 2018). However, challenges remain, such as the risk of "deskilling" when AI takes over cognitive tasks traditionally performed by humans, as seen in the aviation case, where autopilot led to a decline in pilots' opportunities for intervention (Casner & Geven, 2013).

The ethical challenges of "moral outsourcing" add to the complexity, as demonstrated by the COMPAS algorithm in U.S. courts, which has been challenged for reinforcing racial prejudices (Buolamwini & Gebru, 2018). These considerations underscore the need for responsible AI deployment, as exemplified by the UK AI Roadmap and Canada's Pan-Canadian AI Strategy, which focus on transparency, fairness, and interdisciplinary collaboration (AI Council, 2021). Multilateral projects, such as the UN "AI for Good" initiative, contribute to a more inclusive and shared access to AI, which has to be harnessed to avoid that "technology concentrates its benefits in the hands of a few" actors in society (UN, 2021).

Enterprises need to focus on enhancing, not replacing, human intelligence and develop conditions where AI becomes a complement to human judgment. Accenture's "bar raiser" model of recruitment uses AI to challenge biases and encourage less subjective decision—making in a way that organisations would not previously have been able to implement. The real future of human-AI-robotics partnership resides in developing a system that balances innovation with ethics, so technology serves to amplify human potential rather than diminish it. Moreover, as AI and robots integrate into the workplace, organizations must navigate the dynamics of trust, transparency, and inclusivity to ensure that the application of these technologies is ethical, sustainable, and conducive to human flourishing. In the future, business success will depend on how effectively tech augments human intelligence in a manner that embodies ethical integrity and inclusivity.

9. PRACTICAL STRATEGIES FOR INTEGRATING HUMAN INTELLIGENCE WITH AI AND ROBOTICS

The incorporation of AI and robots into the fabric of organizations is not merely a new technological development; it necessitates a fundamental shift in how companies perceive workforce development, leadership, and organizational structure. The conventional human capital model, which has focused heavily on hard technical skills for so long, should shift to encompass a broader range of competencies, including emotional intelligence, ethical reasoning, and critical thinking—abilities that are crucial for navigating an AI-infiltrated world (World Economic Forum, 2023).

As these AI and robotic technologies take over human work roles, the emphasis will need to shift from automation to designing for collaboration, enabling people and intelligent machines to work together and make decisions. This requires new models of leadership, such as adaptive leadership and complexity leadership, that cultivate the leader's ability to work with uncertainty, promote collective intelligence, and emphasize moral decision-making (Heifetz, 1994; Uhl-Bien, 2006). These models highlight the demand for leadership that not only guides the implementation of AI but also actively builds organizational cultures that promote inclusiveness, transparency, and psychological safety, such as the leadership of Sundar Pichai at Google that articulates responsible AI practices alongside ethics boards to steer through the morass of AI's moral hazards (Vincent, 2019).

However, the growing dependence on AI paradoxically upends the old shibboleths of how leadership works. While one might argue that delegating decision-making to AI can make a process more efficient, it can also dilute accountability and responsibility. The ability for employees to question AI output can build trust, but it can also hinder efficiency. One way to address those tensions is by adopting a meta-leadership style — one that combines control with input, allowing strategies to remain adaptive and flexible. Furthermore, they should provide middle management with the tools to connect strategy to execution, an undertaking which is vital in instilling the digital transformation blockade.

This new world requires the creation of collaborative ecosystems between people and AI as partners (not as siloed entities), and to do so, we need to redesign how workflows, communications, and workspaces come together to create synergy. The significance of intuitive and transparent interfaces is emphasized from the perspective of Human-Machine Interaction (HMI), where it is a requirement to avoid automation bias and technology rejection (Brown & Martin, 2015).

IDEO and IBM are two examples of companies that recognize the importance of a collaborative design approach in making AI systems accessible and relevant. Training for this new ecosystem also extends beyond technical skills. It includes mixed-reality games and reflex exercises to prepare individuals for the cognitive and emotional demands of working in tandem with machines. At the international level, the OECD AI Principles and other global movements promote inclusive growth based on transparency, openness, and international collaboration (European Commission, 2021).

Nonetheless, many dangers persist, such as techno-elitism, whereby AI advantages disproportionately accrue to advanced economies, undermining social equity. Equity-focused approaches, including localized training and open access to digital resources, are essential to addressing this divide. Additionally, the cultural dimension must also be considered – AI systems developed in one region may be perceived as alienating users in other regions. Cross-cultural design teams and value-sensitive design approaches can resolve these challenges.

Increasingly, AI robots are being equipped with emotional capabilities, which raises questions about human attachment and empathy, prompting the development of digital well-being guidelines and regulations that respect human dignity to ensure proper human-robot relationships. The realization of AI and robotics is not a technological evolution but a cultural revolution. Our systems need an overhaul, as do our values: organizations should now lead by their values and act with future-thinking and empathy. Ultimately, the success of AI will not be measured by productivity levels, the efficiency of factories, or the cost efficiency of enterprises, but by the success of organizations in striking a balance between technological innovation and ethical integrity, and in placing humanity at the core of progress powered by AI.

10. ETHICAL CONSIDERATIONS AND GOVERNANCE IN AI AND ROBOTICS INTEGRATION

The integration of AI and robotic technologies within organizational structures is a crucial nexus of technological advancement and profound ethical considerations. With AI systems taking over high-stakes decision-making, whether it is hiring or surveillance, issues of autonomy, bias, and accountability arise. Uber's self-driving car fatality in 2018 was a classic case when the responsibility for an accident is not clear-cut, as it can be shared between the autonomous machine, its programmer, and the company that is involved. (Lin, 2018). With reinforcement learning, the situation becomes even more complex as machines develop behaviors not limited by their initial coding, leading to unpredictable moral hazards.

Extensive work has been done in projecting people to measure AI bias, for example, by considering factors such as gender and ethnicity, as well as the distance from the training data means. For instance, the application of facial recognition technology has been found to misidentify individuals with darker skin at higher rates than lighter-skinned individuals, thus fuelling racial disparities (Buolamwini & Gebru, 2018). Equally, AI-powered recruitment algorithms, informed by historical discrimination, continue to play a role in maintaining gender and race-based inequalities, like that seen in the UK's pandemic A-level grading fiasco (BBC, 2020). These biases, when programmed into automated systems, not only become more challenging to surface but are scaled in ways that lack human moral intuitions. Privacy issues compounded the ethical challenges of AI implementation.

Mass usage of smart home and surveillance devices means our data is being hoovered up at an unprecedented pace in the name of efficiency or security. Scandals, such as those involving Cambridge Analytica, illustrate how AI can also be exploited to distort and manipulate democracy, thereby infringing on individual consent and autonomy (Isaak & Hanna, 2018). In the workplace, even when AI systems increase efficiency, they can also make it difficult to distinguish between monitoring productivity and surveillance, eroding trust and lowering morale. To address these ethical challenges, institutions must establish robust governance structures that prioritize transparency, inclusivity, and accountability.

Moral theories, such as deontology, consequentialism, and virtue ethics, provide resources for assessing AI behavior and help ensure that such development directs it towards systems that value human dignity, maximize aggregate societal welfare, and promote good moral character in the builders of AI systems. Microsoft and Google (along with all their controversy) are two companies that have taken internal ethics boards for AI development, with a focus on fairness, transparency, and accountability (Hao, 2020)

However, in practice, such efforts often fall short of the independence and enforcement necessary to check commercial pressures. Therefore, external controls are indispensable. The AI Act of the European Union, which adopts a risk-based approach and prioritizes human oversight, serves as a best practice example of responsible governance (European Commission, 2021). Similarly, the US NIST framework for trustworthy AI emphasizes the importance of explainability and ongoing monitoring to ensure ethical accountability (NIST, 2023).

However, regulation is a two-edged sword. At the same time, excessive regulation may have a chilling effect on innovation, particularly for startups; under-regulation risks deepening discrimination, exploitation, and surveillance that undermine democratic legitimacy. Therefore, a balanced and flexible regulatory approach is crucial, one that can keep AI development in line with agreed-upon principles. Ultimately, for AI to contribute to the public good, governance approaches must be pluralistic and involve multiple stakeholders in their development and implementation.

Mechanisms like Canada's Algorithmic Impact Assessment also advocate for frontier governance, arguing that AI systems should be assessed for transparency and bias reduction (Treasury Board of Canada Secretariat, 2021). Additionally, the diversity of teams developing AI is as important for avoiding ethical blind spots as it is for achieving more equitable AI systems (West et al., 2019). It is not a box-ticking exercise, but an evolving, dynamic process of determining what the technological future holds for society. Properly understood and acted upon, it is the foundation of AI in the service of the common good, upholding human dignity and the common good.

11. IMPACT ON ORGANIZATIONAL CULTURE AND EMPLOYEE EXPERIENCE

The integration of AI and robotics is not just a change in organizational methods; it is a qualitative shift in work roles and social contracts between employers and employees. As machines carry out repetitive tasks and make decisions, human work is shifting to cognitive, relational, and innovative areas. However, this transition is accompanied by massive tensions, as the working and employment population struggles with new types of job crafting and redefined professional identities.

Inspired by the work of Leontiev (Activity Theory), human work has a purpose, not just the work task, but its motives, as AI will be doing the routine data entry and humans can focus on creativity, judgment, and ethical reasoning (Engeström, 1987). Such a shift can be liberating as well: DBS Bank utilizes robotic process automation to free human workers for relationship management and strategic thinking, thereby enhancing customer satisfaction and driving innovation.

However, that transformation has its paradoxes of relevance and dignity. When AI takes over such conventional roles, employees may start to consider how to feel expert, having a sense of purpose that may need to be intentionally crafted and co-created with intelligent partners. This purpose can be intentionally crafted through job crafting and the re-conceptualization of professional identity. According to Herzberg's Two-Factor Theory, AI has the potential to impact both hygiene factors (e.g., compliance) and motivators (e.g., achievement) by enabling employees to spend less time on monotonous tasks and more time doing meaningful work.

The MIT Sloan Management Review and BCG (2020) report that when using AI to augment human roles, employee engagement increases by 20–30%. Likewise, Adobe's AI-driven analytics tools had allowed the staff to concentrate on the creative work they enjoyed and be more connected to it. However, not all stories are good stories. The threat of job loss is overwhelmingly dominating, especially in industries where AI seems to substitute for, rather than augment, human roles. It is estimated that automation could displace as much as 14% of all global jobs, causing worry and push-back, particularly in sectors ranging from manufacturing to logistics (International Labour Organization, 2021). Employees are noticing it too: they become passive resistance, disengagement, or even sabotage, in an environment with poor communication and change management (as demonstrated in a McKinsey 2022 study).

Moreover, the emergence of automation highlights a significant gap in reskilling and upskilling efforts. The focus is often on technical training related to coding and data literacy, but the essential soft skills, such as ethical reflection and collaborative creativity, are often ignored. Such uneven development exacerbates the inequality between older workers and low-wage workers, especially in non-urban areas, who may not have the financial resources needed to acquire new skills. Initiatives such as France's "IA pour Tous" and India's "FutureSkills PRIME" seek to democratize AI education, but the success of initiatives such as these ultimately relies on context-specific deployment and continued investment (MEITY, 2022; French Ministry of Education, 2021)

Leadership is crucial in navigating these cultural trends. Transformational leaders (e.g., Satya Nadella at Microsoft) create inclusive, growth-fostering climates in which AI applications unfold in harmony with human thriving (Bass, 1985). These leaders instill empathy, humility, continual learning, and growth (which are all critical to achieving human-centered AI). Another important factor is the psychological safety in the development of human-AI relationships. Amy Edmondson's research on psychological safety confirms that the highest-performing teams only work when people feel safe to take risks, make mistakes, and question outputs, whether those are produced by AI or human output (Edmondson, 2019).

Firms such as Salesforce have established cross-functional councils to weigh not only the technical risks of A.I. but also their broader cultural implications. Still, there are difficulties. As AI increasingly replaces back-end tasks, there is the danger that human labour becomes invisible or devalued, which could undermine workers' motivation and initiative. However, the paradox of dependency presents itself as well: If we rely too heavily on AI, the capacity to think critically could decrease; creative workers might become little more than validators of machine-generated recommendations. Such dilemmas need continuous reflection and dialogue between technology and the workers/leadership. Indeed, the union of AI and robotic technology is nothing short of a new wave of technology – it represents a shift in culture and a vital being. It is a situation that can lead to empowerment or alienation, depending upon the values, leadership, and collective strategies that shape the transformation process, guaranteeing that automation becomes a vehicle for the expansion rather than the eclipse of human dignity, creativity, and cooperative capacities.

12. FUTURE TRENDS: THE NEXT FRONTIER OF HUMAN-AI-ROBOTICS SYNERGY

The infiltration of AI and robots into businesses is not just a technological shift; it is a fundamental cultural shift, one that calls into question highly valued notions of work, agency, and the inflow of value. As AI evolves from specialized systems to generalizable, learning-based models, it will enable organizations to enhance human decision-making by embedding cognitive scaffolding in complex workflows, as demonstrated by models such as OpenAI's GPT and Google's PaLM 2 (Bommasani et al., 2021).

Robotics is also shifting away from clunky automation to flexible, biomimetic systems that replicate biological dexterity, such as Boston Dynamics' two-legged robots and medical exosuits, which are redefining last-mile delivery and human augmentation. This crossover between AI and robotics is referred to as "bio-digital convergence," in which biological and digital systems co-evolve, complementing one another. It raises fundamental concerns about authorship, responsibility, and the ethical dimensions associated with human-AI co-creativity (World Economic Forum, 2020).

When AI enters strategic spaces, such as making decisions in air traffic control or emergency response, it raises fundamental questions about who owns the insights formulated by AI and the ethics of robots performing emotional labour, as is the case in eldercare. These challenges necessitate a reevaluation of accountability, as exemplified by the 2018 fatality caused by the Uber self-driving car, for which responsibility was attributed to the machine, the programmer, and the company (Lin, 2018).

To overcome this, the CPS organization enables human action decision-making and machine-generated data to coordinate in real time. This new model of complex adaptive systems requires a shift from static capacity and scale to dynamic, real-time adaptability, in which both cross-modal intelligence and ethical trust influence success (Orlikowski, 2007). The convergence also holds the promise of enabling sustainable transformation, as AI can optimize energy use and help eliminate waste in precision manufacturing. At the same time, robotics can tackle hazardous tasks such as waste recycling, thereby reducing human exposure to danger.

AI is integrated into companies' practices. Schneider Electric's low-carbon smart factory and Unilever's AI-informed supply chain are two examples where AI can become an integral part of corporate DNA towards sustainability (Schneider Electric, 2021). However, the dangers of "technosolutionism"—the conviction that technology alone can fix society—are not out of the question, especially in the challenge of inequality in AI access. Furthermore, developing countries are continually being left behind as AI-driven solutions, such as precision farming, remain unattainable for smallholder farmers in the Global South.

To close this gap, inclusive innovation policies, open-source infrastructures, and fair access to AI technologies are needed. As AI becomes more creative, questions of human redundancy have been raised; the psychological value of human-created creativity may decrease in areas such as art or music generation. Organizations, in turn, need to craft cultures that do not prioritize human need-making and relationality over theories of creativity and ethical reflection at the center of innovation. "Augmented intelligence," in which machines supplement rather than replace human judgment, provides a promising model, particularly in areas such as medical diagnosis, where AI complements but does not supplant physician decision-making.

To steer that change, we need interdisciplinary leadership that marries the skills of technologists, ethicists, and sociologists to navigate the ethical complexities of integrating AI. With the ongoing development of AI and robotics, it becomes essential for organizations to embed ethical governance "by design" in all their technological activities, making technology more human-centered, sustainable, and beneficial to society (Yang et al., 2021). Ultimately, the future of human-AI-robotics collaboration hinges not on technological expertise but rather on striking a balance between technological innovation and moral responsibility, as well as human well-being.

13. META FRAMEWORK

Figure 1. Human Intelligence – AI/Robotics Framework

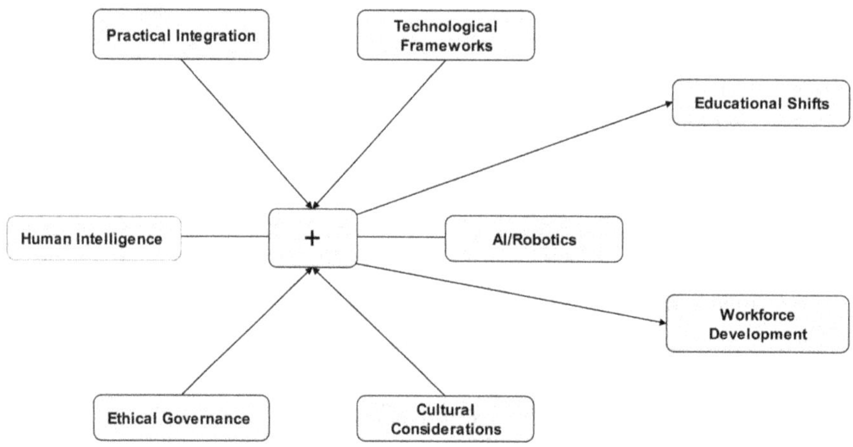

Source: *Authors*

This meta-framework, depicted in Figure 1, illustrates a synergistic perspective on human and artificial intelligence (AI) in various applications, encompassing dimensions such as practical fusion, technological frameworks, educational changes, workforce training, ethical management, and cultural context. Both the intersection of human intelligence with AI/robotics and the combined potential of AI technologies and human cognitive abilities are considered the primary focus. The system demonstrates how human intelligence is integrated into the interpretation and application of AI in real-world scenarios, encompassing ethical governance and cultural risk considerations. Secondly, it emphasizes the importance of pedagogical changes for ensuring the labour force is adequately competent to assimilate AI.

What makes this meta-framework different from existing ones is that it focuses on the collaboration of human intelligence and AI in practical use-cases, and also takes more account of ethics and cultural judgment. The model also moves beyond more technologically oriented or automation-only frameworks to include human-centered factors, acknowledging the necessary roles of education, ethical mindfulness, and cultural competence in the adoption of AI. Its "whole system" consideration of the workforce is more all-encompassing than frameworks such as those that have only considered technological readiness or operational efficiency, allowing for a more complete and integrated view of AI implementation.

14. IMPLICATIONS

The combination of human intelligence with AI and robotics presents several complex theoretical, practical, and managerial challenges that organizations must address. In theory, complications arise from combining traditional models of cognitive labour with the growing autonomy of machines, rendering the issues of intention and liability in reasoning less straightforward. The theory of distributed agency challenges the traditional interpretation of decision-making as exclusive to humans and raises challenging questions about how to distribute responsibilities for machine-autonomous decisions.

In reality, organizations encounter challenges when implementing AI and robotics, such as technology integration, data quality, and compatibility with existing systems. Insufficient data or biased algorithms can lead to adverse outcomes, and the cost of implementation can be prohibitively expensive, particularly for smaller entities. Moreover, AI and robots cannot replace those human roles that involve creativity, empathy, and subtle sense, namely, there is a void in machine power.

From a managerial perspective, the arrival of hybrid intelligence means transitioning from a classical hierarchical leadership style to adaptive models that involve human-machine collaboration. Managers also need to focus on reskilling and upskilling staff so that they can manage more complex tasks and avoid deskilling. Additionally, ethical leadership is another area where managers must ensure that AI is used responsibly, addressing concerns such as algorithmic bias, data privacy, and fairness. In conclusion, AI and robotics hold significant potential for organizational growth. However, this potential must be harnessed appropriately to address the challenges of these technologies, including hypothetical, realistic, and management issues, for a practical, ethically grounded, and sustainable cooperation between human and machine intelligence.

15. CONCLUSION

The implementation of AI and robotics in organizing is not merely a technological change, but an epistemological and cultural re-creative moment that repositions how value is created, the role of humans, and the practice of organizing. The more machines perform precise, repetitive activities, the more human work becomes creative, ethical, and empathetic. However, this reformatting is not a straightforward transformation; instead, it introduces a variety of complex questions related to identity, individual freedom, and estrangement from the world of work.

When human-machine symbiosis is deliberate and fair, evidence-based research has shown that it accelerates innovation, job enrichment, and resilience (Brynjolfsson et al., 2023), as well as improving decision-making by combining the analytical capabilities of AI and human judgment. The new models of adaptive learning, where both AI systems and human feedback shape each other, are crucial in driving continuous organizational evolution toward that ideal, aligning with the theory of double-loop learning by Argyris and Schön (1978).

However, despite these potentials, issues remain, including algorithmic bias, ethical black-boxing, and skill obsolescence. Deploying AI without appropriate ethical frameworks is how systemic inequalities become ingrained. Conversely, an over-dependence on automation can lead to degraded human judgment, as evidenced by catastrophic decision errors in high-risk domains. To address these nuances, businesses will need to invest in comprehensive capability ecosystems, which encompass not only technical training but also metacognition, digital ethics, and adaptive leadership. Efforts such as Singapore's SkillsFuture model highlight the need for lifelong, inclusive education that responds to the evolving landscape of AI and robotics.

Leadership, in addition, should evolve from hierarchical to adaptive principles, focusing on psychological safety, moral dialogue, and transparency. In this new terrain, Ethical AI governance is necessary to maintain transparency, accountability, and inclusivity in the deployment process, as seen in frameworks developed by Microsoft and the European Commission (2021). Organizational design must also change, shifting from silos to agile, cross-functional teams where humans and machines collaborate. These departments should adopt explainable AI (XAI) technologies to make AI outputs more transparent and to allow humans to oversee the final results.

Additionally, performance metrics should reward learning, creativity, and ethical alignment, rather than just throughput or compliance, and incentivize employees to challenge the use of AI. The deployment of AI and robotics will depend on re-imagining organizational life, creating a workplace that not only optimizes tasks but also enhances human capabilities, and building a culture of dignity, collaboration, and shared agency. This transition necessitates a continuous conversation among technology, human wholeness, and sustainability, where the future of work is not defined by technological brilliance, but rather by how organizations strike a balance between innovation, ethical responsibility, and human well-being.

REFERENCES

Argyris, C., & Schön, D. A. (1978). *Organizational Learning: A Theory of Action Perspective*. Addison-Wesley.

Autor, D., Mindell, D., & Reynolds, E. (2020). *The Work of the Future: Building Better Jobs in an Age of Intelligent Machines*. MIT Task Force.

Bank, D. B. S. (2022). *Annual Sustainability Report*. https://www.dbs.com/sustainability

Bass, B. M. (1985). *Leadership and Performance Beyond Expectations*. Free Press.

BBC. (2020, August 17). A-level results: What went wrong with the algorithm? *BBC News*. https://www.bbc.com/news/education-53805156

Bessen, J. E. (2019). *AI and Jobs: The Role of Demand*. NBER Working Paper No. 24235. DOI: 10.3386/w24235

Brown, T., & Martin, R. (2015). Action design. *Harvard Business Review*, *93*(9), 56–64.

Brynjolfsson, E., Li, D., & Raymond, L. (2023). Can AI improve worker performance? Evidence from a call center experiment. *Harvard Business School Working Paper*. https://www.hbs.edu

Brynjolfsson, E., Li, D., & Raymond, L. (2023). Can AI improve worker performance? Evidence from a call center experiment. *Harvard Business School Working Paper*. https://www.hbs.edu

Buolamwini, J., & Gebru, T. (2018). Gender shades: Intersectional accuracy disparities in commercial gender classification. *Proceedings of the 1st Conference on Fairness, Accountability and Transparency*, 77–91.

Casner, S. M., & Geven, R. W. (2013). The trouble with automation: Learnings from aviation. *Human Factors*, *55*(2), 345–353. DOI: 10.1177/0018720812461372

Clark, A., & Chalmers, D. J. (1998). The extended mind. *Analysis*, *58*(1), 7–19. DOI: 10.1093/analys/58.1.7

Council, A. I. (2021). *AI Roadmap 2021: Making the UK a global center for AI excellence and innovation*. https://www.gov.uk/government/publications/artificial-intelligence-roadmap

Daugherty, P. R., & Wilson, H. J. (2018). *Human + Machine: Reimagining Work in the Age of AI*. Harvard Business Review Press.

Dellermann, D., Ebel, P., Söllner, M., & Leimeister, J. M. (2019). Hybrid intelligence. *Business & Information Systems Engineering, 61*(5), 637–643. DOI: 10.1007/s12599-019-00595-2

Dellermann, D., Reck, F., Weyrich, M., & Ebel, P. (2021). Hybrid intelligence. *Business & Information Systems Engineering, 63*(5), 435–447. DOI: 10.1007/s12599-021-00695-5

Edmondson, A. C. (2019). *The Fearless Organization: Creating Psychological Safety in the Workplace for Learning, Innovation, and Growth.* Wiley.

Engeström, Y. (1987). *Learning by Expanding: An Activity-Theoretical Approach to Developmental Research.* Orienta-Konsultit.

European Commission. (2020). *AI White Paper: A European approach to excellence and trust.* https://ec.europa.eu/info/publications/white-paper-artificial-intelligence-european-approach-excellence-and-trust_en

European Commission. (2020). *Ethics guidelines for trustworthy AI.* https://digital-strategy.ec.europa.eu/en/library/ethics-guidelines-trustworthy-ai

European Commission. (2021). *Coordinated Plan on Artificial Intelligence 2021 Review.* https://digital-strategy.ec.europa.eu/en/library/coordinated-plan-artificial-intelligence-2021-review

European Commission. (2021). *Proposal for a Regulation Laying Down Harmonised Rules on Artificial Intelligence (AI Act).* https://eur-lex.europa.eu/legal-content/EN/TXT/?uri=CELEX%3A52021PC0206

European Commission. (2021). *Digital Europe Programme: 2021–2027.* https://digital-strategy.ec.europa.eu/en/activities/digital-programme

European Commission. (2021). *Proposal for a Regulation Laying Down Harmonised Rules on Artificial Intelligence (Artificial Intelligence Act).* https://eur-lex.europa.eu/legal-content/EN/TXT/?uri=CELEX%3A52021PC0206

French Ministry of Education. (2021). *AI for Everyone: Promoting Digital Literacy.* https://www.education.gouv.fr

Frey, C. B., & Osborne, M. A. (2017). The future of employment: How susceptible are jobs to computerization? *Technological Forecasting and Social Change, 114,* 254–280. DOI: 10.1016/j.techfore.2016.08.019

Garvin, D. A., Wagonfeld, A. B., & Kind, L. (2013). Google's Project Oxygen: Do Managers Matter? Harvard Business School Case No. 313-110.

Gershman, S. J., Horvitz, E. J., & Tenenbaum, J. B. (2015). Computational rationality: A converging paradigm for intelligence in brains, minds, and machines. *Science, 349*(6245), 273–278. DOI: 10.1126/science.aac6076 PMID: 26185246

Goleman, D. (1995). *Emotional Intelligence: Why It Can Matter More Than IQ*. Bantam Books.

Hao, K. (2020, December 4). We read the paper that led to Timnit Gebru's departure from Google. Here is what it says. *MIT Technology Review*. https://www.technologyreview.com/2020/12/04/1013294/

Haselberger, D. (2020). Symbiotic Intelligence: Merging AI and Human Capabilities for Better Decision-Making. *Journal of Artificial Intelligence Research, 69*, 45–62.

Heifetz, R., & Linsky, M. (2002). *Leadership on the Line: Staying Alive Through the Dangers of Leading*. Harvard Business Press.

Heifetz, R., & Linsky, M. (2002). *Leadership on the Line: Staying Alive through the Dangers of Leading*. Harvard Business Review Press.

Heifetz, R. A. (1994). *Leadership Without Easy Answers*. Harvard University Press. DOI: 10.4159/9780674038479

Herzberg, F. (1966). *Work and the Nature of Man*. World Publishing.

IFR. (2022). World [International Federation of Robotics. https://ifr.org]. *Robotics Reports, 2022*, •••.

Isaak, J., & Hanna, M. J. (2018). User data privacy: Facebook, Cambridge Analytica, and privacy protection. *Computer, 51*(8), 56–59. DOI: 10.1109/MC.2018.3191268

Kahneman, D., & Tversky, A. (1979). Prospect theory: An analysis of decision under risk. *Econometrica, 47*(2), 263–291. DOI: 10.2307/1914185

Latour, B. (2005). *Reassembling the Social: An Introduction to Actor-Network-Theory*. Oxford University Press. DOI: 10.1093/oso/9780199256044.001.0001

Lin, P. (2018). The ethics of autonomous cars. *The Atlantic*. https://www.theatlantic.com/technology/archive/2013/10/the-ethics-of-autonomous-cars/280360/

McKinsey & Company. (2020). *The State of AI in 2020*. https://www.mckinsey.com/business-functions/mckinsey-analytics/our-insights

McKinsey & Company. (2022). *The State of AI in 2022*. https://www.mckinsey.com

McKinsey Global Institute. (2021). *The state of AI in 2021*. https://www.mckinsey.com

MEITY. (2022). *FutureSkills PRIME: Empowering India's Workforce for the Digital Economy*. https://futureskillsprime.in

Microsoft. (2019). *Microsoft AI Principles*. https://www.microsoft.com/en-us/ai/responsible-ai

MIT Sloan Management Review & BCG. (2020). *The Cultural Benefits of Artificial Intelligence in the Enterprise*. https://sloanreview.mit.edu/projects/the-cultural-benefits-of-artificial-intelligence-in-the-enterprises

MIT Sloan Management Review & BCG. (2020). *The Cultural Benefits of Artificial Intelligence in the Enterprise*. https://sloanreview.mit.edu

NIST. (2023). *AI Risk Management Framework 1.0*. U.S. Department of Commerce. https://www.nist.gov/itl/ai-risk-management-framework

NIST. (2023). *AI Risk Management Framework (AI RMF 1.0)*. U.S. Department of Commerce. https://www.nist.gov/itl/ai-risk-management-framework

Norman, D. A. (2013). *The Design of Everyday Things* (Revised ed.). Basic Books.

O'Reilly, C. A.III, & Tushman, M. L. (2013). Organizational Ambidexterity: Past, Present, and Future. *The Academy of Management Perspectives*, *27*(4), 324–338. DOI: 10.5465/amp.2013.0025

OECD. (2020). *The Future of Education and Skills 2030*. https://www.oecd.org/education/2030/

Orlikowski, W. J. (2007). Sociomaterial practices: Exploring technology at work. *Organization Studies*, *28*(9), 1435–1448. DOI: 10.1177/0170840607081138

Pasquale, F. (2015). *The Black Box Society: The Secret Algorithms That Control Money and Information*. Harvard University Press. DOI: 10.4159/harvard.9780674736061

Robertson, J. (2018). *Robo Sapiens Japanicus: Robots, Gender, Family, and the Japanese Nation*. University of California Press.

Schneider, R. J., Lyons, J. B., & Khazon, S. (2015). Emotional intelligence and personality: Predicting transformational, transactional, and laissez-faire leadership styles. *Leadership and Organization Development Journal*, *36*(3), 396–414.

Simon, H. A. (1957). *Models of Man: Social and Rational*. Wiley.

Singapore SkillsFuture. (2021). *Annual Report on Skills Transformation and Workforce Development*. https://www.skillsfuture.gov.sg

Smart Nation and Digital Government Office. (2019). *National AI Strategy of Singapore*. https://www.smartnation.gov.sg

Somashekhar, S. P., Sepúlveda, M. J., Puglielli, S., Norden, A. D., Rauthan, A., Arun, K. N., & Raghunath, S. (2018). Watson for oncology and breast cancer treatment recommendations: Agreement with an expert multidisciplinary tumor board. *Annals of Oncology : Official Journal of the European Society for Medical Oncology*, *29*(2), 418–423. DOI: 10.1093/annonc/mdx781 PMID: 29324970

Sternberg, R. J. (1985). *Beyond IQ: A Triarchic Theory of Human Intelligence*. Cambridge University Press.

Treasury Board of Canada Secretariat. (2021). *Algorithmic Impact Assessment Tool*. https://www.canada.ca/en/government/system/digital-government/digital-government-innovations/responsible-use-ai/algorithmic-impact-assessment.html

Trist, E. L., & Emery, F. E. (1973). *Towards a Social Ecology: Contextual Appreciations of the Future in the Present*. Plenum Press.

Uhl-Bien, M., Marion, R., & McKelvey, B. (2006). Complexity leadership theory: Shifting leadership from the industrial age to the knowledge era. *The Leadership Quarterly*, *18*(4), 298–318. DOI: 10.1016/j.leaqua.2007.04.002

Vincent, J. (2019). Google dissolves AI ethics board after controversy. *The Verge*. https://www.theverge.com/2019/4/4/18295310

West, S. M., Whittaker, M., & Crawford, K. (2019). Discriminating systems: Gender, race, and power in AI. *AI Now Institute*. https://ainowinstitute.org/discriminatingsystems.html

Wilson, S. (2020). Pandemic leadership: Lessons from New Zealand's approach to COVID-19. *Leadership*, *16*(3), 279–293. DOI: 10.1177/1742715020929151

World Economic Forum. (2023). *Future of Jobs Report 2023*. https://www.weforum.org/reports/future-of-jobs-report-2023

Yang, G.-Z., Bellingham, J., Dupont, P. E., Fischer, P., Floridi, L., Full, R., & Wood, R. (2021). The grand challenges of Science Robotics. *Science Robotics*, *6*(50), eabf0821. Advance online publication. DOI: 10.1126/scirobotics.abf0821 PMID: 34043587

Yang, G. Z., Cambias, J., Cleary, K., Daimler, E., Drake, J., Dupont, P. E., Hata, N., Kazanzides, P., Martel, S., Patel, R. V., Santos, V. J., & Taylor, R. H. (2017). Medical robotics—Regulatory, ethical, and legal considerations for increasing levels of autonomy. *Science Robotics*, *2*(4), eaam8638. Advance online publication. DOI: 10.1126/scirobotics.aam8638 PMID: 33157870

KEY TERMS AND DEFINITIONS

Computational Rationality: Computational rationality, as proposed by Gershman et al. (2015), is a theoretical framework for decision-making that incorporates computational models into the decision-making process.

Distributed Agency: Distributed agency, as described in Actor-Network Theory (Latour, 2005), refers to the notion that agency—the capacity to act and influence outcomes—is not confined to human actors but is spread across both humans and non-human entities, such as AI systems and robots.

Ethical AI Governance: Ethical AI governance entails establishing principles, policies, and structures that guide the development and deployment of AI technologies in a manner that is transparent, fair, and accountable.

Extended Cognition: Extended cognition, introduced by Clark and Chalmers (1998), argues that cognition is not confined to the human mind but extends into tools, technologies, and external environments.

Human-AI Synergy: Human-AI synergy refers to the collaborative interaction between human cognitive and emotional intelligence and artificial intelligence systems, where both complement each other to achieve superior outcomes than either could achieve independently.

Human-Computer Interaction (HCI): Human-Computer Interaction (HCI) is a field of study that focuses on the design, evaluation, and implementation of interactive computing systems for human use, to enhance usability and user experience (Norman, 2013).

Hybrid Intelligence: Hybrid intelligence is the integration of human and machine intelligence in a complementary manner, where both human and AI capabilities enhance each other.

Robotic Process Automation (RPA): Robotic Process Automation (RPA) refers to the use of software robots or AI-driven automation to perform routine, repetitive tasks that humans previously handled. These tasks often involve data entry, transaction processing, and rule-based decision-making.

Chapter 4

Human-Robot Collaboration and Organizational Transformation: Leveraging AI and Robotics for Future-Ready Enterprises

Mohit Sharma
https://orcid.org/0009-0007-2280-8077
Maharshi Dayanand University, India

Rishi Chaudhry
Maharshi Dayanand University, India

Nitika Malik
https://orcid.org/0009-0006-3991-9526
Maharshi Dayanand University, India

Raj Kumar
Institute of Management Studies and Research, India

Amit Kumar
https://orcid.org/0000-0003-4606-7289
Gurugram University, India

ABSTRACT

This chapter explores the symbiotic relationship between humans and robots within organizational contexts. Using case studies from various industries—including

DOI: 10.4018/979-8-3373-3404-2.ch004

manufacturing, healthcare, and public administration—this research highlights successful HRC implementations, key barriers to adoption, and best practices for creating AI-augmented workplaces. By leveraging AI-driven workforce transformation, companies can position themselves for long-term competitiveness while fostering a sustainable and inclusive work environment.

INTRODUCTION

The Fourth Industrial Revolution has ushered in a new era of technological convergence, where artificial intelligence (AI), robotics, and digital platforms are rapidly transforming the way organizations operate. Central to this transformation is Human-Robot Collaboration (HRC)—a synergistic partnership between humans and intelligent machines that combines human cognitive skills with robotic precision and automation. HRC is not merely a technological innovation; it represents a fundamental reconfiguration of organizational structures, work processes, and human roles in the enterprise ecosystem. As organizations worldwide seek to enhance productivity, optimize operations, and remain agile in the face of global uncertainties, the integration of AI and robotics into the workplace has become a strategic imperative. In contrast to the earlier waves of automation that often replaced human labor with machines, the emerging model of collaborative automation emphasizes cooperation rather than substitution. Human-robot teams can achieve complex tasks more efficiently, with robots handling repetitive, hazardous, or precision-oriented tasks, while humans contribute creativity, judgment, and empathy (Villani et al., 2018). The adoption of AI and robotics across industries—from manufacturing and healthcare to logistics and public administration—has been accompanied by measurable gains in efficiency, accuracy, and service delivery. For instance, in manufacturing, collaborative robots or "cobots" have been deployed to assist human workers in assembly lines, reducing fatigue and improving output (Bogue, 2016). In healthcare, robotic assistants have been used for surgical procedures, elder care, and diagnostics, augmenting the capabilities of medical professionals while ensuring patient safety (Yang et al., 2017). Similarly, public sector organizations are using AI-powered chatbots and robotic process automation (RPA) to streamline citizen services and improve administrative efficiency (OECD, 2021). However, the transformative potential of HRC comes with a host of organizational, ethical, and socio-technical challenges. A primary concern is the impact on employment and workforce dynamics. While AI and robotics can create new jobs and enhance existing roles, they also pose a risk of job displacement, particularly for routine and low-skilled positions (Frey & Osborne, 2017). The shift in job profiles necessitates reskilling and upskilling of employees, prompting organizations to invest in continuous learning and digital literacy programs (World Economic Forum, 2020). Resistance to change, fear of obsolescence, and a lack of trust in machines further complicate the smooth integration of HRC systems into traditional work environments (Davenport & Ronanki, 2018). Ethical considerations also loom large in the discourse on human-robot interaction. Issues such as decision-making accountability, bias in AI algorithms, and the transparency of autonomous systems raise important questions about governance and responsibility. Furthermore, as robots take on more interactive and decision-making roles, the need for clear ethical guidelines and regulatory frameworks becomes essential (Lin et al., 2011). Organizational cultures must evolve to support inclusivity, transparency, and fairness in technologically mediated workplaces. In response to these challenges, a number of organizations have begun to develop adaptive strategies that align technological adoption with human-centric values. These strategies include participatory design processes where employees are involved in shaping the role of robotics in their work, pilot programs to evaluate the effectiveness of HRC before full-scale implementation, and policies that prioritize worker safety and well-being (De Stefano, 2020). Companies like BMW, Amazon, and Siemens have been at the forefront of integrating robotics with human labor in ways that enhance both productivity and employee satisfaction . Academic and industry research on HRC also points to the importance of organizational readiness. This includes having a digitally mature infrastructure, leadership commitment to innovation, and a workforce culture that embraces change. According to a study by McKinsey (2021), organizations that invest in digital infrastructure and workforce engagement are more likely to achieve successful outcomes from automation initiatives. Furthermore, the interdisciplinary nature of HRC—involving fields like engineering, psychology, management, and ethics—necessitates a collaborative approach to research and implementation. This chapter delves into the evolving landscape of Human-Robot Collaboration by examining its strategic implications for organizational transformation. Drawing on case studies across key industries, it highlights both successful implementations and the barriers that hinder wider adoption. For example, in the automotive industry, the deployment of collaborative robots has reduced manufacturing errors and improved worker ergonomics (Krüger et al., 2017). In the healthcare sector, HRC has contributed to more personalized and responsive care systems, especially in aging societies. These real-world examples underscore the transformative power of AI and robotics when applied in alignment with organizational goals and human values. Moreover, the chapter explores best practices for fostering AI-augmented workplaces, including change management strategies, inclusive technology design, and human-centered leadership. It argues that the future of work lies not in replacing humans with machines, but in redefining roles and workflows to create a harmonious coexistence. In doing so, it supports the broader vision of Industry 5.0, which emphasizes the integration of human creativity with machine intelligence to achieve resilient, sustainable, and human-centric enterprises (European Commission, 2021). The research methodology employed in this chapter includes a qualitative review of academic literature, industry reports, and organizational case studies. By synthesizing insights from diverse sources, the chapter provides a holistic view of how HRC is reshaping the organizational landscape. It also offers recommendations for policymakers and practitioners seeking to leverage AI and robotics for long-term competitiveness and social inclusion.

In conclusion, Human-Robot Collaboration is not just a technological trend—it is a catalyst for profound organizational transformation. It challenges conventional notions of labor, leadership, and value creation, while offering new opportunities for innovation and growth. As organizations navigate the complex terrain of digital transformation, embracing HRC with a strategic, ethical, and human-centered lens will be critical for building future-ready enterprises.

BACKGROUND OF THE STUDY

The rapid digitalization of economies across the globe has fundamentally altered the way organizations operate, compete, and create value. This transformation is driven largely by emerging technologies such as artificial intelligence (AI), robotics, machine learning, and automation systems that enable organizations to increase efficiency, agility, and decision-making capabilities (Bughin et al., 2018). Among these innovations, Human-Robot Collaboration (HRC) has emerged as a key component in the evolution of future-ready enterprises. Unlike earlier forms of automation that were designed to replace human labor, HRC emphasizes cooperation between humans and robots, where both parties contribute their unique strengths to shared tasks (Villani et al., 2018). Historically, automation in industrial settings has been characterized by rigid, isolated robotic systems designed to perform repetitive tasks. These systems operated independently and often required physical barriers to separate them from human workers due to safety concerns (Colgate et al., 1996). However, advancements in sensor technologies, AI, and machine vision have enabled the development of collaborative robots (cobots)—machines capable of working side-by-side with humans without physical separation. Cobots can respond to human gestures, adapt to changing environments, and learn from human behavior, making them ideal for modern work environments where flexibility, safety, and adaptability are critical (Bogue, 2016). The evolution of HRC aligns with broader technological trends shaping the Fourth Industrial Revolution (Industry 4.0), where cyber-physical systems, the Internet of Things (IoT), and cloud computing are converging to create "smart" factories and digitally interconnected enterprises (Schwab, 2017). Within this context, HRC enables seamless integration of human creativity, adaptability, and judgment with robotic precision, strength, and computational speed. This synergy enhances not only operational efficiency but also innovation, customization, and employee satisfaction (Pedersen et al., 2016). In manufacturing, HRC has revolutionized production lines. Companies like BMW and Ford have successfully integrated cobots to support assembly workers, reduce ergonomically challenging tasks, and ensure consistent quality (Zielinska, 2019). In logistics and warehousing, firms such as Amazon use collaborative robotics for inventory management, order picking, and real-time delivery optimization (Wurman et al., 2008). The healthcare sector has also seen the benefits of HRC, particularly with robotic-assisted surgery, telepresence robots, and eldercare automation, allowing healthcare professionals to deliver more precise, accessible, and personalized care (Yang et al., 2017). Despite these advancements, the deployment of HRC presents substantial organizational and societal challenges. One major concern is the potential displacement of workers. According to Frey and Osborne (2017), nearly 47% of U.S. jobs were at risk of automation, sparking debates about job security and the future of work. While collaborative robots are intended to augment rather than replace human labor, their integration inevitably alters job roles, task structures, and required skill sets. As a result, organizations must invest in reskilling and upskilling programs to prepare workers for digitally augmented work environments (World Economic Forum, 2020). The ethical and psychological dimensions of HRC also warrant serious attention. Human workers may experience anxiety, fear, or resistance when working alongside robots, particularly when their roles are poorly defined or perceived as inferior (de Graaf & Allouch, 2013). Furthermore, the opacity of AI decision-making processes—often referred to as the "black box" problem—raises concerns about accountability, transparency, and bias in automated systems (Burrell, 2016). Ethical frameworks must guide the design and deployment of collaborative systems to ensure fairness, respect for human dignity, and the prevention of harm. Organizational readiness plays a pivotal role in the success or failure of HRC initiatives. Research indicates that companies with strong digital leadership, a culture of innovation, and flexible organizational structures are better positioned to adopt and benefit from HRC technologies (KPMG, 2022). Conversely, firms lacking technological infrastructure, strategic vision, or employee engagement may struggle with implementation. Effective HRC adoption thus requires not just investment in hardware and software, but also in change management strategies, inclusive communication, and employee empowerment (Davenport & Ronanki, 2018). Another key consideration is regulatory and policy support. In many countries, labor laws, safety regulations, and data privacy standards have not kept pace with the rapid evolution of HRC technologies. For example, collaborative robots must adhere to ISO 10218 safety standards, yet the practical enforcement and interpretation of these standards vary widely across jurisdictions (International Organization for Standardization, 2011). National governments, industry associations, and international bodies must work collaboratively to establish comprehensive guidelines that promote innovation while safeguarding workers' rights. The COVID-19 pandemic further accelerated the adoption of AI and robotics in organizations, highlighting the critical role of automation in maintaining operations during crises. Robots were deployed for disinfection, delivery, and contactless interaction in hospitals and public spaces, reducing human exposure and enhancing safety (Yang et al., 2020). In the post-pandemic era, businesses are increasingly exploring HRC not only for operational resilience but also for business continuity, health safety, and digital competitiveness. Within the theoretical framework of organizational transformation, HRC represents both a technological and socio-cultural shift. Traditional organizational models based on hierarchy, centralization, and standardized workflows are giving way to adaptive, decentralized, and agile structures where humans and intelligent agents co-create value. As emphasized by the European Commission (2021), the emerging concept of Industry 5.0 builds upon Industry 4.0 by placing humans back at the center of technological advancement. Industry 5.0 promotes human-centric, sustainable, and resilient innovation, recognizing that long-term competitiveness depends on aligning automation with societal well-being and environmental goals. Moreover, as organizations seek to leverage data-driven decision-making through AI, they are also faced with the need to build digital trust among employees, customers, and stakeholders. Trust is essential in human-robot teams, especially when robots are making or influencing decisions. Studies show that transparent communication, participatory design, and shared goals can significantly enhance trust in collaborative systems (Hancock et al., 2011). Companies must foster a culture where humans and robots are viewed as complementary collaborators, not competitors. In light of these developments, this study is motivated by the need to explore the strategic, operational, and human dimensions of Human-Robot Collaboration in the context of organizational transformation. While the technical capabilities of robots continue to advance rapidly, the human and managerial aspects of HRC remain underexplored. Questions about leadership, culture, ethics, inclusivity, and long-term sustainability are central to understanding how AI and robotics can be leveraged for positive transformation rather than disruption. This chapter aims to fill this gap by analyzing real-world applications, challenges, and success factors in the implementation of HRC across multiple sectors. By doing so, it contributes to a growing body of knowledge on how organizations can effectively harness AI and robotics not just for productivity gains, but for creating inclusive, ethical, and future-ready enterprises.

RESEARCH METHODOLOGY

This study adopts a qualitative, exploratory research design to investigate the integration of Human-Robot Collaboration (HRC) within organizational transformation frameworks. Given the evolving nature of AI and robotics, a qualitative approach is suitable for capturing nuanced insights, contextual factors, and the socio-technical interplay between humans and intelligent systems. The research methodology includes a multiple case study analysis of organizations across three key sectors: manufacturing, healthcare, and public administration. These industries were selected due to their diverse applications of HRC and varying degrees of technological maturity. Case data were gathered through secondary sources, including peer-reviewed journal articles, industry reports, white papers, and company documents published between 2017 and 2024. Selection criteria focused on organizations that have demonstrably implemented HRC technologies and have undergone measurable organizational change. Additionally, the study utilizes thematic analysis to identify recurring patterns, challenges, enablers, and best practices associated with HRC implementation. A coding framework was developed to categorize data under major themes such as operational efficiency, workforce transformation, ethical considerations, and leadership adaptability. The analysis also draws from existing conceptual models of digital transformation and socio-technical systems to interpret findings within an established theoretical lens. By combining sector-specific insights with cross-industry comparisons, this methodology aims to generate actionable recommendations for policymakers, business leaders, and technology developers seeking to foster human-centric, AI-augmented workplaces. The research ensures validity through data triangulation and peer-reviewed sources, and emphasizes practical relevance by grounding findings in real-world applications.

FINDINGS OF THE STUDY

The findings from this study present a comprehensive understanding of how Human-Robot Collaboration (HRC) is transforming organizational structures, workforce dynamics, and business models across diverse industries. Based on in-depth case study analyses and thematic interpretation, five key themes emerged: (1) Enhanced Operational Efficiency and Productivity, (2) Workforce Transformation and Skill Development, (3) Ethical and Cultural Challenges, (4) Leadership and Strategic Readiness, and (5) Sector-Specific HRC Applications and Outcomes.

1. Enhanced Operational Efficiency and Productivity One of the most prominent findings of the study is that HRC significantly improves operational perfor-

mance. Across all three sectors examined—manufacturing, healthcare, and public administration—organizations reported noticeable improvements in task accuracy, process speed, and cost optimization. In manufacturing, collaborative robots (cobots) were found to streamline assembly lines, reduce error rates, and work safely alongside human operators without requiring physical separation (Bogue, 2018). For instance, in the case of a German automotive manufacturer, the integration of cobots resulted in a 25% increase in production output while maintaining quality standards. In healthcare, robots were used to perform repetitive or high-risk tasks such as sterilization, medication delivery, and even surgical assistance. Hospitals employing AI-powered robotic assistants like TUG and da Vinci reported improved patient outcomes and reduced staff fatigue (Yang et al., 2017). Meanwhile, in public administration, robotic process automation (RPA) enabled government agencies to automate routine administrative tasks, resulting in faster service delivery and enhanced citizen satisfaction. These findings illustrate that the combination of human decision-making and robotic precision creates a hybrid workforce capable of handling complexity with greater agility. The ability of AI systems to process large volumes of data in real-time allows human workers to focus on judgment-based and creative tasks, thereby amplifying overall productivity (Daugherty & Wilson, 2018).

2. Workforce Transformation and Skill Development The integration of AI and robotics is also catalyzing a shift in workforce roles and required competencies. As organizations automate repetitive or hazardous tasks, human workers are increasingly required to take on supervisory, analytical, and interpersonal roles that machines cannot easily replicate. This transition is particularly visible in manufacturing, where operators are being retrained to oversee and maintain robotic systems rather than engage in manual labor. In one case, a leading electronics manufacturer in Japan reported that over 60% of its workforce had undergone reskilling programs focused on human-machine interaction, AI literacy, and digital troubleshooting. In the healthcare sector, the role of nurses and medical technicians is also evolving. With robotic assistance managing logistics and routine monitoring, healthcare professionals are able to devote more time to patient care and emotional support—functions that remain distinctly human. Likewise, public sector employees are being encouraged to develop competencies in data analytics, cybersecurity, and process redesign to support AI-enabled government functions. Despite these positive transformations, the study found that not all organizations are equally prepared for workforce reskilling. In many cases, a lack of structured training programs and resistance to change among employees hindered effective adaptation. This underlines the importance of proactive change management strategies and inclusive HR policies that promote

continuous learning and employee empowerment (Brynjolfsson & McAfee, 2014).

3. Ethical and Cultural Challenges in Human-Robot Collaboration Despite the promising benefits of HRC, organizations are facing complex ethical and cultural challenges as they integrate AI and robotics into their operations. A major finding from the study was the divergent perceptions of trust and transparency among employees. In multiple case studies, workers expressed concerns over opaque decision-making processes when AI systems were introduced. In public administration, for example, employees questioned the rationale behind certain automated decisions—such as the denial of benefits or allocation of public resources—raising ethical concerns about accountability and bias in algorithmic governance (Eubanks, 2018). In manufacturing, fears of surveillance and loss of privacy emerged, particularly with AI systems monitoring worker productivity and behavior in real-time. While some employees recognized the performance benefits, others felt alienated by the lack of human oversight and context. These findings highlight the importance of transparency and explainability in AI implementation. Organizations that proactively communicated the role and limitations of robotic systems fostered higher acceptance and trust among staff (Floridi et al., 2018). Another key cultural issue is the perception of robots as threats to job security. Across sectors, automation anxiety was a common theme—especially among older employees and those with lower levels of digital literacy. In a healthcare facility studied in the UK, support staff voiced concerns that robots delivering medication or performing cleaning duties could gradually replace them. This fear was compounded by a lack of consultation during implementation, leading to resistance and morale issues. The study also revealed gendered and socio-economic dimensions in HRC adoption. In manufacturing sites where robotic integration was led by male-dominated engineering teams, female workers often felt excluded from upskilling initiatives and decision-making processes. Similarly, lower-income public sector employees had limited access to digital literacy programs, further widening the skills gap. To address these issues, organizations that prioritized ethical design frameworks, inclusive workforce development, and participatory governance models saw smoother transitions and higher engagement. The development of AI ethics committees, employee feedback loops, and AI explainability tools contributed to a more responsible and human-centric integration of robotics (Jobin, Ienca & Vayena, 2019).
4. Leadership and Strategic Readiness A critical success factor in HRC implementation across industries was leadership commitment and strategic foresight. Organizations that excelled in integrating AI and robotics were those where leadership viewed digital transformation not just as a technological upgrade

but as an organizational renewal process. These leaders embraced uncertainty, fostered a culture of experimentation, and aligned their strategic vision with long-term workforce empowerment. In the case of a Scandinavian logistics company, executive leadership initiated a "human-tech partnership" framework that encouraged all departments to identify opportunities where AI could assist rather than replace human efforts. This collaborative approach led to the creation of AI copilots for route optimization, reducing delivery times by 18% without affecting driver employment. Leadership was also instrumental in driving interdepartmental coordination. In many public sector organizations, robotic process automation was initially piloted by IT departments without collaboration with human resource or operations teams. These siloed implementations often led to inefficient or redundant processes. However, when cross-functional teams were established, the outcomes were more sustainable and aligned with organizational goals. A related finding was the importance of adaptive leadership skills. As HRC introduces dynamic and non-linear change, traditional hierarchical management approaches proved insufficient. Organizations led by individuals who embraced agility, empathy, and data-driven decision-making demonstrated higher levels of HRC integration success. Leaders who facilitated open communication about the impact of robotics and involved staff in co-creation processes experienced fewer implementation challenges (Westerman, Bonnet & McAfee, 2014). Moreover, organizations that invested in strategic readiness assessments—including workforce readiness, technology infrastructure, and process redesign—achieved smoother transitions. These assessments helped in identifying gaps, prioritizing investments, and aligning HRC adoption with business objectives. Strategic foresight enabled firms to plan beyond short-term automation goals and envision AI as a long-term partner in innovation and organizational growth.

5. Sector-Specific Applications and Outcomes of HRC The study's findings underscored distinct patterns in how HRC was adopted and operationalized across manufacturing, healthcare, and public administration. Each sector displayed unique drivers, barriers, and performance outcomes, shaped by their respective regulatory environments, labor structures, and service priorities. 5.1 Manufacturing Sector In the manufacturing domain, HRC was largely driven by the need to enhance productivity, reduce operational costs, and mitigate labor shortages. Collaborative robots (cobots) were widely deployed for tasks like welding, assembly, packaging, and quality control. A German automobile parts manufacturer integrated cobots on its production line to assist workers with repetitive lifting tasks, resulting in reduced worker fatigue and a 30% decrease in workplace injuries. A notable trend was the use of AI-powered predictive maintenance, where sensors on robotic arms provided real-time data on wear-

and-tear, enabling proactive servicing and reducing downtime. Manufacturers also used simulation tools and digital twins to train human workers in safe, virtual environments before actual HRC deployment, enhancing both safety and confidence (Lu et al., 2020). However, the sector faced challenges related to workforce resistance, especially from skilled technicians wary of being replaced by intelligent machines. Companies that addressed these concerns through transparent communication and retraining programs found higher levels of acceptance and productivity. Importantly, firms that framed HRC as a tool to assist rather than replace workers were more successful in achieving organizational buy-in. 5.2 Healthcare Sector In healthcare, the adoption of HRC was primarily motivated by the need to improve patient care quality and operational efficiency amid rising demand and workforce shortages. Robots such as da Vinci (for surgical assistance), TUG (for delivery of medical supplies), and Pepper (for patient engagement) were used in combination with human professionals.

Findings revealed that HRC improved both patient satisfaction and clinical efficiency. At a hospital in Singapore, the use of surgical robots led to shorter recovery times and fewer postoperative complications. Nurses were supported by robots in routine tasks like medication delivery, allowing them to focus on critical and compassionate care functions.

Yet, emotional intelligence remains a limitation for robots in healthcare. While machines excel at precision and logistics, human empathy, intuition, and adaptability continue to be indispensable. In facilities where HRC was positioned as augmenting rather than replacing care, patient trust and staff morale were significantly higher.

Further, healthcare professionals exhibited greater openness to robotic assistance when involved early in the planning and training stages. Hospitals that conducted interdisciplinary training sessions—pairing clinicians with technologists—reported smoother transitions and higher user satisfaction.

Public Administration Sector

In public administration, Human-Robot Collaboration is emerging as a transformative force for improving service delivery, citizen engagement, and internal efficiency. The use of Robotic Process Automation (RPA), chatbots, and AI-powered decision-support systems has allowed governments to streamline operations, reduce response times, and optimize resource allocation. For instance, in Estonia—considered a digital governance pioneer—AI systems now help in managing tax declarations, filing legal documents, and providing real-time responses to citizen queries.

A key finding in this sector was the positive impact on administrative workload reduction. Clerical staff who were previously burdened with data entry, document processing, and form validations found significant relief through automation, allowing them to engage more meaningfully in citizen interaction and policy implementation. A case study from a municipality in Canada showed that the deployment of RPA to process social benefit applications reduced approval time by 40%, while errors dropped by nearly 70%.

Another benefit observed was data transparency and decision traceability. AI systems used for budgeting and planning enabled data-driven decisions with traceable logic paths—enhancing accountability and public trust. However, the implementation of AI in governance also raised critical concerns regarding fairness, algorithmic bias, and the opacity of AI recommendations. In one example, a city in the United States had to roll back its AI-based predictive policing system after it was found to disproportionately target marginalized communities.

Moreover, many public institutions lacked the technological infrastructure and digital talent to scale AI-based solutions. Limited funding, bureaucratic resistance, and data privacy concerns further slowed down adoption. However, agencies that partnered with private tech firms and research institutions saw greater success in overcoming these barriers. In particular, public-private innovation labs emerged as effective spaces for experimentation, testing, and co-design of AI-powered public services.

The study also found that citizen perception played a critical role in shaping HRC adoption in the public sector. Citizens were more accepting of automation in back-office functions than in decision-making roles, such as eligibility assessments for welfare programs or judicial support systems. This underlines the need for ethical AI governance frameworks, public education campaigns, and mechanisms for human oversight to ensure transparency and public confidence.

CROSS-SECTOR COMPARATIVE INSIGHTS

By comparing findings across the manufacturing, healthcare, and public administration sectors, the study identified several cross-sectoral patterns and unique differences in HRC adoption:

- Operational Gains: All three sectors reported gains in efficiency, accuracy, and cost savings, though the scale and nature of these gains varied. Manufacturing emphasized physical task automation, healthcare focused on augmenting care delivery, while public administration leveraged process automation.
- Human-Centric Roles: In all sectors, the human role shifted from execution to supervision, requiring new skill sets. Emotional intelligence, ethical reasoning, and creative problem-solving became increasingly important.

• Technology Acceptance: Acceptance of robots and AI was higher when framed as assistive tools rather than replacements. Participation of employees in design and implementation improved adoption rates.
• Challenges: Key challenges were shared across sectors—namely, workforce displacement fears, ethical concerns, and lack of readiness. However, sector-specific issues also emerged, such as the high regulatory burden in healthcare and the public scrutiny in governance.
• Leadership and Policy: Successful HRC implementation hinged on strong leadership, strategic vision, and inclusive policy-making. Organizations with clear digital roadmaps and continuous learning cultures outperformed those with fragmented approaches.

BEST PRACTICES FOR AI-AUGMENTED WORKPLACES

Based on the analysis of multiple case studies and sectoral implementations, the study proposes the following best practices for effectively managing Human-Robot Collaboration in organizational settings:

1. Co-Design with End Users

Involve employees early in the design and deployment process. Co-design fosters ownership, reduces resistance, and ensures that AI solutions address actual pain points.

2. Invest in Digital Literacy and Reskilling

Organizations must build robust learning ecosystems that support continuous reskilling, especially in digital competencies, ethical reasoning, and human-machine interaction.

3. Ensure Transparency and Explainability

Use AI systems that offer traceability and interpretability. Employees and stakeholders should understand how decisions are made, especially in high-stakes scenarios.

4. Create AI-Ethics Committees

Establish cross-functional teams to monitor the ethical implications of AI use and ensure alignment with organizational values and societal expectations.

5. Adopt Agile Change Management

Use iterative, feedback-driven implementation approaches that allow for flexibility and course correction. Agile models enhance responsiveness to user needs.

6. Measure Impact Beyond Efficiency

Track social and emotional outcomes such as employee satisfaction, trust, and work-life balance—not just productivity and cost metrics.

7. Foster Leadership Vision and Commitment

Executive leaders must champion HRC as part of a broader organizational transformation strategy and model the behaviors expected in a digitally empowered workforce.

8. Encourage Interdisciplinary Collaboration

Create cross-departmental innovation teams to facilitate the integration of AI with existing workflows, ensuring seamless human-machine synergy.

Implications of the Study

The integration of artificial intelligence (AI) and robotics through Human-Robot Collaboration (HRC) is not merely a technological shift; it represents a profound transformation in organizational processes, human resource strategies, policy considerations, and societal interactions. The findings of this study provide insights that have both theoretical significance and practical managerial applications.

Theoretical Implications

This study contributes to the theoretical understanding of HRC by emphasizing the multi-dimensional nature of human-machine interaction. Traditional models of workplace automation largely focused on efficiency gains and task replacement. The research highlights the importance of considering emotional, ethical, and social dimensions in HRC, suggesting that future theories should incorporate variables such as employee engagement, trust, and cultural adaptation alongside technological performance. The findings underscore that AI and robotics are not mere operational tools but strategic enablers. From a theoretical perspective, this supports models of digital transformation that view technology as a catalyst for organizational agility, innovation, and competitive advantage. It calls for refinement of existing strategic management theories to account for AI as a dynamic capability that influences long-term organizational outcomes. The study also advances theoretical discussions around workforce development by proposing that competency-based models are more effective than role-based approaches in AI-augmented environments. Incorporating adaptability, digital literacy, emotional intelligence, and ethical decision-making into human capital theories expands the understanding of skills required in technologically mediated workplaces.

By highlighting ethical concerns and cultural shifts in HRC, the study contributes to theoretical frameworks connecting technology adoption with organizational behavior, leadership, and trust. It emphasizes the need for theories that integrate ethical AI governance, human-machine synergy, and inclusive culture as core components of technology adoption models. The research extends theory beyond the organizational boundary, offering a societal lens to HRC. It suggests that theories of technology diffusion, labor economics, and public policy must consider not only productivity outcomes but also social equity, workforce inclusion, and public trust in automation technologies.

Managerial Implications

For managers, the study highlights the need to embed HRC into organizational strategy. AI and robotics should be treated as strategic assets, not just operational tools. Managers must develop enterprise-wide digital roadmaps that align AI adoption with organizational goals, vision, and values. Clear objectives, measurable milestones, and inclusive leadership are crucial for ensuring successful implementation and adoption. The research points to significant managerial responsibilities in reskilling and upskilling the workforce. Managers should invest in learning ecosystems that facilitate continuous professional development, support career transitions, and foster adaptability. Performance management systems must also evolve to recognize contributions in human-machine collaborations, including emerging roles such as AI supervisors and robot coordinators. Managers play a pivotal role in shaping organizational culture in AI-integrated workplaces. Leaders must promote openness to technology, encourage experimentation, and embed ethical practices in decision-making processes. Traditional hierarchical leadership models are insufficient; managers must adopt digital fluency and collaborative approaches to lead hybrid human-machine teams effectively. Managers must also ensure that AI deployment adheres to ethical standards and regulatory requirements. This includes addressing concerns around algorithmic bias, data privacy, and workforce impact. Establishing AI ethics committees, transparent governance structures, and compliance protocols is critical to fostering trust and social legitimacy. The study emphasizes the importance of co-creation and continuous engagement with technology providers. Managers should advocate for human-centric design, inclusive interfaces, and adaptable robotic systems. Engaging end users throughout the development and implementation phases enhances adoption, reduces resistance, and ensures that technologies augment rather than replace human capabilities.

Finally, managerial implications extend beyond the organization. Managers must participate in public discourse, promote AI literacy, and implement socially responsible practices to mitigate risks of inequality and dehumanization. Organizations that actively engage with civil society, educational institutions, and policymakers can enhance both societal impact and organizational legitimacy.

CONCLUSION

The unprecedented rise of artificial intelligence (AI), robotics, and intelligent automation marks the dawn of a new era in organizational transformation. The integration of these technologies into core business operations through Human-Robot Collaboration (HRC) represents not merely a technological upgrade but a fundamental shift in how work is conceived, performed, and managed. This chapter has explored HRC as a multifaceted phenomenon—technological, strategic, social, and ethical—and emphasized its potential to reshape enterprise operations across sectors such as manufacturing, healthcare, and public administration.

Drawing from case studies, theoretical frameworks, and empirical research, this study has provided a holistic analysis of how HRC is enabling the evolution of the workplace into a hybrid ecosystem where humans and intelligent machines coexist, cooperate, and co-create value. The following key themes summarize the overarching conclusions of the study.

1. HRC as a Strategic Imperative in the Digital Age

One of the central conclusions drawn is that Human-Robot Collaboration is no longer a speculative or futuristic endeavor—it is a strategic necessity. Organizations that leverage HRC are better positioned to respond to volatility, complexity, and hyper-competition that characterize modern markets. Automation of routine tasks, real-time data analytics, and predictive decision-making empower firms to increase productivity, reduce operational costs, and enhance agility.

However, the implementation of HRC must be driven not merely by technological feasibility but by strategic alignment with organizational goals, customer expectations, and social responsibility. Companies must develop comprehensive digital transformation roadmaps where HRC plays a pivotal role in achieving sustainable and inclusive growth.

2. Augmenting Human Capabilities, Not Replacing Them

A recurring theme throughout this study is that AI and robotics should be viewed as collaborators, not competitors. When designed and deployed thoughtfully, robotic systems can complement human abilities, taking over physically demanding, dangerous, or repetitive tasks while freeing humans to focus on complex, creative, and interpersonal responsibilities.

The notion of augmentation over automation forms the ethical and strategic bedrock of successful HRC models. For instance, in healthcare settings, surgical robots assist doctors in delivering precision-based procedures while relying on human judgment and empathy. In manufacturing, collaborative robots (cobots) work alongside humans to enhance safety, speed, and accuracy.

This reimagining of roles also shifts the organizational focus from task execution to value creation, with employees seen as innovators, problem solvers, and change agents in AI-integrated environments.

3. The Centrality of Human-Centered Design and Change Management

The study underscores the importance of adopting a human-centric approach to HRC. Successful integration of robots into the workplace is not just a matter of engineering—it requires cultural, behavioral, and structural adaptation. Workers must feel psychologically safe, empowered, and equipped to work alongside intelligent machines.

Human-centered design ensures that robotic systems are intuitive, inclusive, and adaptable. Organizations must actively involve end-users in the co-creation process, solicit feedback, and iterate designs based on usability and ethical considerations. Additionally, the implementation must be supported by comprehensive change management strategies, including transparent communication, inclusive leadership, and mechanisms to address resistance.

Change should be framed not as disruption but as evolution—an opportunity for employees to grow with technology, rather than be displaced by it.

4. Reinventing Workforce Development for the AI Era

Another significant conclusion pertains to the transformation of workforce capabilities. The future of work in AI-augmented environments requires continuous learning and the development of a hybrid skillset combining digital literacy, emotional intelligence, ethical reasoning, and domain expertise. Technical reskilling must go hand in hand with soft skills development, especially in roles that require collaboration between humans and machines.

This calls for a systemic overhaul of educational, vocational, and corporate training frameworks. Organizations must partner with academic institutions, ed-tech platforms, and government agencies to create lifelong learning ecosystems that support skill mobility and employment resilience.

Moreover, the roles of HR professionals and learning leaders are being redefined—they are no longer just administrators of training but architects of workforce transformation.

5. Ethical, Legal, and Social Implications of HRC

HRC introduces a spectrum of ethical and regulatory questions that cannot be overlooked. Algorithmic bias, job displacement, privacy concerns, and autonomy of machines are pressing issues that demand interdisciplinary and cross-sectoral responses. Organizations must go beyond compliance and adopt responsible AI principles that prioritize fairness, transparency, and accountability.

This study found that trust in AI and robots is contingent on clarity in decision-making processes, explainability of algorithms, and ethical safeguards. Regulatory bodies must work collaboratively with industry leaders, technologists, and ethicists to design adaptive governance frameworks that keep pace with innovation while protecting public interest.

The social contract of work is being redefined, and ensuring that the future of HRC is inclusive, equitable, and sustainable must remain a priority.

6. Leadership in the Age of Human-Robot Teams

The rise of HRC calls for a new model of leadership—one that is digitally fluent, emotionally intelligent, and ethically grounded. Traditional top-down approaches are insufficient in environments characterized by decentralization, rapid change, and hybrid workforces. Leaders must demonstrate the ability to orchestrate human-robot teams, resolve techno-social tensions, and inspire innovation.

Moreover, leadership must set the tone for cultural transformation by embodying values of trust, learning, collaboration, and inclusion. As the boundary between human and machine work continues to blur, leaders must champion human dignity, well-being, and purposeful work in all strategic decisions.

The study advocates for transformational and inclusive leadership as a cornerstone of successful HRC implementation.

7. Sector-Specific Applications and Lessons Learned

Throughout this study, case studies from different sectors illuminated how HRC can be tailored to industry-specific needs and constraints:

- Manufacturing: Cobots increased productivity, improved worker safety, and enabled flexible production lines. Yet, success depended on worker training and ergonomic design.
- Healthcare: Robotic surgery, automated diagnostics, and AI-powered triage systems enhanced patient outcomes and efficiency. However, ethical concerns around decision-making and data use remain.
- Public Administration: Chatbots and service robots improved public service delivery, but digital inclusion, user trust, and bureaucratic inertia posed challenges.

These examples reinforce that while the technological components of HRC are transferable, contextual adaptation is key. Best practices must be industry-specific and sensitive to cultural, regulatory, and economic environments.

8. Towards a Vision of AI-Augmented Workplaces

Ultimately, this study envisions the future enterprise as a cyber-physical-social system, where humans, machines, data, and processes converge in a dynamic, intelligent, and adaptive ecosystem. This is not a utopian fantasy, but a tangible possibility already unfolding in pilot projects and innovation clusters across the globe.

For organizations willing to embrace transformation, HRC offers a pathway to resilient business models, empowered workforces, and enhanced customer value. However, this requires not only investment in technology but also in people, governance, and values.

The journey toward AI-augmented workplaces must be guided by four interlinked principles:

1. Human Empowerment – Technology should amplify human potential, not diminish it.
2. Ethical Innovation – AI and robotics must serve the greater good and uphold social justice.
3. Inclusive Growth – All workers, regardless of age, background, or skill level, must have access to opportunity.
4. Collaborative Governance – The future of work is a shared responsibility among business, government, academia, and civil society.

Limitations of the Study

This study, while providing valuable insights into Human-Robot Collaboration (HRC) and AI integration, has several limitations. Its focus on select sectors such as manufacturing, healthcare, and public administration limits the generalizability of findings to other industries with different work structures. The study relied primarily on qualitative data and case-based analyses, which, although rich in context, may not fully represent broader organizational populations, highlighting the need for quantitative validation. Rapid technological advancements in AI and robotics also pose temporal constraints, as practices and tools may evolve quickly. Cultural, regional, and organizational variations in leadership, employee attitudes, and digital literacy may further influence adoption, limiting cross-context applicability. Additionally, reliance on self-reported data may introduce biases, and ethical, social, and policy implications were explored conceptually rather than empirically. Despite these limitations, the study offers important theoretical and managerial insights for guiding HRC implementation.

Recommendations for Future Studies

Future research on Human-Robot Collaboration (HRC) and AI integration should expand across diverse industries beyond manufacturing, healthcare, and public administration to enhance generalizability and capture sector-specific dynamics. Longitudinal studies are recommended to examine the evolving impact of HRC over time, particularly regarding workforce adaptation, organizational performance, and technology adoption. Incorporating quantitative methods, such as large-scale surveys and performance metrics, alongside qualitative approaches, can provide stronger empirical validation and richer insights. Cross-cultural studies are also needed to explore how regional, societal, and organizational differences influence the acceptance and effectiveness of AI-robot systems. Additionally, more focused investigations into ethical, social, and policy-related implications—such as algorithmic bias, workforce displacement, and inclusive technology design—would provide deeper guidance for responsible implementation. Research on co-creation strategies between employees and technology developers could further illuminate best practices for enhancing human-machine synergy.

REFERENCES

Bogue, R. (2016). *Deleuze's way: Essays in transverse ethics and aesthetics*. Routledge. DOI: 10.4324/9781315576312

Brynjolfsson, E., & McAfee, A. (2014). *The second machine age: Work, progress, and prosperity in a time of brilliant technologies*. WW Norton & company.

Bughin, J., Catlin, T., Hirt, M., & Willmott, P. (2018). Why digital strategies fail. *The McKinsey Quarterly*, *1*(1), 14–25.

Burrell, J. (2016). How the machine 'thinks': Understanding opacity in machine learning algorithms. *Big Data & Society*, *3*(1), 2053951715622512. DOI: 10.1177/2053951715622512

Colgate, J. E., Wannasuphoprasit, W., & Peshkin, M. A. (1996, November). Cobots: Robots for collaboration with human operators. In *ASME international mechanical engineering congress and exposition* (Vol. 15281, pp. 433-439). American Society of Mechanical Engineers.

Davenport, T. H., & Ronanki, R. (2018). Artificial intelligence for the real world. *Harvard Business Review*, *96*(1), 108–116.

De Graaf, M. M., & Allouch, S. B. (2013). Exploring influencing variables for the acceptance of social robots. *Robotics and Autonomous Systems*, *61*(12), 1476–1486. DOI: 10.1016/j.robot.2013.07.007

Eubanks, V. (2018). *Automating inequality: How high-tech tools profile, police, and punish the poor*. Macmillan+ ORM.

Floridi, L., Cowls, J., Beltrametti, M., Chatila, R., Chazerand, P., Dignum, V., Luetge, C., Madelin, R., Pagallo, U., Rossi, F., Schafer, B., Valcke, P., & Vayena, E. (2018). AI4People—An ethical framework for a good AI society: Opportunities, risks, principles, and recommendations. *Minds and Machines*, *28*(4), 689–707. DOI: 10.1007/s11023-018-9482-5 PMID: 30930541

Frey, C. B., & Osborne, M. A. (2017). The future of employment: How susceptible are jobs to computerisation? *Technological Forecasting and Social Change*, *114*, 254–280. DOI: 10.1016/j.techfore.2016.08.019

Hancock, P. A., Billings, D. R., Schaefer, K. E., Chen, J. Y., De Visser, E. J., & Parasuraman, R. (2011). A meta-analysis of factors affecting trust in human-robot interaction. *Human Factors*, *53*(5), 517–527. DOI: 10.1177/0018720811417254 PMID: 22046724

Jobin, A., Ienca, M., & Vayena, E. (2019). The global landscape of AI ethics guidelines. *Nature Machine Intelligence*, *1*(9), 389–399. DOI: 10.1038/s42256-019-0088-2

Krüger, R., Klucken, J., Weiss, D., Tönges, L., Kolber, P., Unterecker, S., Lorrain, M., Baas, H., Müller, T., & Riederer, P. (2017). Classification of advanced stages of Parkinson's disease: Translation into stratified treatments. *Journal of Neural Transmission (Vienna, Austria)*, *124*(8), 1015–1027. DOI: 10.1007/s00702-017-1707-x PMID: 28342083

Lin, F. R., Ferrucci, L., Metter, E. J., An, Y., Zonderman, A. B., & Resnick, S. M. (2011). Hearing loss and cognition in the Baltimore Longitudinal Study of Aging. *Neuropsychology*, *25*(6), 763–770. DOI: 10.1037/a0024238 PMID: 21728425

Lu, R., Zhao, X., Li, J., Niu, P., Yang, B., Wu, H., Wang, W., Song, H., Huang, B., Zhu, N., Bi, Y., Ma, X., Zhan, F., Wang, L., Hu, T., Zhou, H., Hu, Z., Zhou, W., Zhao, L., & Tan, W. (2020). Genomic characterisation and epidemiology of 2019 novel coronavirus: Implications for virus origins and receptor binding. *Lancet*, *395*(10224), 565–574. DOI: 10.1016/S0140-6736(20)30251-8 PMID: 32007145

Pedersen, H. K., Gudmundsdottir, V., Nielsen, H. B., Hyotylainen, T., Nielsen, T., Jensen, B. A., Forslund, K., Hildebrand, F., Prifti, E., Falony, G., Le Chatelier, E., Levenez, F., Doré, J., Mattila, I., Plichta, D. R., Pöhö, P., Hellgren, L. I., Arumugam, M., Sunagawa, S., & Pedersen, O. (2016). Human gut microbes impact host serum metabolome and insulin sensitivity. *Nature*, *535*(7612), 376–381. DOI: 10.1038/nature18646 PMID: 27409811

Schwab, K., Samans, R., Zahidi, S., Leopold, T. A., Ratcheva, V., Hausmann, R., & Tyson, L. D. (2017, November). The global gender gap report 2017. World Economic Forum.

Villani, V., Pini, F., Leali, F., & Secchi, C. (2018). Survey on human–robot collaboration in industrial settings: Safety, intuitive interfaces and applications. *Mechatronics*, *55*, 248–266. DOI: 10.1016/j.mechatronics.2018.02.009

Westerman, G., Bonnet, D., & McAfee, A. (2014). *Leading digital: Turning technology into business transformation*. Harvard Business Press.

Wilson, H. J., & Daugherty, P. R. (2018). Collaborative intelligence: Humans and AI are joining forces. *Harvard Business Review*, *96*(4), 114–123.

Wurman, P. R., D'Andrea, R., & Mountz, M. (2008). Coordinating hundreds of cooperative, autonomous vehicles in warehouses. *AI Magazine*, *29*(1), 9–9.

Yang, B., Fu, X., Sidiropoulos, N. D., & Hong, M. (2017, July). Towards k-means-friendly spaces: Simultaneous deep learning and clustering. In *international conference on machine learning* (pp. 3861-3870). PMLR.

Yang, Y., Lu, Q. B., Liu, M. J., Wang, Y. X., Zhang, A. R., Jalali, N., . . . Fang, L. Q. (2020). Epidemiological and clinical features of the 2019 novel coronavirus outbreak in China. medrxiv, 2020-02. DOI: 10.1101/2020.02.10.20021675

Chapter 5
Human-Robot Collaboration for Organizational Transformation: A Bibliometric Analysis

Muhamad Bagas Firmansyah
https://orcid.org/0009-0008-0029-0745
University of Muhammadiyah Malang, Indonesia

Ilyas Masudin
https://orcid.org/0000-0002-0153-4560
University of Muhammadiyah Malang, Indonesia

Fien Zulfikarijah
https://orcid.org/0000-0001-7521-2065
University of Muhammadiyah Malang, Indonesia

Uci Yuliati
University of Muhammadiyah Malang, Indonesia

Aniek Rumijati
University of Muhammadiyah Malang, Indonesia

ABSTRACT

This review explores the transformative role of Human-Robot Collaboration in driving organizational transformation, particularly within the framework of Industry 4.0. The rise of advanced technologies has revolutionized organizational dynamics

DOI: 10.4018/979-8-3373-3404-2.ch005

by fostering enhanced productivity, innovation, and adaptability. Human-Robot Collaboration emerges as a critical enabler of adaptive and resilient work environments, empowering human workers while optimizing operations across various industries. Recognizing the urgency of understanding Human-Robot Collaboration impact, this study employs a systematic bibliometric analysis of peer-reviewed articles from the Scopus database, focusing on publications from 2020–2025. Through rigorous screening and eligibility criteria, 233 articles were selected to provide a comprehensive foundation for examining the implications of Human-Robot Collaboration. By synthesizing insights from existing literature, this review underscores the pivotal role of Human-Robot Collaboration in organizational transformation and its potential to shape the future of work.

INTRODUCTION

The advent of advanced technologies, particularly robotics and artificial intelligence (AI), has fundamentally altered the dynamics of modern organizations (Stasevych & Zvarych, 2023). These technological innovations have paved the way for transformative shifts in organizational processes, enabling businesses to enhance productivity, streamline operations, and gain a competitive edge. Among these developments, Human-Robot Collaboration has emerged as a critical area of focus, offering unprecedented opportunities for integrating human ingenuity with robotic precision (Khoa et al., 2023). This synergy has the potential to drive organizational transformation, fostering adaptive and resilient work environments.

Human-Robot Collaboration represents a paradigm shift in how humans and machines interact within workplaces (Mia & Shuford, 2024). Unlike traditional automation, where robots operate independently, Human-Robot Collaboration involves direct collaboration between humans and robots to achieve shared objectives (Chand & Lu, 2023). This collaboration is characterized by dynamic task allocation, mutual adaptation, and real-time interaction, allowing organizations to leverage the complementary strengths of humans and robots. Humans contribute cognitive flexibility, creativity, and decision-making capabilities, while robots provide precision, consistency, and efficiency.

In the era of Industry 4.0, Human-Robot Collaboration is reshaping various industries by enabling the implementation of collaborative robots (cobots) and other advanced robotic systems that are designed to work seamlessly alongside humans (Carissoli et al., 2023). For example, in the manufacturing sector, cobots are transforming assembly lines by automating repetitive tasks while allowing human workers to focus on more complex and value-added activities (Abeliansky et al., 2024). Similarly, in the healthcare sector, robotic systems are being deployed to assist in surgeries, rehabilitation, and patient care, improving outcomes and operational efficiency (Mukherjee et al., 2022).

The successful implementation of Human-Robot Collaboration requires a multidimensional approach that addresses technical, organizational, and human factors (K. J. Wang et al., 2023). On the technical front, the design of collaborative robots must prioritize safety, user-friendliness, and adaptability (Fraboni et al., 2023). Sensors, machine learning algorithms, and advanced interfaces play a pivotal role in enabling robots to understand and respond to human actions in real time (Toupas et al., 2023). From an organizational perspective, integrating HRC involves reengineering workflows, redefining roles, and fostering an environment of innovation. Human factors, such as trust, communication, and employee willingness to engage with robotic systems, also play a significant role in ensuring the success of HRC initiatives (Erro-Garcés & Aramendia-Muneta, 2023).

The benefits of Human-Robot Collaboration extend beyond operational efficiency (Li et al., 2021). By enhancing flexibility and responsiveness, Human-Robot Collaboration enables organizations to adapt to rapidly changing market demands and technological advancements. Additionally, it fosters a culture of collaboration and innovation, empowering employees to work more effectively and creatively. As industries continue to evolve, Human-Robot Collaboration has the potential to redefine the future of work, creating hybrid teams where humans and robots coexist in symbiotic harmony (Baltrusch et al., 2022).

Despite the growing interest in Human-Robot Collaboration, there remains a need for comprehensive research to explore its full potential and address emerging challenges. Key areas for future investigation include the development of more intuitive robotic systems, strategies for optimizing human-robot interaction, and the ethical implications of widespread Human-Robot Collaboration adoption. Bibliometric analysis offers a robust methodological approach for examining these topics, as it enables researchers to identify patterns, trends, and gaps in the academic literature. By synthesizing existing knowledge and highlighting new directions, bibliometric studies can contribute significantly to advancing the understanding of Human-Robot Collaboration and its transformative impact on modern organizations.

METHODOLOGY

Based on a systematic review on the theme of Human-Robot Collaboration, articles were retrieved from Scopus databases using specific search criteria. In the Scopus database, the search query focused on Title, Abstract, and Keywords fields with terms such as "human robot collaboration" and "Human Resources". The topic field was used, encompassing Title, Abstract, and Author Keywords, with the same search terms. This initial search yielded 525 articles from Scopus.

During the screening phase, all authors participated by examining random sets of articles through their titles and abstracts. Articles were included if they were written in English, publish at least 5 years (2020-2025), journal, final publication stage, open access and specifically addressed human–robot collaboration and human robot collaboration. Of the 525 documents screened, 292 articles were excluded for not meeting these criteria, leaving 233 articles for the eligibility phase.

In the eligibility phase, the full texts of the remaining articles were reviewed to ensure compliance with the criteria. As a result, 233 articles for the final analysis. This rigorous process underscores the systematic approach to selecting high-quality, peer-reviewed studies on human–robot collaboration and human resources in organizational environments.

RESULTS AND DISCUSSION

This research was conducted to analyze the literature related to keywords "Human Robot Collaboration" and "Human Resources" or from Scopus databases journal and proceeding range from 2020 to 2025. Co-authorship, co-accuracy, and citations are among the three aspects examined. The VOSviewer software was used to per form the bibliometric analysis. The researcher employs the minimal occurrence limit 1 when providing co-authorship, minimal 10 keywords of co-occurrence, and citation data. The data can be seen in Figure below

Figure 1. Co-Authorship based on author name

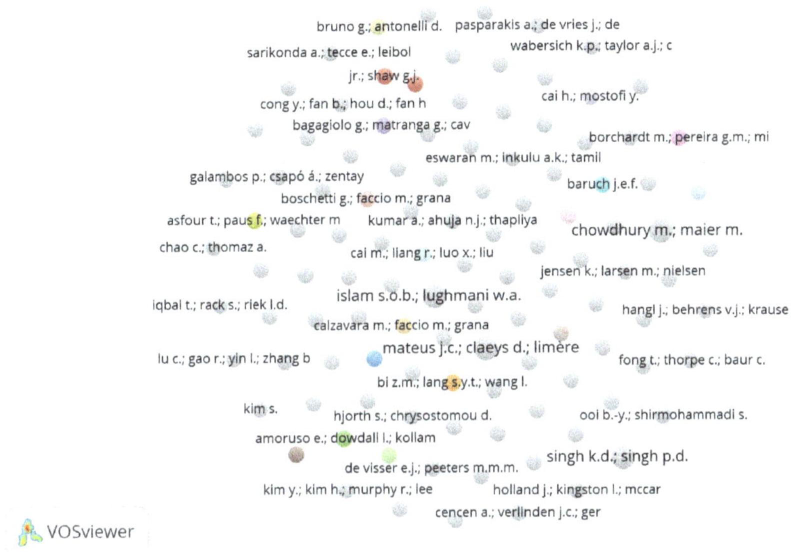

Source: VosViewer data processed (2025)

Based on figure 1 the image represents a co-author network visualization created using VOSviewer, showcasing the collaborative relationships among researchers in a particular research field. Minimum of 1 Document including authors with at least one publication ensures that the analysis captures all potential contributors to the research field. This is particularly important for identifying emerging researchers or those who may only have a few but significant publications. Limiting to authors with no more than 25 documents helps maintain focus on active contributors while avoiding over-representation by prolific authors who might dominate the network. The size of the nodes generally reflects the number of documents an author has contributed to, whereas the colors indicate different clusters or groups of closely collaborating authors. These clusters often emerge due to shared research interests, institutional affiliations, or geographical proximity.

One of the most prominent clusters in the network is centered around the authors is Mateus et al., (2019). Their central positioning and dense interconnections suggest that they are key contributors within their group, potentially serving as leading figures in a particular subfield. Their frequent collaborations indicate strong academic ties and potentially coordinated research agendas. Similarly, another significant cluster can be observed around is Islam & Lughmani, (2023), who also occupy central positions within their respective networks. These authors appear to act as focal points of collaboration, facilitating joint research efforts and possibly influencing the development of their research area.

Other clusters, display a more localized pattern of collaboration, likely reflecting focused research communities or specialized areas of study. The presence of these clusters demonstrates how academic collaborations are often concentrated within specific circles, though some authors act as bridges between groups. The proximity between Singh & Singh, (2023)., though in separate clusters, implies a potential bridging role that could connect distinct research communities.

The network also highlights a degree of fragmentation, with several authors appearing on the periphery of the visualization. These peripheral authors are either weakly connected or entirely isolated, which may suggest limited collaboration or recent entry into the research domain. Despite this, their presence indicates ongoing expansion and diversity within the academic field.

In summary, the co-authorship network provides a comprehensive view of the structural patterns of academic collaboration in the selected domain. It identifies central figures, reveals collaborative hubs, and uncovers potential gaps or opportunities for new partnerships. By analyzing such a network, researchers can better understand the dynamics of their field, identify influential authors, and strategically seek collaboration that enhances the visibility and impact of their work.

Figure 2. Co Occurance based on keywords

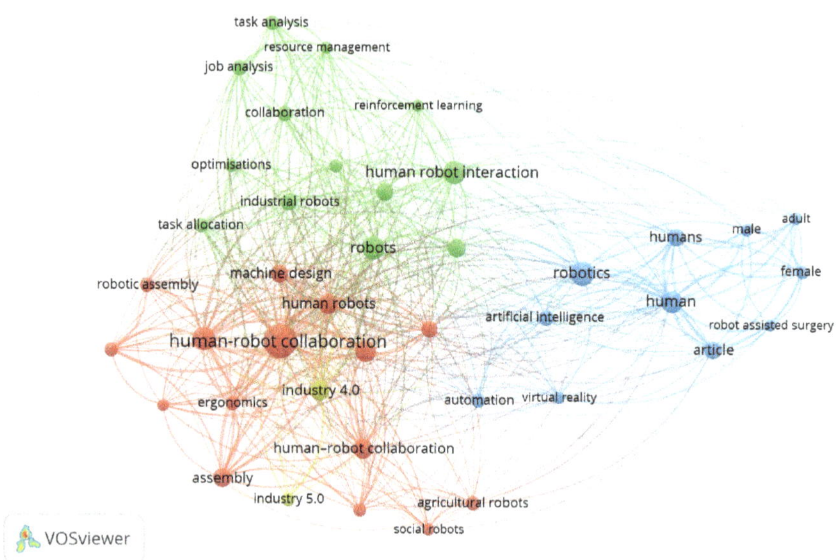

Source: VosViewer data processed (2025)

Based on figure 2 the bibliometric visualization generated by VOSviewer reveals three major clusters that reflect the evolving landscape of Human-Robot Collaboration and its role in driving organizational transformation. These clusters offer a structured understanding of how Human-Robot Collaboration is integrated into operational processes, task management, and human-centered technology interfaces within modern organizations.

The first cluster centers on technological implementation and organizational integration, with "human-robot collaboration" as the dominant keyword, indicating its foundational role in current industrial discourse. Closely associated terms such as industry 4.0, industry 5.0, robotic assembly, machine design, social robots, human robots, ergonomics, and assembly reflect a multidimensional focus on both technological advancement and human-centered design. This cluster illustrates how Human-Robot Collaboration is deeply embedded in the ongoing evolution of industrial systems, where the synergy between human intelligence and robotic precision enhances efficiency, safety, and innovation. The inclusion of ergonomics highlights attention to human well-being and work adaptability in increasingly automated environments. Moreover, the convergence of social and collaborative robotics signals a shift toward systems that not only perform tasks but also interact meaningfully with human co-workers. This cluster emphasizes that Human-Robot Collaboration is not merely a technological upgrade but a comprehensive organizational transformation that redefines workflows, workplace dynamics, and production capabilities through structured, technology-driven integration.

The second cluster emphasizes task management, optimization, and operational efficiency, with "human-robot interaction" and "robots" as its central themes. Surrounding keywords such as task allocation, task analysis, job analysis, resource management, collaboration, reinforcement learning, industrial robots, and optimisations illustrate a detailed focus on how Human-Robot Collaboration enhances the operational dimension of organizational transformation. This cluster captures the evolving dynamics of work distribution, where robotic systems take on repetitive or high-precision tasks, enabling humans to concentrate on decision-making, supervision, and creative problem-solving. Reinforcement learning and intelligent algorithms allow robots to adapt and improve task execution over time, leading to continuous performance optimization. The integration of Human-Robot Collaboration into resource and task management systems also enables more effective scheduling and workload distribution, reducing inefficiencies and bottlenecks. Overall, this cluster highlights a micro-level transformation that reshapes job design, enhances collaboration, and leverages intelligent automation to achieve strategic operational goals in increasingly complex organizational environments.

The third cluster emphasizes the intersection of human factors and technological interface, focusing on the integration of advanced technologies with human-centric design. Central keywords such as "human" and "robotics" are accompanied by related terms including artificial intelligence (AI), automation, virtual reality (VR), and robot-assisted surgery, along with demographic descriptors like male, female, and adult. While the cluster often aligns with healthcare and cognitive applications, its implications extend into broader organizational transformation. The emphasis on human-centered technology highlights the need for adaptive leadership, employee retraining, and inclusive design practices that consider diverse user profiles. This cluster reflects a cultural and social shift in how organizations approach digital integration not only through infrastructure upgrades but also through changes in mindset, communication, and team dynamics. The adoption of AI and VR in particular requires a reevaluation of ethical frameworks, privacy concerns, and decision-making processes. As such, this cluster underscores how technological innovation must be balanced with empathy, usability, and human well-being in organizational contexts.

The integration of Human-Robot Collaboration with organizational transformation is articulated through three interconnected thematic pathways. The first pathway, represented by the red cluster, emphasizes the redesign of physical and operational systems, including robotics implementation, machine design, and ergonomic adaptation, which redefines how workspaces and production lines function. The second pathway, captured in the green cluster, focuses on the intelligent distribution of tasks and decision-making, where robotic systems complement human labor through task allocation, resource management, and optimization strategies, enhancing operational efficiency. The third pathway, reflected in the blue cluster, addresses the sociotechnical alignment between humans and machines, emphasizing human factors, AI, virtual reality, and ethical interfaces. This includes rethinking organizational culture, leadership, workforce capabilities, and emotional well-being. Taken together, these clusters illustrate that Human-Robot Collaboration is more than a technological advancement; it serves as a strategic lever for comprehensive organizational transformation reshaping structures, processes, behaviors, and performance to meet the demands of a rapidly evolving, technology-driven environment..

Table 1. HRC Related to Management Fields

Topics	Source
Human-Robot Collaboration & Operations Management	(Ram et al., 2024); (Yu, Lu, et al., 2024); (Korivand et al., 2024); (Lu et al., 2024)' (Lario et al., 2025); (Eswar Balachandar & Bhaskar Reddy, 2024); (Yi et al., 2023); (Eswaran et al., 2024); (Jia et al., 2025); (He et al., 2025); (Feddoul et al., 2023); (Chung & Tan, 2025); (Elendu et al., 2024); (Huang et al., 2024); (Bolla et al., 2023); (Kavvathas et al., 2024); (Hsu et al., 2024); (Kim et al., 2024); (Kermenov et al., 2025); (Wang et al., 2024); (Ciano et al., 2025); (Rakholia et al., 2024); (Liu et al., 2022); (Coronado et al., 2024); (Inkulu & Bahubalendruni, 2024); (Hjorth & Chrysostomou, 2022); (Y. Kim et al., 2022); (Mbunge et al., 2021); (Holland et al., 2021)
Human-Robot Collaboration & Human Resources Management	(Chuang, 2024); (Shahid et al., 2025); (Gervasi et al., 2024); (Cao & Chen, 2024); (Keshvarparast et al., 2024); (Rastgoftar, 2024); (D'souza & Tapas, 2024); (Calzavara et al., 2024); (J.-M. Li, Liu, et al., 2024); (Dagkinis et al., 2025); (D. Wang et al., 2024); (Liau & Ryu, 2024); (Wu & Zhang, 2024); (Tian et al., 2024); (Shirakura et al., 2025); (Alenjareghi et al., 2024); (Uğuz Arsu & Sipahi Döngül, 2023); (Almokdad & Lee, 2024)
Human-Robot Collaboration & Marketing Management	(Ji et al., 2024); (Li, Zhang, et al., 2024); (Wang et al., 2024)
Human-Robot Collaboration & Finance Management	(Peng & Wu, 2021); (Sarikonda et al., 2024); (Ietto et al., 2024)

Source: VosViewer data processed (2025)

Human-Robot Collaboration has become an integral part of modern organizational transformation across various management domains. In Operations Management, Human-Robot Collaboration enhances process efficiency through collaborative automation, increased precision, and reduced exposure to hazardous manual tasks. Collaborative robots (cobots) allow organizations to automate repetitive work without replacing humans, but rather reposition them toward higher-value activities such as decision-making and quality control. This accelerates the transition toward lean and agile operational models.

Table 2. Impact of HRC on Operations Management

Impact of HRC	Source
Industry 4.0	(Ciano et al., 2025); (Rakholia et al., 2024); (Inkulu & Bahubalendruni, 2024)
Industry 5.0	(Yu, Lu C., et al., 2024); (Feddoul et al., 2023); (Chung & Tan, 2025); (Bolla et al., 2023); (B. Wang et al., 2024); (Coronado et al., 2024)
Healthcare Sector	(Ram et al., 2024); (Elendu et al., 2024); (Koç, 2023); (Mbunge et al., 2021); (Holland et al., 2024)
Quality Assurance	(Lario et al., 2025)
Optimization	(Yi et al., 2023); (Eswar Balachandar & Bhaskar Reddy, 2024)

continued on following page

Table 2. Continued

Impact of HRC	Source
Scheduling	(He et al., 2025); (Kavvathas et al., 2024)
Task Allocation	(Yi et al., 2023); (Eswaran et al., 2024)
Circular Economy	(Eswaran et al., 2024); (Y. H. Kim et al., 2024).
Coal Mining and High Risk Sector	(Huang et al., 2024)

Source: VosViewer data processed (2025)

Based on the image provided (Table 2), Human-Robot Collaboration (HRC) has a broad and transformative impact across various domains within Operations Management. In the context of Industry 4.0 and 5.0, HRC facilitates the integration of intelligent systems, enabling real-time responsiveness, customization, and sustainable practices. In the healthcare sector, HRC supports the delivery of precise and efficient care through robotic-assisted interventions and automation of routine tasks. In terms of quality assurance and optimization, robots enhance accuracy, reduce errors, and maintain consistent standards. HRC also improves scheduling and task allocation through data-driven analysis and predictive modeling, ensuring optimal use of human and robotic resources. Furthermore, its application in the circular economy and high-risk sectors like coal mining improves safety and environmental sustainability. Collectively, these impacts demonstrate that HRC not only enhances productivity and safety but also fosters innovation, adaptability, and systemic resilience within modern operations management frameworks.

As a core component of Industry 4.0, Human-Robot Collaboration bridges the gap between cyber-physical systems and the human workforce, creating a synergistic environment where technology and human intelligence coexist. According to Ciano et al., (2025); Rakholia et al., (2024); and Inkulu & Bahubalendruni, (2024) Human-Robot Collaboration enables the seamless integration of robotics, Internet of Things, big data analytics, and machine learning into daily operational processes, enhancing both precision and efficiency. Through real-time data insights and predictive maintenance, organizations can reduce downtime, optimize resource usage, and prevent equipment failures before they occur. This connectivity facilitates agile production models that can quickly adapt to shifting market demands and support mass customization. Furthermore, Human-Robot Collaboration empowers smart manufacturing by enabling decentralized decision-making and autonomous systems that still align with human oversight. The result is a flexible, responsive, and sustainable production ecosystem that embodies the goals of Industry 4.0. Ultimately, HRC not only transforms operational performance but also redefines the future of work within industrial settings.

According to Yu, Lu C., et al., (2024); Feddoul et al., (2023); Chung & Tan, (2025); Bolla et al., (2023); B. Wang et al., (2024); and Coronado et al., (2024) Human-Robot Collaboration is a pivotal enabler of the Industry 5.0 paradigm, where the goal extends beyond efficiency to include human-centric, resilient, and sustainable operations. Human-Robot Collaboration optimizes workflows by enabling robots to perform repetitive, hazardous, or precision-intensive tasks, thereby reducing human exposure to risks and operational errors. Meanwhile, human workers concentrate on oversight, creative problem-solving, strategic planning, and ethical decision-making, fostering a balanced division of labor. This collaboration increases operational flexibility and responsiveness, especially in dynamic production environments. Moreover, Human-Robot Collaboration supports advanced scheduling, real-time task allocation, and predictive maintenance, which collectively improve productivity and reduce downtime. It also contributes to quality assurance by ensuring consistency in output while allowing humans to oversee critical control points. As a result, operations become not only more efficient but also more adaptive to the demands of sustainability and personalized production.

According to Ram et al., (2024); Elendu et al., (2024); Koç, (2023); Mbunge et al., (2021); and Holland et al., (2021) in the healthcare sector, Human-Robot Collaboration is increasingly recognized as a key driver of operational excellence, particularly in enhancing quality assurance and overall patient care. The integration of intelligent robots into clinical workflows enables healthcare providers to streamline routine tasks, reduce the likelihood of human error, and ensure a higher standard of service delivery. For example, nursing robots can assist in administering medications, monitoring vital signs, and transporting medical supplies activities that, when automated, improve efficiency and reduce the burden on medical staff. This allows healthcare professionals to dedicate more time to complex decision-making and patient interaction, thereby improving the overall quality of care.

From a quality assurance perspective, Human-Robot Collaboration contributes to the standardization of procedures and the reliability of service execution (Lario et al., 2025). Robotic systems are capable of performing repetitive tasks with high precision and consistency, minimizing variability that often arises from human factors such as fatigue or stress. Moreover, these systems can capture real-time data on patient health indicators and procedural outcomes, creating a feedback loop that supports continuous quality improvement. This data can be used for predictive analytics, early detection of anomalies, and evidence-based clinical decisions, all of which are essential components of a robust quality assurance framework.

In addition, Human-Robot Collaboration enhances safety and compliance by ensuring that tasks are performed in accordance with established protocols, thereby reducing the incidence of medical errors and adverse events. Robots can also assist in maintaining hygiene standards through automated disinfection and sterile handling of equipment, further contributing to patient safety. As healthcare systems worldwide face increasing pressure due to aging populations and resource constraints, the strategic deployment of collaborative robots presents a sustainable solution. By optimizing task allocation and supporting real-time communication between human and robotic agents, Human-Robot Collaboration fosters a more responsive, efficient, and accountable healthcare environment aligned with the principles of quality assurance (Yi et al., 2023).

According to Yi et al., (2023); and Eswaran et al., (2024), Human-Robot Collaboration significantly enhances operational performance by driving optimization through intelligent task allocation and adaptive scheduling systems. In modern operational environments, efficiency and responsiveness are critical, and Human-Robot Collaboration introduces a transformative framework to achieve both. Optimization begins with the strategic deployment of robotic systems integrated with artificial intelligence (AI), which enables the automatic assignment of repetitive, time-sensitive, or physically demanding tasks to robots. These systems analyze variables such as task priority, worker availability, and real-time performance data to determine the most efficient allocation of responsibilities.

Task allocation in Human-Robot Collaboration also promotes optimal utilization of both human and machine capabilities. Robots handle routine, hazardous, or high-precision tasks with consistency and speed, while humans are tasked with roles that require emotional intelligence, adaptive thinking, and complex decision-making. This division not only preserves human energy and motivation but also boosts overall system throughput. It enhances employee satisfaction by reducing physical strain and cognitive overload, while maintaining high standards of operational output. Overall, the integration of intelligent task allocation and adaptive scheduling through Human-Robot Collaboration fosters a data-driven, responsive, and highly optimized operational ecosystem. It aligns with the core objectives of operations management efficiency, agility, and resource maximization ultimately leading to increased productivity, improved quality, and long-term organizational resilience.

According to He et al., (2025); and Kavvathas et al., (2024) scheduling, is a traditionally rigid and manual process is revolutionized through Human-Robot Collaboration. Adaptive scheduling systems that integrate both human and robotic capacities allow for dynamic adjustments in response to workflow disruptions, demand fluctuations, or equipment failure. For example, in a manufacturing setting, if a robot encounters an unexpected delay, the system can instantly reassign the pending task to an available human worker or another robot, minimizing downtime. This flexibility ensures smoother operations, better lead-time adherence, and reduced production delays factors crucial in highly competitive markets.

In addition to mitigating disruptions, adaptive scheduling systems facilitated by Human-Robot Collaboration (HRC) optimize resource allocation and enhance productivity across various industries. By leveraging real-time data and predictive analytics, these systems can forecast workload patterns, identify potential bottlenecks, and proactively adjust task assignments to maintain operational efficiency. This predictive capability also enables organizations to maximize the utilization of both human and robotic resources, ensuring tasks are allocated based on strengths, availability, and efficiency. For instance, complex problem-solving tasks can be assigned to human workers, while repetitive or precision-demanding tasks are delegated to robots.

Within the context of the circular economy, Human-Robot Collaboration serves as a key enabler of sustainable and efficient production systems (Hjorth & Chrysostomou, 2022); and (Y. H. Kim et al., 2024). By integrating robots into critical processes such as material sorting, product disassembly, and component recycling, organizations can significantly reduce waste and enhance the recovery of valuable resources. Robotic systems are particularly effective in handling complex tasks that require precision, such as the separation of materials in e-waste or the disassembly of multi-layered products, which may be time-consuming or hazardous for human workers. Through this collaboration, operations gain improved traceability, allowing companies to monitor the life cycle of materials and ensure compliance with environmental regulations. Moreover, Human-Robot Collaboration facilitates the transition from linear to closed-loop production models by increasing throughput and consistency in circular processes. It also supports the implementation of real-time monitoring systems, ensuring that circular strategies are executed efficiently and transparently. In doing so, Human-Robot Collaboration reinforces a culture of sustainability within operations by aligning technological innovation with environmental stewardship. The combination of human oversight and robotic precision allows organizations to optimize resource use while maintaining high standards of safety, quality, and productivity. Ultimately, Human-Robot Collaboration contributes to long-term value creation by embedding sustainable principles into the very core of operational practices.

In high-risk sectors like coal mining, Human-Robot Collaboration significantly improves safety, efficiency, and regulatory compliance (Huang et al., 2024). Robots equipped with sensors and AI can perform hazardous tasks such as exploration, drilling, and inspection, reducing human exposure to dangerous conditions. Real-time data from these robots allows human operators to monitor and manage operations remotely, enabling quicker and safer decision-making. This shift minimizes downtime, enhances predictive maintenance, and ensures better adherence to safety standards. By reallocating workers from risky manual tasks to supervisory roles, Human-Robot Collaboration not only protects human health but also optimizes workforce efficiency. As mining transitions into Industry 4.0 and 5.0, Human-Robot

Collaboration fosters sustainable, human-centered operations without compromising productivity or operational control.

Table 3. Impact of HRC on Human Resources Management

Impact of HRC	Source
Transformation	(Chuang, 2024)
Change Readiness	(Shahid et al., 2025)
Mental Health	(Gervasi et al., 2024); and (D. Wang et al., 2024).
Organizational Support	(Almokdad & Lee, 2024)
Shared Workplace	(Keshvarparast et al., 2024); and (Shirakura et al., 2025)
Decision Making	(Rastgoftar, 2024)
Innovation	(D'souza & Tapas, 2024)
Productivity and Peformance	(Calzavara et al., 2024); and (Tan et al., 2025)
Workload	(J.-M. Li, Zhang, et al., 2024); and (Tian et al., 2024), (Mao et al., 2025)
Reward and Punishment	(Cao & Chen, 2024)
Motivation	(Uğuz Arsu & Sipahi Döngül, 2023).
Knowledge Management	(Liau & Ryu, 2024)
Workplace Safety	(Alenjareghi et al., 2024)

Source: VosViewer data processed (2025)

Based on the image provided (Table 3) Human-Robot Collaboration in the field of Human Resources Management demonstrates a significant and multidimensional impact. Human-Robot Collaboration compels organizations to overhaul their human capital strategies, emphasizing not only transformation and change readiness but also mental health, innovation, and workplace safety. The implementation of Human-Robot Collaboration requires targeted reskilling and upskilling programs to prepare employees to work effectively alongside robotic systems. Moreover, HRM plays a vital role in fostering a technology-adaptive organizational culture, managing workload, supporting motivation, and reconfiguring fair reward and punishment mechanisms. Human Resources also contributes to shared decision-making, supports knowledge management, and strengthens employee engagement in digitally transforming environments. This transformation demands a shift in mindset, as well as the development of digital competencies across all organizational levels. Thus, Human-Robot Collaboration within HRM is not merely a technological integration it is a strategic, holistic approach to enhancing performance, resilience, and innovation in the evolving future of work.), Human-Robot Collaboration has a broad and transformative impact across various domains within Operations Management. In the context of Industry 4.0 and 5.0, Human-Robot Collaboration facilitates the integration of intelligent systems, enabling real-time responsiveness, customization, and sustainable practices. In the healthcare sector, Human-Robot Collaboration supports the delivery of precise and efficient care through robotic-assisted interventions and automation of routine tasks. In terms of quality assurance and optimization, robots enhance accuracy, reduce errors, and maintain consistent standards. Human-Robot Collaboration also improves scheduling and task allocation through data-driven analysis and predictive modeling, ensuring optimal use of human and robotic resources. Furthermore, its application in the circular economy and high-risk sectors like coal mining improves safety and environmental sustainability. Collectively, these impacts demonstrate that Human-Robot Collaboration not only enhances productivity and safety but also fosters innovation, adaptability, and systemic resilience within modern operations management frameworks.

The implementation of Human-Robot Collaboration necessitates a profound and multidimensional organizational transformation (Chuang, 2024). This transformation extends beyond technological upgrades, requiring a strategic reconfiguration of human capital, workflows, and organizational culture. Organizations must redesign their talent management approaches, emphasizing the development of digital competencies, cross-functional collaboration, and adaptive thinking. Employees are no longer solely responsible for task execution but must learn to work alongside intelligent robotic systems in a shared environment. As such, the workforce must be reskilled and upskilled to meet the demands of a hybrid human-robot ecosystem. This transformation also calls for leaders to promote agility, continuous learning, and open-mindedness across all levels of the organization.

Within this process, change readiness emerges as a critical success factor. Organizational readiness for change encompasses both the psychological preparedness of employees and the structural capacity to support change initiatives (Shahid et al., 2025). Human Resources plays a pivotal role in managing this readiness by fostering transparent communication, building trust, and engaging employees in the transformation journey. Proactive engagement reduces uncertainty and helps mitigate resistance to change, which often arises from fear of job displacement, role ambiguity, or a lack of technological confidence. Human Resources must also implement support systems such as mental health resources, peer mentoring, and feedback mechanisms to ensure that employees feel valued and supported throughout the transition.

Without adequate change readiness, the introduction of Human-Robot Collaboration may encounter significant internal resistance, creating friction that undermines strategic goals. Conversely, a well-prepared organization can leverage Human-Robot Collaboration as a catalyst for innovation, improved productivity, and enhanced employee engagement. Therefore, organizational transformation and change readiness must progress in tandem, with careful alignment between technological adoption and human adaptation. Only through this integrated approach can organizations realize the full potential of Human-Robot Collaboration and maintain long-term competitiveness in an increasingly automated and dynamic business environment.

As robots increasingly become co-workers in modern workplaces, employees are confronted with significant psychological challenges that can impact their mental health and overall well-being (Gervasi et al., 2024); and (D. Wang et al., 2024). The integration of robotic systems often leads to concerns about job security, especially when automation replaces or transforms traditional roles. Employees may experience heightened anxiety, stress, and uncertainty, particularly when the boundaries between human and robotic responsibilities are not clearly defined. This role ambiguity can result in a perceived loss of control and purpose, which are critical factors in workplace stress.

Human Resources departments play a crucial role in mitigating these mental health risks by proactively addressing the emotional impact of technological transitions. Providing access to mental health resources, such as counseling services, stress management workshops, and peer support groups, can help employees cope with the psychological demands of change. Transparent communication is equally important; when organizations clearly articulate the role of technology and its intended purpose such as augmenting rather than replacing human labor employees are more likely to feel secure and valued. Furthermore, involving employees in the decision-making process related to technological adoption fosters a sense of ownership and reduces feelings of exclusion. Encouraging open dialogue and feedback allows workers to voice their concerns, ask questions, and participate in shaping the future of their roles. Human Resources should also emphasize emotional safety and psychological resilience by cultivating a supportive and inclusive workplace culture.

Human-Robot Collaboration thrives in environments where strong organizational support is present (Almokdad & Lee, 2024). This support involves more than just providing technological infrastructure it encompasses leadership commitment, policy alignment, and the cultivation of a collaborative culture between humans and intelligent systems. Organizations must ensure that employees receive adequate training, clear communication, and psychological safety during transitions involving robotic integration. Supportive practices such as transparent change management, accessible resources, and inclusive decision-making foster trust and acceptance among the workforce. Moreover, recognizing employee contributions in hybrid work settings where tasks are shared between humans and robots reinforces motivation and role clarity. Effective organizational support also includes establishing ethical guidelines, promoting well-being, and ensuring that robotic implementation aligns with the organization's long-term goals. Without such holistic support, the benefits of Human-Robot Collaboration may be undermined by resistance, confusion, or operational inefficiencies. Therefore, organizational support is a cornerstone for sustainable, productive, and human-centered Human-Robot Collaboration adoption across industries.

Human-Robot Collaboration transforms the traditional workplace into a hybrid environment, where humans and robots work side by side to achieve shared objectives (Keshvarparast et al., 2024); and (Shirakura et al., 2025). This integration requires significant behavioral adaptation from employees, as they must learn to interact with intelligent machines that possess varying degrees of autonomy (Wu & Zhang, 2024). In shared workplaces, effective communication and mutual understanding between humans and robots become essential for maintaining seamless workflows and preventing operational errors. As such, employee behavior must evolve to include openness to technology, problem-solving in real-time, and proactive engagement with robotic systems.

Organizations must recognize that this transformation goes beyond technical skills it necessitates a cultural shift that values adaptability, cooperation, and continuous learning. Human Resources departments play a critical role in this process by fostering behavioral flexibility and encouraging collaborative mindsets. This includes providing comprehensive training on ethical considerations, machine limitations, and protocols for safe interaction with robotic systems. Additionally, Human Resources should design work environments that support trust, inclusion, and psychological safety, especially in scenarios where humans might feel displaced or threatened by automation. By cultivating a culture of mutual respect and teamwork, organizations can ensure that the human-robot workplace remains productive, safe, and emotionally sustainable for all participants involved.

In environments where humans and robots coexist, decision-making evolves into a more collaborative and data-driven process that integrates artificial intelligence with human insight (Rastgoftar, 2024). Robots, equipped with advanced sensors and machine learning algorithms, can rapidly analyze large volumes of data, detect anomalies, and provide recommendations for operational and strategic decisions. This real-time analytical capability significantly enhances the accuracy, consistency, and speed of decision-making processes across various organizational levels. However, despite these technological advantages, human input remains essential particularly in areas requiring contextual judgment, emotional intelligence, ethical reasoning, and the evaluation of complex trade-offs that algorithms alone cannot address.

The fusion of robotic efficiency and human discernment fosters a fertile ground for innovation (D'souza & Tapas, 2024). By relieving employees of routine, repetitive tasks, Human-Robot Collaboration allows them to concentrate on creative, value-adding activities such as product development, process improvement, and strategic planning. This shift not only drives continuous innovation but also cultivates a culture of experimentation and adaptability. To ensure that this synergy is effective, organizations must establish clear governance frameworks that define the boundaries and protocols of human-robot interaction. HR departments and leadership teams play a critical role in shaping these frameworks, aligning them with organizational goals, values, and compliance standards. Ultimately, well-structured collaboration between humans and robots can transform decision-making into a strategic advantage and position innovation at the core of organizational growth.

Human-Robot Collaboration holds significant promise for enhancing organizational productivity and performance by automating repetitive, time-consuming, and low-value tasks (Calzavara et al., 2024); and (Tan et al., 2025). Through the strategic deployment of collaborative robots (cobots), companies can streamline operations, reduce process inefficiencies, and increase output without compromising quality. This allows human employees to shift their focus toward more complex, creative, and decision-oriented responsibilities that contribute directly to organizational growth and innovation. As a result, performance metrics such as speed, accuracy, and customer satisfaction can improve substantially, making HRC a valuable asset in the pursuit of operational excellence.

According to Li et al., (2024); Tian et al., (2024); and Mao et al., (2025) the integration of Human-Robot Collaboration also brings challenges related to workload distribution. While robots can alleviate certain burdens, there is a risk of uneven task allocation if job roles are not redesigned with clear consideration for the division of labor between humans and machines. For example, as robots take over routine tasks, employees might face increased expectations to deliver high-level results under tighter deadlines, potentially leading to physical and cognitive overload. This imbalance can result in burnout, decreased morale, and diminished overall performance.

To mitigate such risks, Human Resource Management must play a proactive role in redefining job descriptions and implementing fair work allocation strategies. A comprehensive task analysis should be conducted to assess which responsibilities are best suited for automation and which require human judgment and flexibility. Additionally, continuous monitoring of workload, employee feedback, and performance outcomes is essential to ensure that the benefits of Human-Robot Collaboration are equitably realized across the workforce. Training and upskilling programs must also accompany Human-Robot Collaboration adoption to prepare employees for evolving job demands and foster a culture of adaptability. Ultimately, when thoughtfully managed, Human-Robot Collaboration can lead to a harmonious balance between productivity gains and sustainable employee performance, reinforcing the organization's competitive advantage in an increasingly automated world.

With the emergence of Human-Robot Collaboration and shifting job roles, traditional reward and punishment systems are becoming increasingly obsolete. The integration of robotics into human workflows requires organizations, particularly Human Resources departments, to redesign performance appraisal mechanisms to accommodate new forms of collaborative and hybrid work (Cao & Chen, 2024).

Rather than focusing solely on individual output, performance evaluations must now account for team-based outcomes, human-robot synergy, adaptability, and innovation. For example, metrics should recognize employees who demonstrate effective collaboration with robotic systems, display initiative in learning new technologies, or contribute to process improvements.

In terms of motivation, equitable and transparent reward systems are essential to sustaining employee engagement during periods of technological transition. When employees perceive fairness in how performance is measured and rewarded, their intrinsic motivation and organizational commitment are likely to increase (Uğuz Arsu & Sipahi Döngül, 2023). Conversely, outdated or misaligned punishment mechanisms may lead to frustration, resistance to change, and reduced morale. Therefore, penalties should be reframed not as punitive measures, but as developmental feedback opportunities aimed at improving future performance.

Human-Robot Collaboration generates knowledge through new patterns of interaction, workflow adjustments, and technological integration (Liau & Ryu, 2024). Effective knowledge management becomes essential to capture these insights, codify them, and ensure their accessibility across departments. Human Resources plays a central role in fostering a culture of continuous learning by facilitating training programs, establishing collaborative platforms, and encouraging cross-functional communication. Competency frameworks must be regularly updated to reflect the evolving skillsets required in Human-Robot Collaboration environments. By promoting systematic knowledge sharing and retention, organizations can enhance their agility, preserve institutional memory, and ensure sustainable innovation amid ongoing technological transformation.

Workplace safety becomes increasingly critical in environments where humans and robots operate collaboratively (Alenjareghi et al., 2024). Human Resources must implement comprehensive safety protocols that address both physical and psychological aspects of this interaction. This includes establishing clear operational boundaries, providing specialized training on robot behavior and emergency procedures, and enforcing real-time monitoring systems to prevent accidents. In addition, Human Resources must promote psychological safety by ensuring employees feel supported, valued, and confident while working alongside intelligent systems. Transparent communication, regular feedback mechanisms, and mental health resources are essential in mitigating fear or uncertainty. A culture of safety fosters trust, enhances performance, and ensures long-term human-robot coexistence.

Table 4. Impact of HRC on Marketing Management

Impact of HRC	Source
User and Customer Experience	(Ji et al., 2024); (J.-M. Li, Zhang, et al., 2024); (R. Wang et al., 2024)

Source: VosViewer data processed (2025)

In Marketing Management, Human-Robot Collaboration (HRC) significantly enhances customer experience by enabling hyper-personalized services, intelligent and context-aware interactions, and more efficient data-driven decision-making. Robots are utilized to deliver prompt, consistent, and accurate support, while human employees focus on empathy-driven communication and resolving complex customer issues. This dual approach ensures operational efficiency without compromising emotional intelligence. Furthermore, HRC allows marketers to analyze consumer behavior in real-time, leading to proactive service delivery and higher satisfaction. As customer expectations evolve in the digital age, HRC supports a strategic shift toward data-powered, customer-centric models that enhance brand loyalty and competitive positioning.

According to Ji et al., (2024); J.-M. Li, Zhang, et al., (2024); and R. Wang et al., (2024) Human-Robot Collaboration plays a transformative role in elevating user and customer experience by enabling personalized, efficient, and intelligent service delivery. In the era of digital transformation, where customer expectations continue to rise, Human-Robot Collaboration offers organizations a competitive edge by merging technological precision with human empathy. Robots equipped with artificial intelligence (AI), machine learning, and natural language processing can now interact with users in real-time, delivering rapid and accurate responses to inquiries, processing routine transactions, and recommending products or services based on behavioral analytics. This automation not only enhances operational efficiency but also ensures 24/7 service availability, reducing customer wait times and significantly boosting overall satisfaction levels.

While robots efficiently manage standardized and repetitive tasks, human professionals continue to play a crucial role in emotionally nuanced interactions. Tasks such as conflict resolution, brand storytelling, and handling sensitive or complex complaints require emotional intelligence, ethical judgment, and cultural sensitivity areas where human skills remain irreplaceable. This division of labor allows businesses to allocate resources more effectively, ensuring that customers receive both high-speed service and authentic human engagement where it matters most. Moreover, Human-Robot Collaboration supports the continuous collection and real-time analysis of customer data, providing valuable insights into user behavior, preferences, and feedback. These insights facilitate the development of hyper-personalized marketing strategies, dynamic user interfaces, and interactive digital environments. For instance, smart retail systems and online platforms powered by Human-Robot Collaboration can tailor content, product suggestions, and navigation flows based on individual user profiles, resulting in more intuitive and engaging experiences. Robots can guide users through these interactions seamlessly, while human designers ensure alignment with the organization's branding and customer experience goals.

The synergy between humans and robots in customer-facing roles reflects a broader organizational shift toward customer-centricity where the emphasis is placed not only on service delivery but also on building trust, fostering loyalty, and deepening emotional engagement. By balancing technological efficiency with human-centered values, Human-Robot Collaboration redefines the customer journey, making it more responsive, inclusive, and emotionally resonant across all touchpoints.

Table 5. Impact of HRC on Marketing Management

Impact of HRC	Source
Marginal Cost	(Sarikonda et al., 2024)
Stakeholders	(Peng & Wu, 2021)
SMEs	(Ietto et al., 2024)

Source: VosViewer data processed (2025)

Meanwhile, in Finance Management, Human-Robot Collaboration supports the automation of financial processes such as transaction processing, AI-assisted audits, and operational risk management. This collaboration improves the speed and accuracy of financial reporting, which in turn facilitates real-time strategic decision-making. Thus, organizational transformation is not limited to operations but extends to finance as a central hub for control and planning.

Human-Robot Collaboration significantly contributes to reducing marginal costs, particularly in finance-related functions, by automating repetitive and time-consuming tasks such as transaction processing, invoice handling, and data reconciliation (Sarikonda et al., 2024). By delegating these routine operations to intelligent robotic systems, organizations can minimize the need for extensive human involvement in administrative financial duties, thereby lowering labor costs. This efficiency leads to a reduction in the cost incurred for producing each additional unit of output, which is particularly beneficial in large-scale operations and highly competitive markets. Additionally, the consistency and accuracy provided by robots reduce the likelihood of errors, which can otherwise lead to costly corrections and compliance risks. The time saved through automation allows human employees to focus on strategic financial planning, forecasting, and value-added analysis. Consequently, Human-Robot Collaboration not only enhances productivity but also drives financial agility, enabling firms to reallocate resources more effectively and adapt swiftly to changing economic conditions without inflating operational expenses.

As organizations increasingly adopt intelligent automation to meet digital transformation goals, Human-Robot Collaboration has emerged as a critical factor influencing financial governance and stakeholder engagement (Peng & Wu, 2021). This enhanced accuracy and timeliness of financial data benefit a wide range of stakeholders including investors, regulatory authorities, board members, and internal decision-makers by providing reliable insights into an organization's fiscal performance and compliance. Through automated processes and intelligent analysis, Human-Robot Collaboration reduces the likelihood of human error, fraud, and data manipulation, thereby fostering a greater level of trust among stakeholders. Moreover, this improved transparency supports informed decision-making and strengthens investor confidence, which is critical for long-term capital investment and sustainable growth. Regulatory bodies are also better equipped to assess risk and ensure legal compliance, while internal stakeholders can make proactive, data-driven financial plans. The implementation of Human-Robot Collaboration further encourages stakeholder engagement by ensuring that strategic financial actions are aligned with broader organizational goals and ethical standards. Overall, Human-Robot Collaboration serves as a cornerstone of responsible and responsive financial governance.

For Small and Medium Enterprises (SMEs), Human-Robot Collaboration presents a transformative opportunity to enhance financial and operational capabilities through digitalization (Ietto et al., 2024). SMEs often operate with constrained resources, lacking access to extensive finance teams or sophisticated enterprise software. By integrating robotic process automation and AI-driven systems, SMEs can streamline critical financial tasks such as payroll processing, invoice generation, expense tracking, tax compliance, and financial forecasting. This automation significantly reduces manual errors and administrative burdens, allowing business owners to focus on strategic growth and customer engagement. Furthermore, Human-Robot Collaboration enables real-time data analysis, improving decision-making and responsiveness to changing market dynamics. It also supports marginal cost reduction by increasing efficiency and minimizing labor-intensive processes. With greater accuracy, transparency, and speed in financial reporting, SMEs can enhance investor and stakeholder confidence. Overall, Human-Robot Collaboration empowers SMEs to compete more effectively in a technology-driven marketplace, aligning their operations with broader digital transformation goals while maintaining agility and resilience.

Human-Robot Collaboration, as a cornerstone of Industry 4.0 and 5.0, revolutionizes operations management by enhancing productivity, workplace safety, and sustainability through the seamless integration of human intelligence and robotic precision. In human resource management, it drives workforce optimization by necessitating reskilling, fostering a technology-adaptive culture, and ensuring equitable work distribution, while also prioritizing psychological and physical safety. Marketing benefits significantly from this synergy, as AI-driven automation enables personalized and efficient customer engagement, while human emotional intelligence addresses complex interactions to build trust and loyalty. In the finance domain, Human-Robot Collaboration automates routine tasks, reduces costs, improves accuracy, and supports real-time decision-making, fostering transparency and agility. The successful implementation of Human-Robot Collaboration requires comprehensive organizational transformation, encompassing innovative decision-making frameworks, inclusive reward systems, and effective change management, ultimately creating resilient and responsive environments across sectors.

Overall, integrating Human-Robot Collaboration across the four management areas drives comprehensive Organizational Transformation, impacting processes, structures, culture, and strategic direction. HRC serves as a catalyst for shaping future-ready organizations that are more responsive, efficient, and innovative in facing the challenges of the Industry 5.0 era.

CONCLUSION

Human-Robot Collaboration has emerged as a transformative force in modern organizations, redefining how humans and machines interact across various management fields. Human-Robot Collaboration is not merely a technological innovation but also a strategic driver of organizational transformation. Through bibliometric analysis, three key thematic clusters have been identified: technological implementation and integration, task optimization and efficiency, and human-centered interfaces. These clusters highlight Human-Robot Collaboration impact on both macro-level structural changes and micro-level operational processes.

In operations management, Human-Robot Collaboration contributes to improved productivity and workplace safety. In human resource management, it necessitates workforce reskilling, cultural adaptation, and the development of digital competencies. Marketing benefits from Human-Robot Collaboration through enhanced personalization and data-driven customer engagement, while the finance domain experiences improved accuracy and faster decision-making processes. By integrating Human-Robot Collaboration across these domains, organizations can foster a more responsive, agile, and innovative environment.

Moreover, the application of Human-Robot Collaboration introduces broader adaptability to complex business needs in the modern era. In operations management, adaptive scheduling systems enable dynamic adjustments to workflow disruptions, demand fluctuations, or equipment failures. For instance, if a robot encounters a delay, the system can promptly reassign tasks to an available human worker or another robot, minimizing downtime. This approach not only enhances efficiency but also ensures adherence to critical production deadlines in highly competitive markets.

At a strategic level, HRC represents a new paradigm of hybrid collaboration between humans and machines. By combining technological capabilities, such as robotics and artificial intelligence (AI), with human strengths like creativity and empathy, organizations can holistically optimize their resources. This transformation aligns with the vision of Industry 5.0, which emphasizes a balance between operational efficiency, adaptability, and human well-being.

Overall, HRC not only enhances operational efficiency but also steers organizations toward a more inclusive, adaptive, and future-ready workplace. By leveraging the strengths of human-machine collaboration, organizations can effectively address the complexities of the modern era, create sustainable value, and build a competitive edge in an ever-evolving business landscape.

REFERENCES

Abeliansky, A. L., Beulmann, M., & Prettner, K. (2024). Are they coming for us? Industrial robots and the mental health of workers. *Research Policy*, *53*(3), 104956. Advance online publication. DOI: 10.1016/j.respol.2024.104956

Alenjareghi, M. J., Keivanpour, S., Chinniah, Y. A., Jocelyn, S., & Oulmane, A. (2024). Safe human-robot collaboration: A systematic review of risk assessment methods with AI integration and standardization considerations. *International Journal of Advanced Manufacturing Technology*, *133*(9-10), 4077–4110. Advance online publication. DOI: 10.1007/s00170-024-13948-3

Almokdad, E., & Lee, C. H. (2024). *Service Robots in the Workplace: Fostering Sustainable Collaboration by Alleviating Perceived Burdensomeness*. Sustainability., DOI: 10.3390/su16219518

Baltrusch, S. J., Krause, F., de Vries, A. W., van Dijk, W., & de Looze, M. P. (2022). What about the human in human robot collaboration?: A literature review on HRC's effects on aspects of job quality. In *Ergonomics* (Vol. 65, Issue 5). DOI: 10.1080/00140139.2021.1984585

Bolla, R., Bruschi, R., Davoli, F., Lombardo, C., Mohammadpour, A., Trivisonno, R., & Poe, W. Y. (2023). Adaptive Reliability for the Automated Control of Human-Robot Collaboration in Beyond-5G Networks. *IEEE Transactions on Network and Service Management*, *20*(3), 2489–2503. Advance online publication. DOI: 10.1109/TNSM.2023.3298171

Calzavara, M., Faccio, M., Granata, I., & Trevisani, A. (2024). Achieving productivity and operator well-being: A dynamic task allocation strategy for collaborative assembly systems in Industry 5.0. *International Journal of Advanced Manufacturing Technology*. Advance online publication. DOI: 10.1007/s00170-024-14302-3

Cao, J., & Chen, N. (2024). The Influence of Robots' Fairness on Humans' Reward-Punishment Behaviors and Trust in Human-Robot Cooperative Teams. *Human Factors*, *66*(4), 1103–1117. Advance online publication. DOI: 10.1177/00187208221133272 PMID: 36218282

Carissoli, C., Negri, L., Bassi, M., Storm, F. A., & Delle Fave, A. (2023). Mental Workload and Human-Robot Interaction in Collaborative Tasks: A Scoping Review. *International Journal of Human-Computer Interaction*. Advance online publication. DOI: 10.1080/10447318.2023.2254639

Chand, S., & Lu, Y. (2023). Dual task scheduling strategy for personalized multi-objective optimization of cycle time and fatigue in human-robot collaboration. *Manufacturing Letters*, *35*, 88–95. Advance online publication. DOI: 10.1016/j.mfglet.2023.08.064

Chuang, S. (2024). Indispensable skills for human employees in the age of robots and AI. *European Journal of Training and Development*, *48*(1/2), 179–195. Advance online publication. DOI: 10.1108/EJTD-06-2022-0062

Chung, K. C., & Tan, P. J. B. (2025). Artificial intelligence and internet of things to improve smart hospitality services. *Internet of Things (The Netherlands)*. DOI: 10.1016/j.iot.2025.101544

Ciano, M. P., Peron, M., Panza, L., & Pozzi, R. (2025). Industry 4.0 technologies in support of circular Economy: A 10R-based integration framework. *Computers & Industrial Engineering*, *201*, 110867. Advance online publication. DOI: 10.1016/j.cie.2025.110867

Coronado, E., Ueshiba, T., & Ramirez-Alpizar, I. G. (2024). A Path to Industry 5.0 Digital Twins for Human–Robot Collaboration by Bridging NEP+ and ROS. *Robotics (Basel, Switzerland)*, *13*(2), 28. Advance online publication. DOI: 10.3390/robotics13020028

D'souza, C. P., & Tapas, P. (2024). Diversity 5.0 framework: Managing innovation in Industry 5.0 through diversity and inclusion. *European Journal of Innovation Management*. Advance online publication. DOI: 10.1108/EJIM-02-2024-0139

Dagkinis, I. K., Spyrou, S., Georgantis, G. K., Psomas, P. M., Platis, A. N., & Tsoulfas, G. (2025). Impact of innovative technologies on quality and safety in surgery. *Journal of Robotic Surgery*. Advance online publication. DOI: 10.1007/s11701-024-02192-2 PMID: 39707082

Elendu, C., Elendu, T. C., & Elendu, I. D. (2024). *5G-enabled smart hospitals: Innovations in patient care and facility management*. Medicine., DOI: 10.1097/MD.0000000000038239

Erro-Garcés, A., & Aramendia-Muneta, M. E. (2023). The role of human resource management practices on the results of digitalisation. From Industry 4.0 to Industry 5.0. *Journal of Organizational Change Management*, *36*(4), 585–602. Advance online publication. DOI: 10.1108/JOCM-11-2021-0354

Eswar Balachandar, G., & Bhaskar Reddy, C. (2024). *Mathematical Model for Multi-Objective Optimization of Facility Allocation in Flexible Manufacturing Systems with Collaborative Robots*. International Research Journal of Multidisciplinary Scope., DOI: 10.47857/irjms.2024.v05i03.0892

Eswaran, M., Inkulu, A. K., Tamilarasan, K., Bahubalendruni, M. V. A. R., Jaideep, R., Faris, M. S., & Jacob, N. (2024). Optimal layout planning for human robot collaborative assembly systems and visualization through immersive technologies. *Expert Systems with Applications*, *241*, 122465. Advance online publication. DOI: 10.1016/j.eswa.2023.122465

Feddoul, Y., Ragot, N., Duval, F., Havard, V., Baudry, D., & Assila, A. (2023). Exploring human-machine collaboration in industry: A systematic literature review of digital twin and robotics interfaced with extended reality technologies. *International Journal of Advanced Manufacturing Technology*, *129*(5-6), 1917–1932. Advance online publication. DOI: 10.1007/s00170-023-12291-3

Fraboni, F., Brendel, H., & Pietrantoni, L. (2023). Evaluating Organizational Guidelines for Enhancing Psychological Well-Being, Safety, and Performance in Technology Integration. *Sustainability (Basel)*, *15*(10), 8113. Advance online publication. DOI: 10.3390/su15108113

Gervasi, R., Capponi, M., Mastrogiacomo, L., & Franceschini, F. (2024). Eye-tracking support for analyzing human factors in human-robot collaboration during repetitive long-duration assembly processes. *Production Engineering*. Advance online publication. DOI: 10.1007/s11740-023-01230-6

He, P., Jiang, X., Wang, Q., & Zhang, B. (2025). Multi-objective Human-robot collaborative batch scheduling in distributed hybrid flowshop via automatic design of local search-reconstruction-feedback algorithm. *Computers & Industrial Engineering*, *203*, 110983. Advance online publication. DOI: 10.1016/j.cie.2025.110983

Hjorth, S., & Chrysostomou, D. (2022). Human–robot collaboration in industrial environments: A literature review on non-destructive disassembly. *Robotics and Computer-integrated Manufacturing*, *73*, 102208. Advance online publication. DOI: 10.1016/j.rcim.2021.102208

Holland, J., Kingston, L., McCarthy, C., Armstrong, E., O'dwyer, P., Merz, F., & McConnell, M. (2021). Service robots in the healthcare sector. *Robotics (Basel, Switzerland)*, *10*(1), 47. Advance online publication. DOI: 10.3390/robotics10010047

Holland, J., Kingston, L., McCarthy, C., Armstrong, E., O'dwyer, P., Merz, F., & McConnell, M. (2024). Service robots in the healthcare sector. *Robotics (Basel, Switzerland)*.

Hsu, C.-H., Liu, J.-C., Cai, X.-Q., Zhang, T.-Y., & Lv, W.-Y. (2024). Enabling Sustainable Diffusion in Supply Chains Through Industry 5.0: An Impact Analysis of Key Enablers for SMEs in Emerging Economies. *Mathematics, 12*(24), 3938. Advance online publication. DOI: 10.3390/math12243938

Huang, Z., Ge, S., He, Y., Wang, D., & Zhang, S. (2024). Research on the Intelligent System Architecture and Control Strategy of Mining Robot Crowds. *Energies, 17*(8), 1834. Advance online publication. DOI: 10.3390/en17081834

Ietto, B., Ancillai, C., Sabatini, A., Carayannis, E. G., & Gregori, G. L. (2024). The Role of External Actors in SMEs' Human-Centered Industry 4.0 Adoption: An Empirical Perspective on Italian Competence Centers. *IEEE Transactions on Engineering Management, 71*, 1057–1072. Advance online publication. DOI: 10.1109/TEM.2022.3144881

Inkulu, A. K., & Bahubalendruni, M. V. A. R. (2024). *Human-robot collaborative task planning for assembly system productivity enhancement*. Robotic Intelligence and Automation., DOI: 10.1108/RIA-05-2023-0067

Islam, S. O. B., & Lughmani, W. A. (2023). A Connective Framework for Safe Human–Robot Collaboration in Cyber-Physical Production Systems. *Arabian Journal for Science and Engineering, 48*(9), 11621–11644. Advance online publication. DOI: 10.1007/s13369-022-07490-1

Ji, H., Li, S., Chen, J., & Zhou, S. (2024). On-site human-robot collaboration for lunar exploration based on shared mixed reality. *Multimedia Tools and Applications*.

Jia, Z., Xie, S., & Zhang, W. (2025). Flexible task assignment and assembly scheduling for human-robot collaboration cell considering uncertainty. *International Journal of Production Research, 63*(16), 6134–6154. Advance online publication. DOI: 10.1080/00207543.2025.2469286

Kavvathas, K., Kampourakis, E., Andronas, D., Fousekis, N., & Makris, S. (2024). A service-oriented orchestration and planning tool for plug and produce manufacturing: A deformable object handling supervision paradigm. *International Journal of Computer Integrated Manufacturing, 37*(12), 1626–1649. Advance online publication. DOI: 10.1080/0951192X.2024.2331533

Kermenov, R., Di Biase, A., Pellicani, I., Longhi, S., & Bonci, A. (2025). Near Time-Optimal Trajectories with ISO Standard Constraints for Human–Robot Collaboration in Fabric Co-Transportation. *Robotics (Basel, Switzerland), 14*(2), 10. Advance online publication. DOI: 10.3390/robotics14020010

Keshvarparast, A., Berti, N., Chand, S., Guidolin, M., Lu, Y., Battaia, O., Xu, X., & Battini, D. (2024). Ergonomic design of Human-Robot collaborative workstation in the Era of Industry 5.0. *Computers & Industrial Engineering*, *198*, 110729. Advance online publication. DOI: 10.1016/j.cie.2024.110729

Khoa, D. T., Gip, H. Q., Guchait, P., & Wang, C. Y. (2023). Competition or collaboration for human–robot relationship: A critical reflection on future cobotics in hospitality. *International Journal of Contemporary Hospitality Management*, *35*(6), 2202–2215. Advance online publication. DOI: 10.1108/IJCHM-04-2022-0434

Kim, Y., Kim, H., Murphy, R., Lee, S., & Ahn, C. R. (2022). Delegation or Collaboration: Understanding Different Construction Stakeholders' Perceptions of Robotization. *Journal of Management Engineering*, *38*(1), 04021084. Advance online publication. DOI: 10.1061/(ASCE)ME.1943-5479.0000994

Kim, Y. H., Ye, W., Kumar, R., Bail, F., Dvorak, J., Tan, Y., May, M. C., Chang, Q., Athinarayanan, R., Lanza, G., Sutherland, J. W., Li, X., & Nath, C. (2024). Unlocking the Potential of Remanufacturing Through Machine Learning and Data-Driven Models—A Survey. *Algorithms*, *17*(12), 562. Advance online publication. DOI: 10.3390/a17120562

Koç, M. (2023). ARTIFICIAL INTELLIGENCE IN GERIATRICS. *Turk Geriatri Dergisi*.

Korivand, S., Galvani, G., Ajoudani, A., Gong, J., & Jalili, N. (2024). Optimizing Human–Robot Teaming Performance through Q-Learning-Based Task Load Adjustment and Physiological Data Analysis. *Sensors (Basel)*, *24*(9), 2817. Advance online publication. DOI: 10.3390/s24092817 PMID: 38732923

Lario, J., Mateos, J., Psarommatis, F., & Ortiz, Á. (2025). Towards Zero Defect and Zero Waste Manufacturing by Implementing Non-Destructive Inspection Technologies. *Journal of Manufacturing and Materials Processing*. DOI: 10.3390/jmmp9020029

Li, J.-M., Liu, K.-X., Xie, J.-F., & Wu, H.-Y. (2024). How Does Human-Robot Collaboration Affect Hotel Employees' Proactive Behavior? *International Journal of Human-Computer Interaction*. Advance online publication. DOI: 10.1080/10447318.2024.2431769

Li, J.-M., Zhang, R.-X., Wu, T.-J., & Mao, M. (2024). How does work autonomy in human-robot collaboration affect hotel employees' work and health outcomes? Role of job insecurity and person-job fit. *International Journal of Hospitality Management*, *117*, 103654. DOI: 10.1016/j.ijhm.2023.103654

Li, S., Wang, R., Zheng, P., & Wang, L. (2021). Towards proactive human–robot collaboration: A foreseeable cognitive manufacturing paradigm. *Journal of Manufacturing Systems*, *60*, 547–552. Advance online publication. DOI: 10.1016/j.jmsy.2021.07.017

Liau, Y. Y., & Ryu, K. (2024). Framework of knowledge management for human–robot collaborative mold assembly using heterogeneous cobots. *Journal of Intelligent Manufacturing*, *35*(8), 3713–3729. DOI: 10.1007/s10845-024-02439-7

Liu, R., Natarajan, M., & Gombolay, M. C. (2022). Coordinating Human-Robot Teams with Dynamic and Stochastic Task Proficiencies. *ACM Transactions on Human-Robot Interaction*, *11*(1), 1–42. Advance online publication. DOI: 10.1145/3477391

Lu, C., Gao, R., Yin, L., & Zhang, B. (2024). Human-Robot Collaborative Scheduling in Energy-Efficient Welding Shop. *IEEE Transactions on Industrial Informatics*, *20*(1), 963–971. Advance online publication. DOI: 10.1109/TII.2023.3271749

Mao, Z., Zhang, J., Sun, Y., Fang, K., & Huang, D. (2025). Balancing parallel assembly lines with human-robot collaboration: Problem definition, mathematical model and tabu search approach. *International Journal of Production Research*, *63*(1), 51–85. Advance online publication. DOI: 10.1080/00207543.2024.2356627

Mateus, J. C., Claeys, D., Limère, V., Cottyn, J., & Aghezzaf, E.-H. (2019). A structured methodology for the design of a human-robot collaborative assembly workplace. *International Journal of Advanced Manufacturing Technology*, *102*(5-8), 2663–2681. Advance online publication. DOI: 10.1007/s00170-019-03356-3

Mbunge, E., Chitungo, I., & Dzinamarira, T. (2021a). *Unbundling the significance of cognitive robots and drones deployed to tackle COVID-19 pandemic: A rapid review to unpack emerging opportunities to improve healthcare in sub-Saharan Africa*. Cognitive Robotics., DOI: 10.1016/j.cogr.2021.11.001

Mbunge, E., Chitungo, I., & Dzinamarira, T. (2021b). *Unbundling the significance of cognitive robots and drones deployed to tackle COVID-19 pandemic: A rapid review to unpack emerging opportunities to improve healthcare in sub-Saharan Africa*. Cognitive Robotics.

Mia, M. R., & Shuford, J. (2024). Exploring the Synergy of Artificial Intelligence and Robotics in Industry 4.0 Applications. *Journal of Artificial Intelligence General Science (JAIGS) ISSN:3006-4023, 1*(1). DOI: 10.60087/jaigs.v1i1.31

Mukherjee, S., Baral, M. M., Pal, S. K., Chittipaka, V., Roy, R., & Alam, K. (2022). Humanoid robot in healthcare: A Systematic Review and Future Research Directions. *2022 International Conference on Machine Learning, Big Data, Cloud and Parallel Computing, COM-IT-CON 2022.* DOI: 10.1109/COM-IT-CON54601.2022.9850577

Peng, L. P., & Wu, Y. T. (2021). Assessing Multiple Stakeholders' Collaborative Network in Promoting Eco-agriculture. *Taiwan Shuili.*

Rakholia, R., Suarez-Cetrulo, A. L., Singh, M., & Simon, C. R. (2024). *Advancing Manufacturing Through Artificial Intelligence: Current Landscape, Perspectives, Best Practices, Challenges, and Future Direction.* IEEE., DOI: 10.1109/ACCESS.2024.3458830

Ram, C., Pitt, J. B., Wallace, M. W., Goldstein, S. D., & Zamora, I. J. (2024). Framework for pediatric robotic surgery program development. *Seminars in Pediatric Surgery, 33*(1), 151389. Advance online publication. DOI: 10.1016/j.sempedsurg.2024.151389 PMID: 38245993

Rastgoftar, H. (2024). Safe Human-UAS Collaboration from High-Level Planning to Low-Level Tracking. *IEEE Transactions on Automation Science and Engineering.* Advance online publication. DOI: 10.1109/TASE.2024.3461726

Sarikonda, A., Tecce, E., Leibold, A., Mansoor, A. D., Thalheimer, S., Heller, J., Prasad, S., Sharan, A., Jallo, J., Harrop, J., Vaccaro, A. R., & Sivaganesan, A. (2024). What is the Marginal Cost of Using Robot Assistance or Navigation for Transforaminal Lumbar Interbody Fusion? A Time-Driven Activity-Based Cost Analysis. *Neurosurgery, 95*(3), 556–565. Advance online publication. DOI: 10.1227/neu.0000000000002899 PMID: 38465927

Shahid, S., Kaur, K., Mohyuddin, S. M., Prikshat, V., & Patel, P. (2025). Revolutionizing HRM: A review of human-robot collaboration in HRM functions and the imperative of change readiness. *Business Process Management Journal, 31*(7), 2892–2928. Advance online publication. DOI: 10.1108/BPMJ-12-2023-0951

Shirakura, N., Maruyama, T., Makihara, K., Ueshiba, T., Itadera, S., Endo, Y., Tada, M., & Domae, Y. (2025). Synergistic effects of robot performance on human–robot mutual assistance systems in manufacturing. *Advanced Robotics, 39*(4), 208–221. Advance online publication. DOI: 10.1080/01691864.2024.2445600

Singh, K. D., & Singh, P. D. (2023). *Machine Learning in Robotics with Fog/Cloud Computing and IoT.* EAI Endorsed Transactions on AI and Robotics., DOI: 10.4108/airo.3621

Stasevych, M., & Zvarych, V. (2023). Innovative Robotic Technologies and Artificial Intelligence in Pharmacy and Medicine: Paving the Way for the Future of Health Care—A Review. In *Big Data and Cognitive Computing* (Vol. 7, Issue 3). DOI: 10.3390/bdcc7030147

Tan, L., Liu, C., Wang, Y., Li, Y., Zhao, J., Wang, S., & Zhong, B. (2025). The effect of human-robot collaboration on frontline employees' service performance: A resource perspective. *International Journal of Hospitality Management, 126*, 104033. Advance online publication. DOI: 10.1016/j.ijhm.2024.104033

Tian, B., Kaul, H., & Janardhanan, M. (2024). Balancing heterogeneous assembly line with multi-skilled human-robot collaboration via Adaptive cooperative co-evolutionary algorithm. *Swarm and Evolutionary Computation, 91*, 101762. Advance online publication. DOI: 10.1016/j.swevo.2024.101762

Toupas, P., Tsamis, G., Giakoumis, D., Votis, K., & Tzovaras, D. (2023). From Detection to Action Recognition: An Edge-Based Pipeline for Robot Human Perception. *2023 5th International Conference on Control and Robotics, ICCR 2023*. DOI: 10.1109/ICCR60000.2023.10444804

Uğuz Arsu, Ş., & Sipahi Döngül, E. (2023). The impact of robots and robotic systems on businesses in the workplace: Systematic review. *Journal of Organizational Change Management, 36*(7), 1193–1221. Advance online publication. DOI: 10.1108/JOCM-11-2022-0340

Wang, B., Chen, S., & Xiao, G. (2024). Advancing healthcare through mobile collaboration: A survey of intelligent nursing robots research. *Frontiers in Public Health, 12*, 1368805. Advance online publication. DOI: 10.3389/fpubh.2024.1368805 PMID: 39659720

Wang, D., Ma, E., & Leung, X. Y. (2024). Working with robots: A job design perspective of hospitality employees' collaboration intentions with service robots. *Journal of Hospitality and Tourism Management, 61*, 66–77. DOI: 10.1016/j.jhtm.2024.09.012

Wang, K. J., Lin, C. J., Tadesse, A. A., & Woldegiorgis, B. H. (2023). Modeling of human–robot collaboration for flexible assembly—A hidden semi-Markov-based simulation approach. *International Journal of Advanced Manufacturing Technology, 126*(11–12), 5371–5389. Advance online publication. DOI: 10.1007/s00170-023-11404-2

Wang, R., Zhao, D., Gupte, A., & Min, B.-C. (2024). Initial Task Allocation in Multi-Human Multi-Robot Teams: An Attention-Enhanced Hierarchical Reinforcement Learning Approach. *IEEE Robotics and Automation Letters*, *9*(4), 3451–3458. Advance online publication. DOI: 10.1109/LRA.2024.3366414

Wu, T.-J., & Zhang, R.-X. (2024). Exploring the impacts of intention towards human-robot collaboration on frontline hotel employees' positive behavior: An integrative model. *International Journal of Hospitality Management*, *123*, 103912. DOI: 10.1016/j.ijhm.2024.103912

Yi, B., Lv, J., Chen, J., Wang, X., & Li, K. (2023). Digital Twin Constructed Spatial Structure for Flexible and Efficient Task Allocation of Drones in Mobile Networks. *IEEE Journal on Selected Areas in Communications*, *41*(11), 3430–3443. Advance online publication. DOI: 10.1109/JSAC.2023.3313193

Yu, F., Lu, C., Yin, L., & Zhang, B. (2024). A Self-Learning Memetic Algorithm for Human-Robot Collaboration Scheduling in Energy-Efficient Distributed Mixed Fuzzy Welding Shop. *IEEE Transactions on Automation Science and Engineering*. Advance online publication. DOI: 10.1109/TASE.2024.3448435

Yu, F., Lu, C., Yin, L., & Zhang, B. (2025). B., Yin, L., & Zhang. (2024). A Self-Learning Memetic Algorithm for Human-Robot Collaboration Scheduling in Energy-Efficient Distributed Mixed Fuzzy Welding Shop. *IEEE Transactions on Automation Science and Engineering*, *22*, 6595–6607. Advance online publication. DOI: 10.1109/TASE.2024.3448435

Chapter 6
Transforming Service Delivery With Self-Service Kiosks:
A Human-Robotic Collaboration Approach to Customer Satisfaction

Heru Susanto
https://orcid.org/0000-0002-1823-357X
National Research and Innovation Agency, Indonesia

Wafiq Lamit
Universiti Teknologi Brunei, Brunei

ABSTRACT

This study investigates the influence of service quality dimensions on customer satisfaction regarding the use of self-service kiosks (SSKs) in Brunei. SSKs are a type of self-service technology that allows customers to perform tasks without the assistance of service employees. The research adopts the SSTQUAL scale, which consists of seven dimensions: functionality, enjoyment, security, design, assurance, convenience, and customization. The study also addresses the challenges and concerns associated with SSKs, such as theft and identity theft. Additional security measures to deter theft could make honest customers feel uncomfortable or distrusted, which could also lead to longer waiting times, negating the benefits of SSKs. The data has been collected from 163 respondents who may or may not have used SSKs. The result of this study has revealed that only 2 constructs have a significant impact towards customer satisfaction towards using SSKs in which are considered the most

DOI: 10.4018/979-8-3373-3404-2.ch006

important. Functionality has been found to be the most statistically significant while customers perceive security to be the most important. These findings offer guidance to Brunei's service sector on where to invest in recent technology to improve customer satisfaction and experience.Along with adding to the body of knowledge on self-service technologies and customer satisfaction, the study is relevant to Brunei, a nation undergoing digital transformation and working toward realizing its 2035 vision as it improves understanding of customer behavior in Brunei, which can be used in comparative studies and to better understand cultural influences on technological adoption. From a business perspective, the study helps firms understand the elements impacting customer happiness, enabling them to improve their SSKs and so promote customer satisfaction and loyalty.

INTRODUCTION

Technology has advanced as we move towards the future, and it has also innovated service delivery, particularly on self-service options, where some delivery processes have been switched from human-to-human interaction to self-service mode of delivery (Campbell et al., 2010). Self-service delivery can be driven by two factors: non-technological-based services such as checkout stands, which were first introduced in the 1880s, and technological-based services known as Self-Service Technologies (SSTs), which began to appear in the 2000s (Acrelec United Kingdom, 2022; Larivière et al., 2017; Poushneh & Vasquez-Parraga, 2018). This study will focus on self-service channels, which are driven by SSTs, where there is a technological contact between the user and the technology but no direct engagement with frontline service employees (Meuter et al., 2000). Some examples of self-service options driven by technology include self-service kiosks in supermarkets, self-order kiosks in fast food restaurants, ATMs in banks, self-check-in at airports or hotels, and vending machines that are popular with consumers.

SSTs were first introduced by Dr Howard Schneider in Price Chopper in Clifton Park, Upstate New York, in 1992 but it was only during the last decade that the use of SSTs started to skyrocket where in 2021. the market was estimated to be worth USD 34.3 billion. (Dittig, 2023; The Independent, 2011; Turkar, 2023). Additionally,the emergence of the pandemic COVID-19 has expedited the usage of SSTs by providing digital transformation, remote interactions, health and safety, efficiency and convenience, and the use of artificial intelligence (AI) (Horowitz et al., 2022). According to Parise et al. (2016) SSTs have been implemented in delivery services as a tool of interaction to aid the front-line employee. As a result, more firms are advocating the use of SSTs because it allows them to reduce service delivery costs while improving customer experience. (Scherer et al., 2015). Furthermore, this enables customers to have access to a modern and convenient services and firms can address customers demand and their satisfaction (Kim & Lee, 2020;McGrath & Astell, 2016; Othman et al., 2020).

Since transactions using cards and phones have increased because they are more convenient for customers who do not need to bring a lot of cash, SSTs are always incorporated with multiple types of transactions, with the use of QR codes, debit cards, and credit cards being the major ones where it uses the principles of tap, scan, and go, but some SSTs are still accepting cash (Anitha et al., 2019; Collier & Kimes, 2012; Hilton et al., 2013;). In the context of self-service, convenient refers to the projected time and effort required to locate and use an SST (Collier & Sherrell 2010). As a result, the current generation of SSTs is associated with electronic cloud-based payment platforms such as Ingenico, which is a firm that creates a payment acceptance ecosystem by connecting smart terminals to a cloud-based platform and offering end-to-end terminal management. (Ingenico, n.d.).

Self-service kiosks (SSKs) are a popular type of SST in the retail business, especially when COVID-19 hits the world and face-to-face interactions are limited (Park et al., 2020). SSKs, sometimes also known as self-payment kiosks, self-checkout kiosks, and self-ordering kiosks, refer to a computer with a touchscreen or an interactive tablet that allows clients to access information or services without speaking to a live person (Kiosk Group, n.d.). SSKs are proving to be an effective tool for dealing with uncertainties in a changing service environment and escalating client expectations, and they are currently incorporated into the service algorithm using the most recent knowledge on consumer behaviour (Vakulenko et al., 2018). Moreover, SSKs are associated with the fourth industry retail transformation, in which it uses technology that encompasses artificial intelligence (AI), the Internet of Things (IoT), cloud computing, big data analytics (BDA), and augmented reality (AR) to fulfil customers' needs (Sakrabani et al., 2019).

SSKs are also developed based on the needs of clients in various locations. Tesco, a UK-based retail behemoth, has had a huge influence in South Korea by creating "virtual stores" in public locations, most notably subways and bus stations with considerable foot traffic(Robson & Pitt, 2018). These virtual stores are considered a type of SSK because they allow customers to finish their purchases independently using their smartphones, eliminating the need for interaction with store staff. This is similar to how clients use SSK in physical stores. Customers use a computer interface to scan QR codes and make purchases, rather than engaging with a physical machine. So, while the format has changed, the premise of allowing customers to serve themselves remains the same. These "virtual stores" operate by utilizing an app to scan the QR codes of the products users wish to purchase and place orders that may be paid online. The rationale for this method is that the average Seoul commuter does not have time to shop at the local brick-and-mortar store due to their lengthy work hours(Kim & Min, 2023).

In Brunei, there appears to be a positive perception toward technologies such as SSK. For example, Progresif, a Bruneian company, established a SSK at their Times Square store, providing consumers with a contactless option to pay bills and top up their prepaid accounts (Amu, 2021). This approach was part of the company's business continuity plan, which also involved employees working from home and changing store procedures to accommodate social distancing measures. The SSK was well-received, and the business intended to expand it to their Seria shop.

Additionally, SSK also helps to contribute to Brunei goals in achieving Wawasan 2035 where it plays a consequential role for all Wawasan 2035 goals (WAWASAN BRUNEI 2035, n.d.). SSK can play a significant role in retail environment where the use SSK requires a certain level of digital literacy. As such, their widespread use could encourage the development of digital skills among the population. Next, as the use of SSK can reduce waiting times and providing more efficient shopping experience,help businesses to reduce costs and increase efficiency, it can contribute to a more dynamic and sustainable economy. The development and maintenance of SSK also require robust infrastructure and advance info-communication technology, which aligns with one of the 13 strategies under Wawasan Brunei 2035 where Brunei is aiming to be a smart nation(Kiffle, 2019).

Regardless of the ingenious interaction that SSKs provide, there are still some problems that both firms and consumers face when dealing with and interacting with such technologies and these can affect customer inclination in using it. For instance, it is easier for people to commit theft when no front-line employees are keeping an eye on customers during transactions, and this creates huge losses for both businesses and customers (Kwak, 2020). There is also part of the population that is afraid of identity theft as fraudulent activities in the digitalized economy have started to increase, and some have suggested biometric authentication as a solution for this issue (Obermeier et al., 2022). These factors can increase consumers tendency in using SSK and some may try to avoid using it as they may have anxiety and do not feel comfortable in using it.

Understanding the adoption of technology such as SSK has been a continuous concern as there are both positive and negative factors in how a person can accept technology which can be affected by internal and external variables (Blut & Schoefer, 2016). As an analogy, Davis introduced the Technology Adoption Model (TAM), which said that the factors that impact technology adoption include perceived usefulness and ease of use, however it only portrays consumer positive responds and fail to mention the negative ones (Davis, 1989; Wang et al., 2018). As there are options for purchaser to adopt SSK, the public will compare different selection, assessing the cost and benefits before choosing their preferable method of checkout (Leung et al., 2021; Zhu et al., 2017).

Moreover, a person will also think about the action they will take, and this will affect how they decide whether to accept an SSK. Before determining whether to utilize a particular piece of technology, each person will compare its benefits (such as usability, value, and convenience of use) against its drawbacks (such as perceived risk, loss of human contact, and lack of comprehension) (Blut & Schoefer, 2016). Furthermore, valence theory has been utilised numerous times to understand user behaviour when introducing new technologies. (Ghasemaghaei, 2020; Hajli, 2019). Valence theory is "a central component of all affective states, including pains, pleasures, emotions, moods, and feelings of desire or repulsion" (Carruthers, 2018). The valence-based approach also affects judgments and decisions regarding the use of SSK (Gelbrich, 2009).

1.1 Problem Statement

As the world moves towards a rapid advancement in technology, more companies have deployed the use of frontline technological interfaces within employee and customer service encounters (Grand View Research, 2020). Industrial operations such as hotels, airports, fast-food restaurants, and supermarkets have implemented SSKs as an option for customers to check in and pay for their stuff (Bogicevic et al., 2017; Jie & Kamsin, 2021; Kaushik & Rahman, 2017; Park et al., 2021). This is due to the benefits of self-service delivery in terms of increased productivity and lower operational costs for the company. For example, a company can cut the number of service employees, reduce the workload of employees, and improve customer satisfaction (Djelassi et al., 2018; Yang & Klassen, 2008). Additionally, SSKs also have a positive impact on the customer as they help reduce time spent at checkout, increase perceived control, and improve ease of use (Fernandes & Pedroso, 2016).

Despite the innovative utility provided by SSKs, it can be observed that the usage of these technologies is still unpredictable (Lawton, 2022; Lee & Lee, 2019). For instance, part of the consumer population in the USA still prefers to use the traditional way of transacting at the supermarket (Commisso, 2022). This is believed to be attributed to customers who still want to have interpersonal interactions with supermarket employees, are afraid of using technology (technophobia), or simply follow the country's culture (Blut & Schoefer, 2016; Bulmer et al., 2018; Tawari, 2019). Since implementing these types of technologies does cost money, firms need to analyse whether the SSKs can give high-quality services to customers and whether they are satisfied in using them, and therefore determine whether it is feasible to use SSKs over traditional cashiers (Purba et al., 2022; Ul Hassan et al, 2020).

Technology has revolutionized the industry in Brunei over the years (Azahari, 2022; Noor, 2022), and this includes SSKs, which have started to bloom in Brunei's retail industries as well as food service industries. It all started when McDonald's Brunei introduced a self-ordering kiosk in its Lambak branch in 2019 (Geekturf, 2019). This was then followed by Jollibee, PCSB, and Hua Ho, which introduced SSKs in 2021 (Tabsquare, 2021; The Bruneian, 2021; Time in Brunei, 2021). Likewise, in 2022, Jaya Hypermart and Milimewah also set up SSKs in their respective supermarkets (Mohamad, 2022). However, the use of SSKs among Brunei consumers is still uncertain as the establishment of these technologies is still new. Limited studies have been made on how the service quality provided by these kiosks impacts customer satisfaction.

While it is widely known that improved service quality leads to increased customer satisfaction, the exact characteristics of service quality that are most significant in the context of Brunei SSKs are not well understood (Shahid Iqbal et al., 2018). Factors such as efficiency, accuracy, reliability, security, and user-friendliness are often cited as important and may play a role in determining Bruneian satisfaction (Shahril et al., 2021). Furthermore, an SSK that is now being deployed around public spaces needs a lot of effort for people to use, and it can be observed that some of the consumers are still hesitant and are still considering using a self-service machine (Vakulenko et al., 2019). Additionally, demographic factors such as age, gender, and technological literacy might influence customer satisfaction when using SSKs. Therefore, the purpose of this study is to determine the relationship between service quality and customer satisfaction in the context of SSKs, with a particular focus on the dimensions of service quality that drive customer satisfaction.

Furthermore, SSK is also vulnerable to fraudulent activities, and security issues such as thefts caused by SSK implementation also increase in some countries (Jie & Kamsin, 2021). Fraudulent activity may cause financial losses, affect customer trust and satisfaction, and damage a firm's reputation. To mitigate these problems, a firm needs to develop a comprehensive operational strategy that includes robust security measures and regular system updates. Additionally, educating customers about potential fraud and theft risks and how to operate SSKs safely may also help. In the context of Brunei, fraud and theft issues could be particularly impactful as the country has a small population and a close-knit community. A single issue regarding SSK could damage a supermarket's reputation, leading to a significant loss of customers. Therefore, it is also crucial for supermarkets in Brunei to take proactive steps to prevent fraud and theft, protecting their customers' well-being.

1.2 Research Objectives

The purpose and direction of the research are as follows:

1. To investigate the relationship between dimensions of service quality and customer satisfaction in the context of (SSKs).
2. To examine how these dimensions interact with each other to influence overall satisfaction.
3. To identify the most important dimension perceived by customers on self-service kiosks in supermarkets and explore potential strategies for optimising it to enhance customer satisfaction.

1.3 Research Question

The aim of this research paper is to answer the following questions:

1. What is the relationship between dimensions of service quality and customer satisfaction in the context of SSK provided in Brunei?
2. How do the various dimensions of service quality on self-service kiosks (SSKs) interact with each other to influence customer satisfaction?
3. How does the most important dimension perceive by customers on self-service kiosks in supermarkets impact customer satisfaction, and what strategy can be implemented to maximise it and improve customer satisfaction?

1.4 Research Hypothesis

The developed hypotheses are based on the theoretical concepts of which what is expected between the variables. The following are the developed hypotheses:

1. Dimensions of service quality of SSK have a significant and positive impact on customer satisfaction.
2. The interaction between different dimensions of service quality on self-service kiosks (SSKs) will not have a significant impact on customer satisfaction.
3. The most important dimension perceived by customers on self-service kiosks in supermarkets will have a significant positive impact on customer satisfaction. By implementing a strategy to optimize this dimension, customer satisfaction may be further enhanced.

1.5 Significance of the Study

The study's relevance is to investigate how technology-based services, specifically SSK, affect consumer satisfaction in Brunei's service sector. The investigation may indicate a significant association between SSK service quality characteristics, both directly and indirectly through customer satisfaction. These findings may help the

Brunei service sector invest in new technology to improve the consumer experience and customer satisfaction. The study also may imply that organizations can improve customer satisfaction by improving the quality of SSK service

LITERATURE REVIEW

2.1 Introduction

Customarily, all service interactions took place in the presence of a service employee and a customer. When a firm can provide good service quality to their customers, ensuring positive customer interaction, this can lead to increased customer satisfaction, loyalty, and major business success (Hoffman & Bateson, 2016). In the case of SSK, since there is no human-to-human interaction, the interface that will be used by customers will represent the quality of the services. Things such as efficiency, accuracy, availability, user-friendliness, security, reliability, aesthetics, and information quality can provide high service quality in SSKs, ensuring a positive user experience and enhancing customer satisfaction (Demirci, Orel & Kara, 2014). After all, a satisfied customer can lead to a range of benefits for both customers and businesses, such as customer retention and increased return on investment. Therefore, this section will focus on literature regarding customer satisfaction with service quality dimensions, specifically for SSK.

Further, because human-to-human interaction is minimal, this may help a corporation cut its costs. This may be analysed using Transaction Cost Theory and Transaction Cost Economics because it involves minimizing transaction costs(Hennart & Verbeke, 2022). Labor-intensive transaction processes are reduced, lowering the need for negotiation between parties involved and thus lowering negotiation costs; lowering information costs, as SSK frequently provides customers with clear and accurate information about product prices, promotions, and discounts; and making it easier for customers to complete transactions. While the implementation of SSK lowers transaction costs, it is crucial to note that the overall impact on transaction costs is determined by several factors, including consumer acceptability, technology reliability, and kiosk maintenance costs. After alllowering transaction costs can lead to a better customer experience, more satisfaction, and better service(Otto et al., 2019).

Ensuring excellence in service quality through SSK can result in better service organization in supermarkets, optimization of service procedures, the elimination of lineups, and increased comfort for both customers and employees. Service quality, which includes information quality and interpersonal service quality, has a favourable impact on user happiness and the intention to continue using technology (Lian, 2018). Good service quality can also generate favourable word-of-mouth and repeat business. As a result, organizations that offer self-service technology must guarantee that their systems are dependable, safe, and simple to use, as well as ensure their customers receive high-quality information and service.

2.2 Service Quality

The extent to which a company meets customer expectations of the services it provides is measured by service quality. Customers buy services to meet specific needs, and they have standards and expectations about how the service provider meets those needs. Services provided by a company with high service quality will meet or exceed customer expectations. To add more, the concept of service quality incorporates procedures that are related to how the service is being conveyed to customers and how their outcomes will affect both firms and the ones that are using those services. The paradigm of service quality has been researched multiple times in previous studies, and it is usually associated with the expectancy-confirmation model, which suggests that the foundation of this model posits that an individual's impression of quality is determined by the extent to which he or she believes the service has met customer expectations (Taylor, 1993; Poister & Thomas, 2011). Service quality determinants have also been conceptualised into having ten determinants in which it circumscribes (1) Access, (2) Communication, (3) Competence, (4) Courtesy, (5) Credibility, (6) Reliability, (7) Responsiveness, (8) Security, (9) Tangibles and (1) Understanding the customer (Parasuraman et al., 1985). In addition, although there are many different measures, SERVQUAL is used most often to evaluate service quality. SERVQUAL is based on a set of five variables that customers consistently consider most important to service quality, regardless of the industry in which they provide the service. These variables are (1) Tangibility, (2), Reliability, (3) Responsiveness, (4) Assurance and (5) Empathy (Murdifin Haming et al., 2019). Three American marketing experts – Parasuraman, Zeithami, and Berry – created this scale in 1988 to assess the size of the service sector where they build it on the core idea of the "service quality gap model," which explains how consumer expectations of service quality differ from the actual impressions (Parasuraman et al., 1988). According to previous literature, SERVQUAL beset with technical, functional, physical, interactive, and corporate quality and therefore, conventional service circumstances, service quality measures have primarily focused on customers' interactions with the organization's employees (human interaction) as well as certain aspects of the marketing mix variables (Shahid Iqbal et al., 2018).

SERVPERF is another scale to which SERVQUAL is frequently compared. It was developed by Cronin and Taylor in 1994, modified based on the SERVQUAL and therefore uses similar categories to judge service quality (Cronin & Taylor, 1994). The difference that distinguishes SERVPERF to SERVQUAL here is that

SERVPERF uses attitude instead of satisfaction to measure service quality where it connects further into satisfaction with further purchase intention (Rodrigues et al., 2011). The SERVPERF scale makes the idea of service quality easier to understand as the service quality notion is explained in a more convergent and discriminant-valid way (Jain & Gupta, 2004). Empirical research analysing the validity, reliability and methodological correctness of the service quality scale clearly shows that the SERVPERF scale is superior (Adil et al., 2013; Sohail & Hasan, 2021). However, the scale lacks diagnostic capabilities (Danjuma et al., 2018). Therefore, the SERVQUAL scale should be employed due to its superior diagnostic capabilities to identify areas connected to service quality problems for future management efforts.

Significant research fields related to technology-enabled services includes SSTs together with SSKs have been made to attain customer related services (Collier & Kimes, 2012; Fernandes & Pedroso, 2016; Ul Hassan et al, 2020). According to Curran and Meuter, Consumer opinions on service quality vary depending on the type of self-service implemented (Curran & Meuter, 2005). SSKs has pilot the population to perceive better service as transactions can be completed more swiftly and easily (Hilton et al., 2013; Vakulenko et al., 2019). As SSK also provides different choice to customers in terms of how they will get their services, retailers start to gradually employ SSK in few of their outlets to replace their traditional means of service delivery. For instance, SSK will provide its user more control over how they make their payment as all the activities are done themselves. According to Wang, SSK can provide service employee involvement independency, co-creation and customization(Chih-Hung Wang, 2012). SSKs are projected to assist service organizations in increasing efficiency and lowering operational costs and therefore it is considered to better manage costs and thereby improve satisfaction (Ku & Chen, 2013; Parise et al., 2016).

Customer preference is also crucial when a firm decides to implement SSK as an option for customers to checkout. While customers often prefer SSK's efficiency and convenience, the quality of the technology, their readiness to use it, and situational factors can exert influence on their perception. According to Fernandes and Pedroso (2016), customers base their perceptions on their control, reliability, ease of use, enjoyment, and the speed of SSK. Therefore, they suggested that firms should emphasize these dimensions to make sure their SSK checkout options offer specific attributes that consumers value. The study found that the younger generation of customers is more likely to have a higher sense of control over SSK. Further, they also identify that the more customers use SSK, the more they find it enjoyable and easier to use, and therefore firms need to make customized service strategies depending on the differences among group customers who perceive SSK. Anyhow, the study is conducted only focusing on re-patronage and satisfaction, and other behaviours should also be included.

It is also important to mention that the quality of service provided by SSK can impact customer preferences for using it. According to Ul Hassan, Iqbal, and Habibah (2020), when an SSK provides a high service quality to the user, this can significantly increase the positive effect of technology trust, which will then increase behavioural intention and loyalty toward using the SSK provided by retailers. When the technology is created with excellent quality, SSK can be an effective instrument for boosting customer service and generating loyalty. They can improve their trust and continue to use it. Customers may become upset and lose trust in technology because of a low-quality SSK, resulting in decreased customer satisfaction and loyalty. Regardless, the study made by Ul Hassan may be subjected to response or social desirability bias as it relies on self-reported data.

Furthermore, Technology Advance Model (TAM) developed by Davis is used to discern how individuals will use and accept a particular technology where it exhibited Perceived Usefulness (PU) and Perceived Ease-of-Use (PEOU) as a factor that influence decision making, attitude and intention towards technology enabled services (Davis, 1989). Another model, Technology Readiness (TR) introduced by Parasuraman aims to deduce people readiness in embracing a new type of technologies for accomplishing their own agenda where in the case of SSK, it reflects their mental preparedness to absorb SSK as a new technology in their repertoire (Parasuraman, 2000). Both models have been integrated by Lin, Shih, and Sher to construct TRAM, a combination of Technology Advance Model (TAM) and Technology Readiness (TR) to further analyse the concept of customer's intentions when utilising various technological services (Lin et al., 2007).

On top of that, the Innovation Diffusion Theory (IDT), popularized by Rogers, is also associated with technological acceptance because it accesses six stages of innovation that the populace has gone through, and this includes different societies and cultures when adopting an unfamiliar technology (Kaminski, 2011). The model comprises five stages dealing with how innovation is adopted: (1) knowledge, (2) persuasion, (3) decision, (4) implementation, and (5) confirmation. Over time, this can lead to changes in user perception as they get more involved in using the technology (Halton, 2023). Additionally, another model that is predominantly used when analysing user intention and ensuing their usage behavior towards a certain technological system is the Unified Theory of Acceptance and Use of Technology (UTAUT), developed by Venkatesh, Morris, Davis, and Davis (Ursavaş, 2022). The model focuses on four different key constructs and targets two different determinants. The constructs include (1) performance expectation, (2) effort expectation, and (3) social influence, where these three foci directly on determining the behaviour and usage intention, while (4) facilitating conditions examine the determinant of the user behaviour.

Previous literatures have uses different measurement scales to quantify the service quality constructs however several of it has its own drawbacks. For instance, SERVQUAL and SERVEPREF does not address the interaction of customer to technological interface as it only investigates the interaction between employee and customer. Scales such as WebQual Index, SITEQUAL, E-S-QUAL and e-SELFQUAL is made for assessing service quality provided by online shopping sites (Barnes & Vidgen, 2001; Yoo & Donthu, 2014; Parasuraman, Zeithaml, and Malhotra, 2005; Ding et al., 2011). However, SSTQUAL scale made by Lin and Hsieh (Figure 1) is specifically made examine service quality provide by SSTs (Lin & Hsieh, 2011). The scale consists of seven dimensions which encompasses (1) Functionality, (2) Enjoyment, (3) Security, (4) Design, (5) Assurance, (6) Convenience, (7) Customization. Therefore, the scale is used as the base model for this research.

Figure 1. Causal Model relating SSTQUAL factors to overall service quality (Lin & Hsieh, 2011)

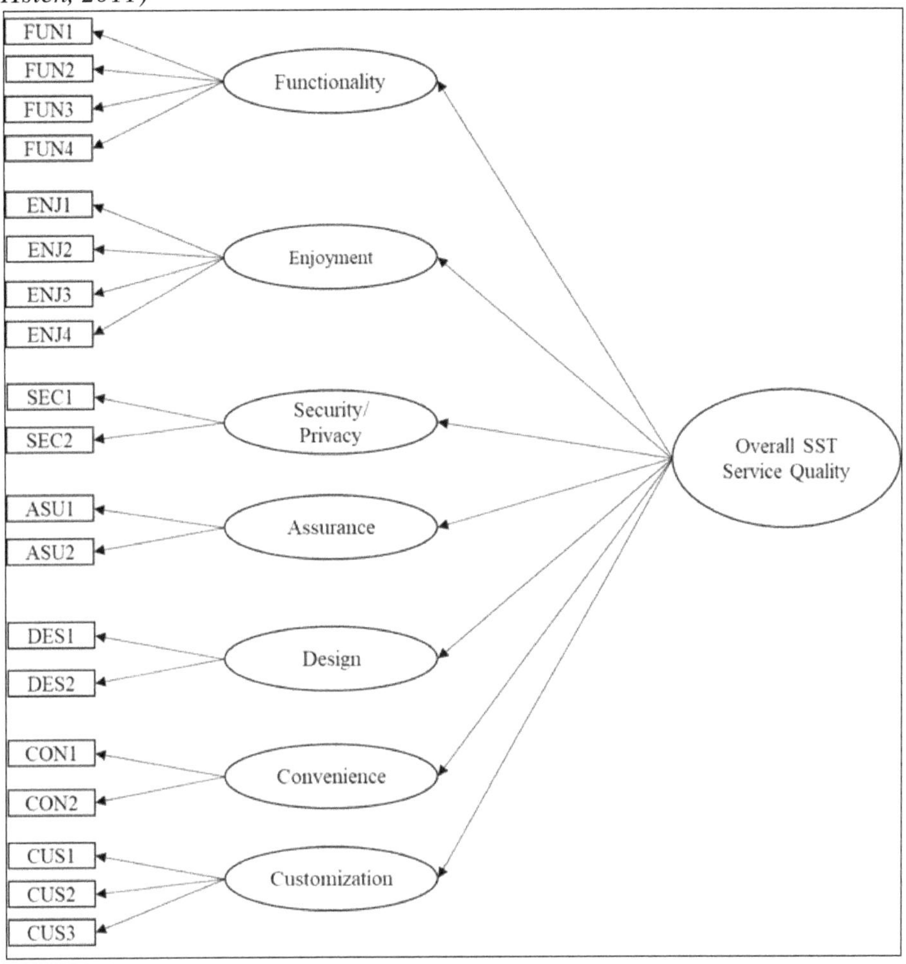

2.3 Dimensions of Service Quality

2.3.1 Functionality

Technology adoption and SST literature have been used to understand the determinants of the actual use of SSKs and the situational factors that influence their use. Since the use of SSKs is meant to give ease to customers, SSKs can reduce customer waiting time in queueing and may increase service quality. Kokkinou and Cranage (2013) mention that the processing speed, number of resources available to customers, number of customers arriving to receive service, and failure rate of SSK influence waiting times and service levels. Service levels here refer to the pre-specified waiting time that a proportion of customers will receive. Further, the impact of SSK on system performance depends on the utilization of the system and accurate demand information. Operation managers need to identify appropriate resources, and this also includes SSK, and service employees needed to meet customer demand. The study also highlights that it is important for managers to concentrate on the processing time of SSK, which can be impacted by the experience of customers in using the kiosk, the design of the SSK, and the number of options that a customer can choose to proceed with the checkout. Despite that, the study is only done for the hotel check-in process, and the study assumes that introducing SSK in this setting would not influence the service employee's processing time, and they also do not consider the attitude of the customers. After all, simplistic customers are more likely to use SSK, while complex customers tend to go for service employees.

SSKs, or SSTs in general, can assist users in completing their services swiftly and efficiently. A study by Considine and Cormican (2016) found that it is important for firms to make SSTs more user-centric and personalized, as this can improve the user experience, and this can be done by improving the design and customization of their SSTs. This can be accomplished by doing user research to determine the user's individual needs and preferences, such as customized interfaces or personalized recommendations. The respondents to the study also suggested that it is important for a firm to have clear policies in place for using SSTs to provide easier guidelines when using them. However, because the study solely focuses on one multi-national financial services organization, more studies in other industries are needed to increase the quality of the results.

SSK can transform the customer experience, and retailers need to provide immediate and personalized content to consumers anytime and anywhere. Therefore, the technology needs to provide relevant information and expertise to shoppers when they need it (Parise et al., 2016). Parise also emphasizes in their writing the need to give consumers a sense of immersion while using SSK and encourage shops to provide a deeper customer experience by delivering content-in-content, or the correct information, at the right time and in the right place. Things such as providing personalized assistance can help customers who are lost, especially when handling complex products or services that require consultation, interactivity, and knowledge sharing.

Since these kiosks are machines that enable customers to carry out tasks without needing assistance from a staff member, there are certain elements that a user might encounter and affect their behavioural intentions to use them (Chao, 2019). Existing research on customer acceptance of SSKs in retail settings indicates that there is a connection between customers' acceptance and the technical and managerial aspects of this technology. According to Kaushik and Rahman (2015), the level of user-friendliness and trustworthy interfaces when using SSKs are important when providing services in terms of consumer innovativeness. They also emphasize that when customers have a high level of trust in using SSKs, they are more likely to adopt them, and therefore, trust plays a significant role in organized retail companies creating favourable attitudes and adoption intentions towards SSKs. Furthermore, the study implies that to increase customer interest in using such technology, firms need to build an SSK interface that is valuable and easy to use, as these factors can encourage SSK adoption. However, the study only considers the impact of beliefs and subjective norms (SN) on consumers' attitudes and intentions toward applying SSK and ignores other factors that can affect the adoption behaviour of a consumer. Further, there is also a possibility of bias in terms of social desirability and response as it only relies on self-reported data.

According to previous empirical articles examined by Blut, culture can also have an impact on the use of SSK (Blut et al., 2016). He divides the different acceptances of SSK based on two factors, which include high-uncertainty avoidance and low-uncertainty avoidance. Customers in a high-uncertainty avoidance culture rely on non-provided information sources such as friends with personal expertise in technology, and this culture values computer playfulness, experience, and subjective norms. The influences of utility, the convenience of use, and the desire for connection are greater in a low-uncertainty avoidance because they are less sceptical and more inclined to trust provider-related information such as interaction with service staff. Nonetheless, the study fails to mention the causes that influence the non-acceptance of SSK.

Understanding how consumers engage with the SSK checkout process is critical for the success of retailers. Bulmer, Elms, and Moore (2018) mention that when utilizing SSK or comparable technologies, users typically have conflicting feelings, with one half feeling confident and happy when they use them successfully and the other half feeling frustrated and inadequate when difficulties arise. Therefore, retailers must also improve the user experience by giving training and help to customers, as well as reconsidering the layout and design of SSK. After all, shoppers are more likely to utilize SSK if it is simple to use. This is possible when a staff member demonstrates how the automated option works, particularly when customers believe the technology is extremely hard to use. Regardless of the findings, Bulmer sample size is relatively small, consisting of only 20 participants, which may limit the representativeness of the findings.

Ghosh (2020) explains that functionality is significant since it pertains to a product's or service's usefulness in terms of function, quality, and dependability. Functionality is one of the seven dimensions that measure overall SSTs service quality in the context of SSK. Customers feel greater value when they encounter functional value, which indicates that the SSK they use are efficient, effective, and dependable. They are more likely to use and promote an SSK if it can complete a service in a short period of time with minimum effort and a clear method. As a result, functionality is an important aspect in influencing customer happiness and SSK adoption, which can lead to a long-term connection between customers and service providers.

Consumer readiness and perceived ease of use are both essential factors to consider while implementing SSK. According to Khalufi and Shah (2022), customers are more likely to adopt SSK if they believe it is simple to use and if they are ready to utilize it. Convenience, security, and attitude or behavioural intention have been identified as important concerns of the retail industry for improving customer experience and satisfaction. Customers like an easy experience since it is convenient and saves them time and effort. This can influence the acceptance of technology by consumers and further build their trust in using self-service technology. As a result, firms must optimize the use of SSK in the retail business and guarantee that the technology is user-friendly and simple to use. This will help to attract and keep clients while also providing a pleasurable experience. However, the study recommends future studies to assess previous knowledge on the factors of social influence, need for human interaction, technological failure, and consumers' intention to use self-service technology in other industries and to look for some environmental and cultural factors that are unable to predict changes in consumers' intentions, this may indicate that the findings are not suitable for other industries or cultural contexts.

2.3.2 Enjoyment

Customers perceive more value in a service when they have a pleasant experience while using SSK (Ahn & Seo, 2018). Furthermore, pleasurable SSK encourages positive affective reactions in clients, such as pleasant sentiments. Also, a hedonically oriented self-service method, such as a fun experience, has a good association with perceived value and positive feelings in consumers. As a result, introducing more engaging features alongside practical benefits into SSK can increase customers' perceived value. Therefore, fun is a crucial aspect in the quality of self-service technology since it can result in positive feelings, higher perceived value, and customer satisfaction.

Customer-perceived usefulness and behavioural intention to use kiosks can also be affected by enjoyment (Joe et al., 2020). In the case of hotel kiosks, enjoyment may operate as a precursor to perceived usefulness, where the new technology's enjoyableness may act as intrinsic motivation, increasing individuals' beliefs that using the technology will improve their performance. Furthermore, enjoyment in adopting a new technology might reduce anxiety, lessening perceived risk. To boost good opinions and kiosk usage, hotels or retailers should aim to identify which functions and features of kiosks might provide a delightful experience for their users. Joe's study, on the other hand, only looks at the impacts of social influence and perceived satisfaction on kiosk adoption in a hotel setting, and the findings may not apply to other locations or technology.

2.3.3 Security

Using SSK as a checkout mechanism can also raise security concerns. technologies used in retail to give customers a more customized shopping experience. Retailers may face challenges in balancing privacy and personalization. Although personalization can result in more affordable advertising and higher customer retention rates, consumers' worries about marketers abusing their personal information have grown dramatically because of recent developments in retail technology (Inman & Nikolova, 2017). According to reports, consumers' top issue is privacy, with the majority feeling that their rights are "under serious threat" because of the abundance of businesses that gather personal data (Johnson et al., 2021; Martin & Palmatier, 2020; Shan et al., 2020). Hence, merchants must figure out how to tailor their offerings to customers' needs while simultaneously making sure that their personal data is not misused. This can be a difficult undertaking because consumers' privacy worries can cause backlash and negatively affect their intentions to make a purchase.

A critical component of Self-Service Technology Quality (SSTQUAL) that emphasizes the security of user data and information is the Security/Privacy dimension (Ghosh, 2020). Trust between customers and service providers is greatly influenced by security and privacy, which are important features that concentrate on the protection of user information and data. When using self-service technologies (SSTs), clients must supply data and personal information to finish transactions. If this data is not safeguarded, it may be exploited and result in identity theft, financial loss, and other undesirable outcomes. Customers should therefore have peace of mind knowing that when they use SSTs, their privacy is safeguarded, and their personal information is secure. Increased use of SSTs could result from this confidence, which would be advantageous to service providers as well as customers. Companies must also abide by data protection rules, as mandated by regulatory agencies; noncompliance can result in fines and other consequences. Therefore, to guarantee client happiness, legal compliance, and commercial success, security, and privacy are essential factors that service providers must give top priority to.

Several different types of theft and fraud activities also happen within the uses of SSK. There are individuals that have done the act of stealing items while using self-checkout lanes Some may uses the "Banana Trick" where a low-priced item, such as a banana, is manually keyed in or scanned while placing another, high-priced item onto the scale or into their bags (Chun, 2019). This act of individuals stealing or giving themselves a hefty unauthorised discount at SSK is known as SWIPERS, which is an acronym for "Seemingly Well-Intentioned Patrons Engaging in Routine Shoplifting" (Taylor, 2016). These activities can lead to substantial losses for retailers and illegal which can result in criminal change.

SWIPERS actions may be impacted by the quality of service, especially about the efficiency of anti-theft measures and the user-friendliness of self-service checkouts. Improved SSK user experience or increased difficulty of stealing could contribute to a decrease in SWIPERS activity. It is noteworthy that these relationships may be impacted by several additional elements, such as the unique attributes of each client, the retail setting, and the retailer's particular initiatives. As a result, merchants must continue to provide excellent customer service and maintain high standards while putting in place strong safeguards against SWIPERS. This can lessen theft-related losses and guarantee that clients have a great shopping experience. Therefore, it is important for customer to feel satisfied as they are less likely to engage in fraudulent activities in self-service checkouts. Conversely, dissatisfied customers may feel justified in manipulating the system, leading to an increase in SWIPERS activities.

According to Jie and Kamsin (2021), even though SSK systems in retail outlets were designed to minimize labor expenses and speed up the customer checkout procedure, merchants who utilize these technologies have a lower annual income. This heavily influenced the increased possibility of SWIPERS due to less monitoring of checkout products when customers use SSK. Therefore, Jie and Kamsin suggested the use of radio frequency identification (RFID) technology to scan goods faster, and more accurately, provide the latest information on the item, and help retailers track the location of inventory and items, especially when they are unpaid. RFID technology here refers to a technology that employs electromagnetic fields to identify and track tags attached to items (Chanchaichujit et al., 2020). However, RFID technology is expensive to implement, especially for small businesses, and it can be affected by another electronic device. Further, RFID technology can also raise privacy concerns as RFID tags can be read from a distance.

2.3.4 Design

Vakulenko suggested that design can become a critical aspect of service quality for SSK as a well design SSK can help customers complete their transactions more quickly and effectively, whereas a poorly designed kiosk might cause frustration and discontent (Vakulenko et al., 2018). A well-designed self-service kiosk can also improve its usability and accessibility, making it more appealing to a larger spectrum of clients. This can boost user happiness and loyalty while also lowering support costs for service providers. Further, Vakulenko also mention that the design of self-service kiosks can also affect their dependability and maintenance. A well-designed kiosk can be easier to maintain and repair, decreasing downtime and enhancing overall network performance.

It is also important to model an SSK based on user-friendly web design, information management, and frequent updates to ensure its effectiveness (Chan & Petrikat, 2022). Setting up a user-friendly SSK can help improve interactions with customers and increase their satisfaction. The study by Chan and Petrikat also emphasizes the need for firms to protect clients' personal information and improve its accessibility. Therefore, a firm needs to ensure the SSK is easy to use, well-designed to meet the unique requirements of the business, provide thorough training to employees on the use of SSK systems, offer a search bar and sorting feature, and ensure the SSK system is well-encrypted to protect the firm's information database. Because SSK will improve over time as strong technologies such as AI and machine learning become available, the study encourages future research to investigate the benefits of these technologies for SSK.

2.3.5 Assurance

Assurance is vital for SSK because it helps consumers develop trust and confidence in the system (Singh, 2018). When users utilize SSK, they must have confidence that the technology is dependable, secure, and simple to use. Clear instructions, user-friendly interfaces, and responsive customer service can help achieve this. Passengers are more likely to use SSK again if they are certain that the technology is working effectively and that they can get assistance if necessary. For service providers, this can result in better efficiency, lower costs, and higher customer happiness.

Tahta Kuswanto and Rahmat Shah (2021) further explain that customers' assurance relates to their level of confidence or trust in the service provider or the self-service technology they are employing. Customers are more likely to trust the service provider or the technology when they have a high level of assurance, which can lead to better levels of satisfaction, trust, and commitment. This, in turn, can improve consumer loyalty. As a result, it is possible to conclude that assurance is a significant component that can influence service quality, particularly in the context of self-service technology.

2.3.6 Convenience

Situational factors may also affect customers' decisions about using SSK. According to Demoulin and Djelassi (2016), these situational factors may include the length of queues at both regular and self-service checkouts, types, sizes, and numbers of items bought, time pressure, the need for interaction with employees, and coupons to redeem. These factors contribute to minimizing the efforts and resources required, which may seem preferable to customers. Since most customers use SSK in a hurry and when they are trying to avoid long queues, SSK should be made more relevant to the actual behaviour of customers, and stores should promote the use of SSK in a way that is pleasant and free from unpleasant situations such as breakdowns and long waiting times. Even so, the study did not consider moderating variables such as customers' assessments of their ability to use SSK or their level of self-efficacy. Scanning coupons can be considered extra work, requiring more effort and time from customers. Lastly, the study did not consider the impact of customers' age, gender, or purchasing motivation, which could be tested empirically in future studies.

Additionally, Park, Lehto, and Lehto argue that security is perceived as relatively less important by customers, who find convenience to be the most critical attribute of an SSK (Park et al., 2021). Customers may prefer convenience because it simplifies the SSK process generally and provides customers with empowerment. Since convenience is highly valued by customers, it is an essential factor for firms when they set up an SSK to provide a positive experience to customers. Considering security is found to be relatively less important in the study, this does not mean that it is not relevant in designing SSK. Service providers consider it an important attribute for assurance, but since it is considered the lowest profit factor, it is not expected to be a profit-generating component. It may not be as critical as other attributes such as convenience, which are highly valued by both customers and service providers. However, it is important to note that their study focuses only on the restaurant industry.

2.3.7 Customization

Self-service technology requires customization since it allows users to adapt their experience to their own requirements and preferences (Shim et al., 2020). Self-service technology can improve client satisfaction and increase the possibility of repeat usage by allowing for customisation. In a restaurant, for example, guests can customize their orders by picking specific ingredients or toppings, resulting in a more delightful and personalized dining experience. Additionally, by ensuring that customers receive exactly what they want, personalization can assist to eliminate errors and improve productivity. Overall, personalization is critical to the success of self-service technology since it can increase customer satisfaction, utilization, and, ultimately, revenue for enterprises.

Figure 2. Research framework using SSTQUAL scale hypothesized conceptual model.

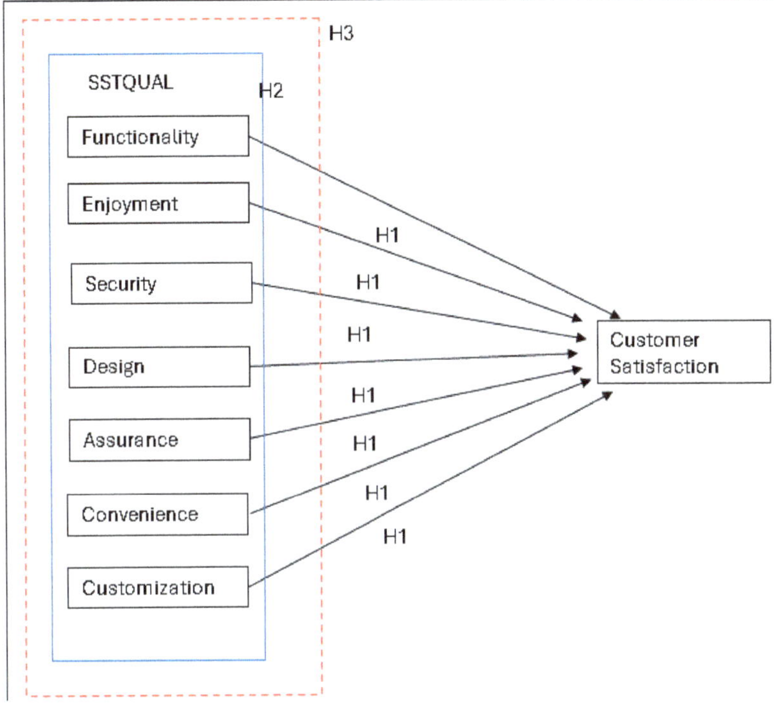

2.4 Customer Satisfaction

SSKs are increasingly changing the way customers interact with businesses to deliver services (Meuter et al., 2000). Therefore, a few literatures emphasize the importance of customer satisfaction when utilizing SSKs. Iqbal, Hassan, and Habibah (2017) mention that to increase customer satisfaction with using SSKs, firms should take steps to increase consumers' technology literacy and provide information about the SSKs they offer. Further, they also suggested that a firm should invest in high-quality and easy-to-use SSKs, as this could provide consumers with significant trust when using such technology. The study also found a significant relationship between service quality, loyalty, and behavioural intentions that have affected customer satisfaction in the sense that if a company provides a high-quality service, it can lead to customer satisfaction, which can increase loyalty and behavioural intentions. If a customer is satisfied with the services they receive, they are more likely to remain loyal to using the technologies provided by the company. In this case, loyalty means the likeliness of customer commitment to continuously using the SSK (Definition of Customer Loyalty, Gartner Marketing Glossary, n.d.). Therefore, it is essential that businesses strive to improve the quality of their services to increase customer satisfaction, loyalty, and intention. Nonetheless, the study has a limited insights in which it only covers major big cities in Pakistan and therefore additional research need to be made to observe the behaviour of customers in another places.

Companies have begun using SSK to reduce costs in a variety of ways, including reducing labour requirements, improving efficiency, reducing support costs, improving resource allocation, and saving money on data entry (Larivière et al., 2017). On the other hand, a customer's impression of SSK may vary based on criteria such as product or service category, customer group, and stage of the customer journey. If customers do not see an increase in their added value, some may be hesitant to cope with SSTs. According to Lariviere, firms must convey the possible experience rewards of appropriate role behaviour through the customer encouragement process so that customers understand the benefits of adopting SSTs. However, because it relies on self-reported data from customers or employees, the study fails to minimize social desirability or recollection bias in its data gathering.

Firms should, however, evaluate the performance of SSKs based on long-term satisfaction and loyalty impacts rather than just cost savings and efficiency advantages, as satisfied customers can increase usage, improve retention, enhance brand image, and increase sales (Djelassi et al., 2018). Additionally, Djelassi also mentions in his article that a positive SSK experience, short waiting times, and satisfaction with the SSK itself increase customer satisfaction when dealing with SSKs. Evaluating customer satisfaction when using SSKs also helps firms understand customer perceptions and interactions with these technologies, which can help them identify areas for improvement and make changes to enhance the customer experience. Additionally, it can help businesses decide which technologies to invest in and which resources to commit to, which can lead to better customer satisfaction, increased adoption rates, and cost savings for the organization. Despite that, the study indirectly analysed retailer satisfaction through waiting time satisfaction and satisfaction with SSKs, while other aspects such as service quality and price were not directly considered. The study is also based solely on self-reported data, which may contain biases and mistakes.

Firms must also understand their consumers' ecosystems to thrive. Gummerus, Lipkin, Dube, and Heinonen (2019) proposed that firms should treat the consumer as the principal actor in service supply, with the firms functioning as the secondary actors. Firms must also understand their customers' processes and activities, even if they are unrelated to the primary service offering. This technique may result in the discovery of new service features that could be implemented into SSK. Further, firms must discover new ways to locate customers' problems and requirements to better assist them in solving them, as customers' problems are not always visible to the organization. Firms will therefore be able to help and serve their clients in the best possible manner, which is critical for firms because delivering better service through SSK can boost customer satisfaction. Nonetheless, the study is only based on conceptual exploration with qualitative insight from several companies, and further research involving empirical studies needs to be done.

Understanding customer value can help businesses make more informed decisions at various phases of contemplating, deploying, and improving SSK networks. Vakulenko, Oghazi, and Hellström (2019) mention that it is important for firms to understand customer value knowledge, which encompasses functional value, epistemic value, emotional value, and social value, as this can help service providers understand what customers want and need from their products or services. In the case of SSK, service providers can build an SSK that meets their customers' wants and expectations, which can lead to higher customer satisfaction, loyalty, and retention. Furthermore, understanding customer value knowledge can assist merchants in identifying areas for development in their SSK networks, leading to improved customer experiences and enhanced profitability. Despite that, Vakulenko fails to present empirical data to back his assertion; hence, more study may be required to validate customer value knowledge in practice.

2.5 Transaction Cost Economics (TCE)

SSK is also directly related to transaction cost economics, where they can significantly reduce transaction costs in various ways. TCE refers to the cost of running an economic system of companies. It is a theory that considers the cost of making an economic exchange, which is incurred during economic trade in a market and does not accrue to any participant in the transaction (Rindfleisch, 2019). Transaction costs include various expenses that are significant factors in business operation and management, and these include expenses such as search and information costs, bargaining and decision costs, and policing and enforcement costs (Chen et al., 2022).

In terms of SSK, supermarkets employ it as a practical application of TCE, where cost transactions are reduced, and efficiency is increased. SSK can lower operational costs, which is one way to achieve this. While the kiosks take orders, serve clients, guide them appropriately, and accept payments, supermarket personnel may focus on more important responsibilities. Next, SSK contributes to greater efficiency and sales by automating transactions that would normally require the assistance of a cashier. Automation can shorten the time required to perform transactions, lowering transaction costs. Furthermore, SSK can result in increased sales and a higher return on investment. SSK can also improve the consumer experience by delivering speed and convenience. Most shoppers who understand how to use self-service checkouts believe they are quick and simple to use (Zaveri et al., 2022). Furthermore, a sense of control over one's shopping might contribute to increased consumer satisfaction and intent to use and repurpose SSK. Implementing SSK also helps to minimize labour costs because the supermarket will need fewer colleagues and reduces mistakes by empowering customers to control the ordering process, lowering the possibility of errors such as wrong orders or overcharging.

2.6 Transaction Cost Theory (TCT)

SSK is likewise meant to deal directly with individual transactions, which makes it closely related to Transaction Cost Theory. TCE and TCT are frequently used interchangeably since they both address the expenses of conducting an economic exchange; however, TCT focuses on the governance of individual transactions (Cuypers et al., 2021).TCT believes that managers and directors may arrange transactions in an opportunistic manner.From a TCT standpoint, SSK in supermarkets can govern individual transactions more efficiently by lowering transaction costs. Customers can scan, bag, and pay for their purchases without the assistance of a cashier because the checkout procedure is fully automated. This can help to lower the costs of hiring, training, and keeping a cashier team.

In addition, SSK can speed up transactions. Customers are not required to wait in line for a cashier and can finish their transactions at their own pace. This can result in shorter lines and reduced waiting time, enhancing the overall shopping experience and therefore enhancing transaction accuracy. The automated technique reduces the possibility of human error, such as wrong pricing or missing items. This can result in more precise inventory management and pricing. Furthermore, SSK gives customers more control over their purchases by allowing them to select, scan, and bag their items exactly as they desire. This can result in higher consumer satisfaction and loyalty.

Managing transaction costs is important for businesses as it directly affects the efficiency and profitability of economic activities (Yousuf, 2017). High transaction costs can be a barrier to trade, which may reduce the potential market transactions leading to inefficiencies in allocating resources. According to Cuypers et al. (2021), when transaction costs are high, it becomes more challenging for parties to engage in trade and economic activities. Additionally, due to high transaction costs, certain economic exchanges encompassing trade relationships, contractual agreements, and governance mechanisms become inefficient within a market-based framework, leading to failure of market transactions to overcome barriers posed by high transaction costs.

In terms of SSK, when customers must deal with long queues, complex procedures, or hidden fees when using SSK, this can contribute to high transaction costs due to the unsmooth customer experience causing customer dissatisfaction. Customers will feel that using SSK is not worth it, causing the perceived value of the transaction to decline. This can lead customers to abandon SSK and prefer to do their transactions using other methods.Therefore, businesses need to utilize different strategies accordingly based on market conditions, such as mean reversion and asset correlations, to determine the most effective approach for managing transaction costs and achieving desired risk-return profiles(Bernoussi & Rockinger, 2022).

METHODOLOGY AND DATA

3.1 Research Design

Various research methods are developed to add new knowledge, gather evidence for generalization, and develop theories, however for this research, a quantitative approach is used, which is an objective, rigorous, formal, deductive approach, and systematic strategy for generating and refining knowledge for problem-solving (Haradhan, 2020). In this case, the phenomenon of the study is how the attributes of seven different dimensions of service quality are affecting customers' experiences and satisfaction during their usage. The research relies on primary data where it is a data that has not been previously published or analysed and is collected specifically for the purpose of the research study (Siva Durga Prasad Nayak & Narayan, 2019). The method for data collection is using a set of questionnaires, as it ensures a faster and more accurate data collection process and facilitate data processing (Roopa & Rani, 2012).

Further, Roopa & Rani define a questionnaire as a research tool consisting of a set of questions or other 'prompts' to collect data from a set of respondents. Questionnaires describe and compare several customer experiences and reveal variable that may be associated with SSK satisfaction outcomes. The survey is design to be flexible and broad to evaluate the different variables while maintaining the economic feasibility. The variables in this study include questionnaire responses such as support, added value, affirmation, ease of use, privacy, and usage outcome. This study will provide a prospective description of customers who were given a chance to use SSK and either they accepted or declines it. Prospective involve forecasting future trends or outcomes based on current data and can be helpful for businesses in making strategic decision (Schweizer et al., 2020).

True experimental studies allow establishment of causal relationship (Eisend & Kuss, 2019). Descriptive research, on the other hand, explains a phenomenon rather than looking for a cause-and-effect relationship ("Descriptive Research," 2021). The goal of this study is to avoid manipulating the environment in which a customer intends to use the SSK. A descriptive study allows for the investigation of an event as it occurs in the retail situation. Customers will be given the option to use traditional or SSK checkout. Self-selection rather than randomization will give customers their philosophical convictions and this will highlight the actual experience of customers who are given the option to use SSK.

Given the current situation of SSK in Brunei, a descriptive study is appropriate to enlighten businesses, customers, and researchers about the variables surrounding satisfaction when using SSK. Researchers will be able to focus on variables and outlines of SSK usage for individuals' practices. Providers and customers will understand the SSK experience and possibly undertake SSK checkout as a good alternative.

3.2 Threats to External Validity and Measures to Control Them

Customers who have previously used SSK on many occasions is focused on the prospective portion of the study. These customers may reflect some bias, as they are likely delighted with their experience with the technology. To avoid this, the questionnaire is designed appropriately, and strategies such as neutral wording, anonymity, single-topic questions, randomizing questions, and answers, limiting scale questions, and placing demographic questions last has been a viable option for obtaining honest, accurate answers for the research. Furthermore, clients who prefer to utilize SSK for the prospective segment will be able to communicate some of their unpleasant experiences with SSK. These strategies increase generalizability for clients who intend to use or learn more about SSK.

Poor survey returns and/or improper responses to any questionnaire might distort the final data analysis and conclusions, lowering the study's repeatability. To improve survey response and avoid loss of documentation, the questionnaire is designed to be completed in about 10 minutes on the internet using Google Forms. Further, the individual question has been made clear, simple and unambiguous to avoid misinterpretation, increase response rate, improve data quality and prevent misleading data.

3.3 Threats to Internal Validity and Measures to Control Them

Threats to internal validity include factors outside the primary variable that could influence the outcome of the study. For instance, an information bias can jeopardize the internal validity of self-service kiosk consumer satisfaction surveys. As only retail industries and fast-food restaurants are considered for this study, airport and hotel segments can also offer SSK to their customers and the way these SSK works might differ from the context of retailers (Anitha et al., 2019; Fernandes & Pedroso, 2016). Customers may have experience using SSK from different industries, but the purpose could be different. Nevertheless, respondents with SSK usage experience from other industries are accepted, considering that there are similarities to some extend across the industries.

Self-selection to using SSK will be studied prospectively, as this will happens in the retail and fast-food restaurant settings. Every customer will have the rights to choose their preferable method of checkouts. Customers who may not prefer to use SSK should not be forced to do so for the sake of research. Likewise, customers who would like to experience SSK should not be prohibited from doing so.

3.4 Sampling Methods

The data were collected through the online survey by distributing the link thorough emails and social media. Simple random sampling and snowball sampling techniques was employed. Simple random sampling is chosen as a sampling strategy as it helps to minimise bias and provide a representative sample of the population, giving an equal chance of collection and provides the simplest form of data collection as it only involves basic observation and recording skills (Bhardwaj, 2019). For snowball sampling, it is a valuable method for a researching people who have specific characteristics that would otherwise be difficult to identify (Bhardwaj, 2019). Participants were asked to help distribute the survey after they finished answering it.

Brunei Darussalam, with a population of around 455,400, has a unique digital landscape (Department of Economic Planning and Statistics - Population, n.d.). To begin, 10% of this population translates to approximately 45,540 people. While data is limited, the gradual acceptance of digital platforms—such as internet usage, social media participation, and mobile connectivity—indicates an increasing interest in technology. To determine the exact number of tech-savvy Bruneians, additional specialized surveys are required. However, a survey from AITI in 2019 indicates that most people do not engage in the Web/Mobile Application activities on a regular basis (AITI, 2019). In terms of convenience, self-service kiosks are becoming increasingly popular. Although accurate data are unavailable, initiatives by institutions such as Bank Islam Brunei Darussalam (BIBD) encourage cashless transactions via kiosks at supermarkets and fast-food restaurants. The population's willingness to accept these kiosks remains an unsettled topic. For a robust survey, a minimum sample size of ~200 (163) respondents are obtained.

Additionally, during the data collection process, ethical issued were well addressed by assuring that the respondents are answering the survey questions anonymously. Consent form is handed at the start of the survey where it stresses out anonymity where personal information will not be captured and shared with anyone inside and outside of the university.

3.5 Data Collection

Table 1 provides relevant information about the respondent's characteristics. The total number of respondents were 163 of which 72 (44.2%) were male respondents while 91 (55.8% were female respondents. Most of the respondents were in between 25-34 years old (47.9%), and most were from Brunei-Muara district (77.3%). Moreover, marital status indicates that 71.8% of the respondent are single, 27.6% are married while 6% prefer not to disclose their status. Additionally, 50.9% of the respondents are the working population, 43.6% are student while 3.1% and 2.5% are unemployed and retired individuals respectively. Lastly, the monthly income of the respondents is mostly below $500 (44.2%). 53.4% of respondents are having monthly income in between $500 to $3000 and 2.5% of respondents have income above $3000. Statistical research was conducted to validate the hypothesized association in the purposed model after the data was collected. The study's findings have significant practical and management implications for understanding customer's attitudes towards SSK. Additionally, to protect the anonymity of the respondents, questions that could potentially identify an individual, such as name, exact address, and email address is avoided

3.6 Measurement Scales

SSTQUAL has been applied to approach the determinants of service quality, and the conceptual model is presented in (Lin & Hsieh, 2011). Seven dimensions of service quality, which encompass functionality, enjoyment, security, design, assurance, convenience, and customization, will be included as independent variables, while customer satisfaction will be the dependent variable. However, since the study focuses only on SSK, the question has been altered towards the experience on using SSK. Additionally, a revised SSTQUAL made by Radomir and Nistor (2014) as it gives better construct reliability. Moreover, some questions have been added to respective variables to increase reliability and understanding of the variables. To measure customer satisfaction, a scale has been adopted from the three-item American Customer Satisfaction Index (ACSI) (Fornell et al., 1996). All items are measured using a six-point Likert-type scale.

Table 1. Demographic Characteristics

Characteristics	Frequency	Percentage (%)
Gender		
Male	72	44.2
Female	91	55.8
Age		
18-24	46	28.2
25-34	78	47.9
35-44	25	15.3
45-54	9	5.5
55 years old & above	5	3.1
District		
Brunei-Muara	126	77.3
Tutong	13	8.0
Belait	19	11.7
Temburong	5	3.1
Marital Status		
Single	117	71.8
Married	45	27.6
Others	1	6
Occupation		
Government	35	21.5
Private	38	23.3
Student	71	43.6
Self Employed	10	6.1
Unemployed	5	3.1
Retired	4	2.5
Monthly Income		
Below $500	72	44.2
$500-$1000	16	9.8
$1001-$1500	17	10.4
$1501-$2000	27	16.6
$2001-$2500	21	12.9
$2501-$3000	6	3.7
Above $3000	4	2.5

Table 2. Public Awareness of SSK in Brunei

Items	Frequency	Percentage (%)
Do you know there are self-service kiosks such as self-checkout kiosks and self-ordering kiosks in Brunei?		
Yes	160	98.2
No	3	1.8
Have you ever used self-service kiosks?		
Yes	154	94.5
No	9	5.5
If yes, have you ever encountered any form of information security breaches when using self-service kiosks?		
Yes	1	0.6
No	152	93.3
How often do you utilize Self Service Kiosk?		
Always	6	3.7
Often	54	33.1
Once in a while	94	57.7
Never	9	5.5

3.7 Pilot Study

A pilot study is an essential step in research as it helps to assess the feasibility of the study, refine research methods, test measurements, refine sample size estimation, cost-effectiveness and ethical considerations as the procedures are tested first on a small scale (Thabane et al., 2010). Generally, 10%-20% of the main sample size is considered a reasonable number for conducting a pilot study (Hazzi & Maldaon, 2015). Since the minimum targeted respondents for this study is 200 respondents, the first pilot study has been done for a total of 30 respondents. Additionally, Johanson and Brooks (2009) also stated that 30 responses are reasonable to test the validity of the survey as it is practical, simple, effortless calculations and sufficient to test the hypothesis.

Reliability test has been done for the first pilot study and it has been found that the design and customization construct have a Cronbach alpha of less than 0.6 which is poor (Johnson, 2021). Therefore, the questions for Design 3 (DES3) have been rephrase as respondents shows confusion in answering the question and the option of neutral has been removed from the Likert scale to reduced social desirability bias.

A second pilot study has been conducted for 20 more respondents after they made changes to test the reliability of the scale. Table 2 summarise the Cronbach alpha for this set of study has shown that all the constructs have a value of more than 0.7 thus ensuring that the scale to be reliable (Bonett & Wright, 2014; Johnson, 2021).

Table 3. Reliability test of the scale

Dimensions		Independent Variables							Dependent Variable
		FUN	ENJ	SEC	ASS	DES	CON	CUSTOM	Customer Satisfaction
Pilot Study	30	0.661	0.612	0.768	0.767	0.470	0.745	0.571	0.793
	50	0.824	0.711	0.876	0.884	0.808	0.883	0.777	0.842
Final Value	163	0.852	0.722	0.803	0.929	0.671	0.936	0.726	0.619

DATA ANALYSIS AND RESULTS

Data analysis has been done using 2 different statistical software. Smart-PLS 4 which has 2 purposes. First, it is used to analyse the measurement and scale's validity alongside its reliability (Measurement Model Assessment). Second, it is used to evaluate the framework model and test the hypothesis proposed (Structural Model Assessment). Both model assessment is executed using structural equation modelling (SEM) which analyses data from 163 respondents. The other software is IBM SPSS used to analyse various descriptive statistics, hypothesis testing, correlations, regressions, t-tests and ANOVA.

4.1 Preliminary test

Before delving into the core analyses to answer the research questions, some preliminary data exploration techniques were employed. These techniques help to ensure the data is suitable for further analysis and provide a basic understanding of the variables involved.

4.1.1 Normality test

Statistical procedures such as t-tests, ANOVA and regression analysis depends on the assumption of normality of the data being analysed. This is due to when data is normally distributed, statistical test provide a more reliable estimates of population parameters such as means and variances and a more accurate p-values. Furthermore, normality allows us to generalise findings from the sample data to the larger population it represents. Based on the central limit theorem, when the sample size reaches 25, the distribution begins to conform to the normal curve and becomes normally distributed at 30 (Chang et al, 2008; Krithikadatta, 2014). Therefore, since the sample gather for this research is 163, the data is already normally distributed.

4.1.2 Descriptive Analysis

According to Loeb et al. (2017) descriptive analysis is a strategy for simplifying data in a systematic and orderly manner. It entails investigating the linkages between variables within a given population or sample group. Furthermore, descriptive statistics are used to explain either a single variable or numerous variables, with the goal of describing their relationship and identifying the key properties of a dataset. Kaliyadan and Kulkarni (2019) mentions that the mean, mode, and median are the most common descriptive statistics methods.

From table 4, it shows that ASS has the highest mean value (5.04) while SEC has the lowest mean value (3.60). Generally, FUN, ENJ, ASS, and CON are above the midpoint which could imply positive perception of the construct while SEC, CUSTOM and DES are below midpoint suggesting a negative perception of the construct. CUSTSTS on the other hand is around the midpoint, indicating average satisfaction. Nevertheless, mode and median are approximately close within the range of 4 and 5. However the median value which are higher than the mean may hint at a skewed distribution. The standard deviation value on the other hand, indicates the variability of customer responses in each category with ASS showing the highest variation and SEC the lowest among the constructs.

Table 4. Descriptive analysis

	FUN	ENJ	SEC	ASS	DES	CON	CUSTOM	CUST STS
N	163	163	163	163	163	163	163	163
Missing	0	0	0	0	0	0	0	0
Mean	4.45	4.36	3.60	5.04	3.74	4.93	3.84	3.99
Median	5	4	4	5	4	5	4	4
Mode	5.00	4.00	4.00	5.00	4.00	5.00	4.00	4.00
Std Dev	0.876	0.792	0.726	0.962	0.813	0.810	0.816	0.797
Min	2	2	2	2	2	3	2	2
Max	6	6	6	6	6	6	6	6
Skewness	-0.359	-0.291	0.203	-0.918	0.855	-0.511	0.0952	-0.137
Std. error skewness	0.190	0.190	0.190	0.190	0.190	0.190	0.190	0.190
Kurtosis	0.0230	0.487	0.176	0.361	0.727	-0.0730	-0.387	0.0927
Std. error kurtosis	0.378	0.378	0.378	0.378	0.378	0.378	0.378	0.378
Shapiro-Wilk (W)	0.880	0.856	0.838	0.827	0.815	0.841	0.871	0.866
Shapiro-Wilk (p)	<.001	<.001	<.001	<.001	<.001	<.001	<.001	<.001

4.1 Relationship Analysis

This section explores on the relationship between dimensions of service quality and customer satisfaction in the context of SSK in Brunei.

4.1.1 Measurement Model Assessment

Measurement model assessment is a crucial step in SEM to ensure the accuracy of the data in representing the framework model. Construct loading was first measured to analyse the strength and direction of the linear relationship between the construct and the indicator variables (Cheah et al., 2023). A high positive loading indicates strong positive relationship, a high negative loading indicates strong negative relationship while a low loading suggests a week relationship between indicator and the construct. The cut off loading of items need to be at least 0.5 where it is considered moderate and acceptable depending on the context while having a cut-off loading of ≥ 0.7 is generally strong and those loadings with < 0.4 must be removed to ensure validity (Henseler et al., 2009). The item ENJ3 and ENJ4 from the variable Enjoyment has been remove due to having low loading (<0.5).

Further, the construct of ENJ has been further analyse and it has been found to have a potential two-factor structure. A rotated component matrix was produced using the Varimax rotation approach to obtain a more interpretable solution for the components underlying the pleasure construct. ENJ3 and ENJ4 have been found to cause the construct to have two component matrixes. After deleting items ENJ3 and ENJ4 with low loadings from the initial analysis, the rotated component matrix revealed a more distinct one-factor structure for the enjoyment construct. This revision improves the construct validity of the enjoyment measure by guaranteeing that all indicators (remaining items) are significantly dependent on a single underlying factor that captures the enjoyment construct (Ghazali et al., 2020; Sreejesh et al., 2013). In the initial study, the inclusion of low-loading items ENJ3 and ENJ4 may have indicated a two-component structure, implying that these items could be used to measure mixed aspects of enjoyment. By removing these items, a more focused and interpretable measure of satisfaction is obtained which contributes to a more consistent with the study's theoretical framework.

Variance inflation factor (VIF) is used to assess the severity of multicollinearity in regression analysis (Liao & Valliant, 2012). Multicollinearity occurs when there is a high degree of correlation between 2 or more independent variables in a regression model. A good model predictor should have a VIF of less than 10 where a range from 5 to 10 indicates moderate correlation and a value of less than 5 indicate a low correlation (Marcoulides & Raykov, 2018). The construct of the data has shown VIF values of <5 which indicates that there is only a low coloration in between the independent variables.

Next, convergent validity. is assessed using Average Variance Extracted (AVE) in which it measures the degree to which the indicators (survey questions/items) truly measure the underlying latent construct (theoretical variable) they are supposed to represent. The value of AVE should be above 0.5 to certify that the latent variable shows at least half of the variance of the indicators.

Further, discriminant validity is assessed using the Fornell and Larcker criterion. The criterion for this analysis is to compare the AVE of each construct to the squared correlations (R-squared) between that construct and the other constructs in the model (Nasution et al., 2020). If the square root of a construct's AVE is greater than all the squared correlations between that construct and other constructs, it suggests a good discriminant validity. Table 6 demonstrates that all the diagonal values, which represent pairs of the same constructs, are higher than the other values suggesting that the constructs are sufficiently distinct.

Table 5. Rotated Component Matrix

	Component	
	1	**2**
ENJ 1	0.786	0.046
ENJ 2	0.893	0.120
ENJ 3	0.280	0.526
ENJ 4	-0.088	0.894
ENJ 5	0.686	0.093

Table 6. Discriminant validity using Fornell-Larcker criterion

	FUN	**ENJ**	**SEC**	**ASS**	**DES**	**CON**	**CUSTOM**	**CUST STS**
FUN	**0.836**							
ENJ	0.760	**0.803**						
SEC	0.526	0.565	**0.709**					
ASS	0.428	0.471	0.568	**0.966**				
DES	0.390	0.483	0.455	0.269	**0.771**			
CON	0.349	0.434	0.497	0.696	0.415	**0.969**		
CUSTOM	0.516	0.538	0.608	0.519	0.465	0.515	**0.804**	
CUSTSTS	0.636	0.612	0.617	0.598	0.462	0.568	0.658	**0.752**

Table 7. Measurement Model Assessment before removal of item with low loadings

Constructs	Loading	VIF	Cronbach's Alpha	CR (rho_c)	AVE
Functionality			0.852	0.901	0.698
FUN1	0.880	2.583			
FUN2	0.885	2.680			
FUN3	0.666	1.364			
FUN4	0.891	2.520			
Enjoyment			0.583	0.732	0.406
ENJ1	0.719	1.677			
ENJ2	0.900	2.175			
ENJ3	0.336	1.061			
ENJ4	0.132	1.015			
ENJ5	0.767	1.404			
Security			0.803	0.855	0.502
SEC1	0.802	1.768			
SEC2	0.544	1.286			
SEC3	0.699	1.542			
SEC4	0.768	1.691			
SEC5	0.551	1.420			
SEC6	0.831	2.024			
Assurance			0.929	0.966	0.934
ASS1	0.964	4.047			
ASS2	0.969	4.047			
Design			0.671	0.809	0.594
DES1	0.883	1.519			
DES2	0.827	1.377			
DES3	0.564	1.212			
Convenience			0.936	0.969	0.940
CON1	0.970	4.399			
CON2	0.969	4.399			
Customization			0.726	0.846	0.647
CUSTOM1	0.853	1.604			
CUSTOM2	0.809	1.460			
CUSTOM3	0.747	1.336			
Customer Satisfaction			0.619	0.789	0.565

continued on following page

Table 7. Continued

Constructs	Loading	VIF	Cronbach's Alpha	CR (rho_c)	AVE
CUSTSTS1	0.904	1.556			
CUSTSTS2	0.540	1.227			
CUSTSTS3	0.765	1.302			

Table 8. Measurement Model Assessment after removal of item with low loadings

Constructs	Loading	VIF	Cronbach's Alpha	CR	AVE
Functionality			0.852	0.901	0.698
FUN1	0.880	2.583			
FUN2	0.885	2.680			
FUN3	0.666	1.364			
FUN4	0.891	2.520			
Enjoyment			0.722	0.843	0.644
ENJ1	0.723	1.660			
ENJ2	0.907	2.155			
ENJ3	-	-			
ENJ4	-	-			
ENJ5	0.767	1.402			
Security			0.803	0.855	0.502
SEC1	0.802	1.768			
SEC2	0.544	1.286			
SEC3	0.699	1.542			
SEC4	0.768	1.691			
SEC5	0.551	1.420			
SEC6	0.831	2.024			
Assurance			0.929	0.966	0.934
ASS1	0.964	4.047			
ASS2	0.969	4.047			
Design			0.671	0.809	0.594
DES1	0.883	1.519			
DES2	0.827	1.377			
DES3	0.564	1.212			

continued on following page

Table 8. Continued

Constructs	Loading	VIF	Cronbach's Alpha	CR	AVE
Convenience			0.936	0.969	0.940
CON1	0.970	4.399			
CON2	0.969	4.399			
Customization			0.726	0.846	0.647
CUSTOM1	0.853	1.604			
CUSTOM2	0.809	1.460			
CUSTOM3	0.747	1.336			
Customer Satisfaction			0.619	0.789	0.565
CUSTSTS1	0.904	1.556			
CUSTSTS2	0.542	1.227			
CUSTSTS3	0.764	1.302			

4.1.2 Structural Model Assessment

For this assessment, the Path Coefficients, R^2 and the Goodness of Fit (GoF) has been used to figure out how well the model framework clarified the relationship between the constructs. The recommended value for a path coefficient is to have values above 0.35, 0.15, and 0.02 which can be regarded as strong, moderate, and weak respectively (Cohen, 2013). However, path coefficients are good if it is less than 1 or greater than -1 as this indicates there are no collinearity problem (Henseler et al., 2016). Since the path coefficients of the constructs towards the dependent variable are within the range of 1 and -1, the structural paths do not have any sorts of collinearity problem.

When looking at the GoF, the value must be at least 0.36 for it to show a significant quality (Alolah et al., 2014). GoF will helps to ensure that the model adequately captures the relationships between variables in the data, provides confidence in inferences and improves model prediction. The GoF value for the model has been found to be 0.653, indicating that the framework model has an acceptable model fit.

Table 9. Structure Model Results showing the Path Coefficients

Structural Paths	Path Coefficients β	Total Effects β
Functionality → Customer Satisfaction	0.281	0.281
Enjoyment → Customer Satisfaction	0.039	0.039
Security → Customer Satisfaction	0.112	0.112
Assurance → Customer Satisfaction	0.160	0.160
Design → Customer Satisfaction	0.073	0.073
Convenience → Customer Satisfaction	0.133	0.133
Customization → Customer Satisfaction	0.239	0.239

Table 10. Goodness of Fit

Constructs	AVE	R^2
FUN	0.698	
ENJ	0.644	
SEC	0.502	
ASS	0.934	
DES	0.594	
CON	0.940	
CUSTOM	0.647	
CUSTSTS	0.565	0.617
Average Score	0.691	
GoF	0.653	

4.1.3 Linear Regression Analysis

Linear regression analysis is used to test or predict the significance of the effect of independent variables towards the dependent variable (Skiera et al., 2021).

Table 11. Liner regression analysis

Model		Unstandardized Coefficients		Standardized Coefficients	t	Sig.	Collinearity Statistics	
		B	Std. Error	Beta			Tolerance	VIF
1	(Constant)	.367	.293		1.254	.212		
	FUN	.161	.073	.187	2.194	.030	.422	2.369
	ENJ	.100	.087	.106	1.154	.250	.367	2.723
	SEC	.155	.079	.145	1.968	.051	.564	1.774
	ASS	.077	.062	.105	1.238	.217	.430	2.328
	DES	.087	.062	.094	1.399	.164	.688	1.453
	CON	.070	.069	.082	1.013	.313	.465	2.152
	CUSTOM	.237	.074	.243	3.187	.002	.531	1.884

a. Dependent Variable: CUSTSTS

4.1.3.1 Coefficients

The unstandardized coefficients (B) indicate the degree of change in Customer Satisfaction (CUSTSTS) for a unit change in the predictor variables. For instance, FUN has a B value of 0.161 meaning for each unit increase in FUN, CUSTSTS increases by 0.161 units. From the table, it can be observed that CUSTSTS changes the most when CUSTOM is affected (0.237).

4.1.3.2 Significance

The significance value denotes the statistical significance of the predictors. A value less than 0.05 indicates a statistically significant impact on CUSTSTS (Alexpoulos, 2010). Among all the constructs, only FUN and CUSTOM have a Sig. value less than 0.05 which may make both as a significant predictor.

4.1.3.3 Collinearity Statistics

Regression analysis also includes VIF which assesses multicollinearity among predictors. Here, all VIF values are below 5, suggesting acceptable levels of collinearity.

4.1.3.4 Structural paths

Table 12. Significance of structural paths of the construct

Structural Paths	Original Sample (o)	Sample mean (M)	STDEV	T Statistics	P Values	f^2
FUN → CUSTSTS	0.281	0.274	0.089	3.159	0.002	0.089
ENJ→CUSTSTS	0.039	0.040	0.094	0.411	0.681	0.001
SEC→CUSTSTS	0.112	0.124	0.083	1.340	0.180	0.016
ASS→CUSTSTS	0.160	0.163	0.083	1.917	0.055	0.029
DES→CUSTSTS	0.073	0.072	0.058	1.267	0.205	0.009
CON→CUSTSTS	0.133	0.124	0.092	1.442	0.149	0.022
CUSTOM→CUSTSTS	0.239	0.239	0.070	3.393	0.001	0.079

Table 12 presents the structural paths between the construct and the dependent variable (CUSTSTS). The purpose of this analysis is to evaluate the significance of the paths, indicating how strongly each function predicts CUSTSTS. Only FUN and CUSTOM are significant as the other constructs have been shown to not affect CUSTSTS. This implies that when SSK performs well and meets the user's functional needs, satisfaction is likely to be higher. Likewise, customization options also can enhance the user experience, leading to greater satisfaction.

Meanwhile the effect size (f^2) indicates the effect of an independent variable on the dependent variable. Selya et al. (2012) mentions that an $f^2 = 0$ indicates no effect size, $0 < f^2 < 0.02$ indicates a negligible effect size, $0.02 \leq f^2 < 0.15$ indicates a small effect size, $0.15 \leq f^2 < 0.35$ indicates a medium effect size while $f^2 \geq 0.35$ presents a large effect size. From table 12 FUN has the largest effect size (0.089) followed by CUSTOM (0.079), ASS (0.029,) and CON (0.022) even though the effect size is only small. In this case, the larger the effect size, the stronger the relationships between the variables.

Furthermore, FUN and CUSTOM also show the highest number of original sample values. Together with path coefficient values from Table 9, low VIF values, and significance, both FUN and CUSTOM may affect customer satisfaction the most. However, the data gathered from the survey indicates that customers perceive that the most important dimension of SSTQUAL is security.

Lastly, the Durbin-Watson value for the regression analysis is 1.67 which is generally considered a substantial value (Abdulhafedh, 2017). This indicates that there is no significant autocorrelation in the residuals of the liner regression model.

Table 13. Customer perspective on the most important dimension

Variables	Frequency	Percentage (%)
Functionality	22	13.5
Enjoyment	11	6.7
Security	63	38.7
Assurance	7	1.8
Design	3	4.3
Convenience	25	15.3
Customization	2	1.2
Missing Values	30	18.4

4.2 Interaction Effects

This section analyses how different dimensions of service quality interact with each other and their collective impact towards customer satisfaction.

4.2.1 Correlation and Collinearity Analysis

Three types of different correlation analysis have been done to ensure correlation consistency in between constructs. Pearson, Spearman and Kendall tau-B have been used to analyse the data as they have their own underlying assumptions about the data. For instance, Pearson's correlations assume continues, normally distributed data and a linear relationship between variable, Spearman's rank correlations are less strict and does not require normality and Kendall's tau-B is closely like Spearman but also applicable to data with ties where they have multiple observations with the same value (El-Hashash & Shiekh, 2022).

For correlation analysis, the range of correlation value in between the constructs is consider a "very strong correlation" "strong correlation", "moderate correlation", "weak correlation" and "negligible correlation" when the value is between 0.90-1.00, 0.70-0.89, 0.40-0.69, 0.10-0.39 and 0.00-0.10 respectively. A strong and very strong correlation may indicate that a linear relationship exists between two or more variables indicating multicollinearity which may cause data problems that affect the reliability of the questionnaire parameters (Alin, 2010).

Correlation analysis from Pearson indicates that there is a positive correlation between FUN and ENJ where the r value is 0.745. The Spearman correlation matrix also indicates a moderate correlation close to 0.7 where the rho value between FUN and ENJ is 0.691. However, for Kendall Tau-B the Tau-B value is 0.579 indicating a moderate correlation. This may suggest that as the FUN factor increases, so does the enjoyment level. Next, the correlation between CON and ASS can also be consider strong. For Pearson, the r value is 0.695, while for Spearman the rho value is 0.653 and Kendall Tau-B value is 0.565. This may suggest that higher convenience is associated with greater assurance.

While strong correlation can increase the risk of multicollinearity, on its own, it is not a definitive indicator. VIF provides a more reliable assessment of multicollinearity by considering how much a variance inflation occurs due to correlated independent variables. When VIF values are low, even with strong correlations, multicollinearity is likely not a significant problem. To ensure there are less chances of collinearity affecting the constructs, VIF value from table 7 and 8 has been analyzed and since the VIF values of the constructs is less than 5, multicollinearity is less likely to happen between the variables (Dormann et al., 2012).

Table 14. Pearson's Correlation

Correlations

		FUN	ENJ	SEC	ASS	DES	CON	CUSTOM	CUSTSTS
FUN	Pearson Correlation	1	.745**	.455**	.419**	.365**	.341**	.497**	.559**
	Sig. (2-tailed)		.000	.000	.000	.000	.000	.000	.000
	N	163	163	163	163	163	163	163	163
ENJ	Pearson Correlation	.745**	1	.515**	.464**	.463**	.430**	.516**	.573**
	Sig. (2-tailed)	.000		.000	.000	.000	.000	.000	.000
	N	163	163	163	163	163	163	163	163
SEC	Pearson Correlation	.455**	.515**	1	.490**	.428**	.426**	.555**	.546**
	Sig. (2-tailed)	.000	.000		.000	.000	.000	.000	.000
	N	163	163	163	163	163	163	163	163
ASS	Pearson Correlation	.419**	.464**	.490**	1	.206**	.695**	.515**	.505**
	Sig. (2-tailed)	.000	.000	.000		.008	.000	.000	.000
	N	163	163	163	163	163	163	163	163
DES	Pearson Correlation	.365**	.463**	.428**	.206**	1	.323**	.406**	.420**
	Sig. (2-tailed)	.000	.000	.000	.008		.000	.000	.000
	N	163	163	163	163	163	163	163	163
CON	Pearson Correlation	.341**	.430**	.426**	.695**	.323**	1	.509**	.480**
	Sig. (2-tailed)	.000	.000	.000	.000	.000		.000	.000
	N	163	163	163	163	163	163	163	163
CUSTOM	Pearson Correlation	.497**	.516**	.555**	.515**	.406**	.509**	1	.605**
	Sig. (2-tailed)	.000	.000	.000	.000	.000	.000		.000
	N	163	163	163	163	163	163	163	163
CUSTSTS	Pearson Correlation	.559**	.573**	.546**	.505**	.420**	.480**	.605**	1
	Sig. (2-tailed)	.000	.000	.000	.000	.000	.000	.000	
	N	163	163	163	163	163	163	163	163

**. Correlation is significant at the 0.01 level (2-tailed).

Table 15. Spearman Correlation

Correlations

		FUN	ENJ	SEC	ASS	DES	CON	CUSTOM	CUSTSTS
FUN	Correlation Coefficient	1.000	.691**	.456**	.459**	.313**	.316**	.507**	.529**
	Sig. (2-tailed)	.	.000	.000	.000	.000	.000	.000	.000
	N	163	163	163	163	163	163	163	163
ENJ	Correlation Coefficient	.691**	1.000	.443**	.480**	.357**	.413**	.513**	.507**
	Sig. (2-tailed)	.000	.	.000	.000	.000	.000	.000	.000
	N	163	163	163	163	163	163	163	163
SEC	Correlation Coefficient	.456**	.443**	1.000	.457**	.342**	.378**	.526**	.545**
	Sig. (2-tailed)	.000	.000	.	.000	.000	.000	.000	.000
	N	163	163	163	163	163	163	163	163
ASS	Correlation Coefficient	.459**	.480**	.457**	1.000	.164*	.653**	.493**	.504**
	Sig. (2-tailed)	.000	.000	.000	.	.037	.000	.000	.000
	N	163	163	163	163	163	163	163	163
DES	Correlation Coefficient	.313**	.357**	.342**	.164*	1.000	.286**	.354**	.346**
	Sig. (2-tailed)	.000	.000	.000	.037	.	.000	.000	.000
	N	163	163	163	163	163	163	163	163
CON	Correlation Coefficient	.316**	.413**	.378**	.653**	.286**	1.000	.470**	.454**
	Sig. (2-tailed)	.000	.000	.000	.000	.000	.	.000	.000
	N	163	163	163	163	163	163	163	163
CUSTOM	Correlation Coefficient	.507**	.513**	.526**	.493**	.354**	.470**	1.000	.598**
	Sig. (2-tailed)	.000	.000	.000	.000	.000	.000	.	.000
	N	163	163	163	163	163	163	163	163
CUSTSTS	Correlation Coefficient	.529**	.507**	.545**	.504**	.346**	.454**	.598**	1.000
	Sig. (2-tailed)	.000	.000	.000	.000	.000	.000	.000	.
	N	163	163	163	163	163	163	163	163

**. Correlation is significant at the 0.01 level (2-tailed).

*. Correlation is significant at the 0.05 level (2-tailed).

Table 16. Kendall Tau-B Correlation

		FUN	ENJ	SEC	ASS	DES	CON	CUSTOM	CUSTSTS
FUN	Correlation Coefficient	1.000	.570**	.337**	.347**	.239**	.250**	.399**	.410**
	Sig. (2-tailed)	.	.000	.000	.000	.000	.000	.000	.000
	N	163	163	163	163	163	163	163	163
ENJ	Correlation Coefficient	.570**	1.000	.338**	.379**	.279**	.339**	.408**	.401**
	Sig. (2-tailed)	.000	.	.000	.000	.000	.000	.000	.000
	N	163	163	163	163	163	163	163	163
SEC	Correlation Coefficient	.337**	.338**	1.000	.367**	.262**	.302**	.406**	.419**
	Sig. (2-tailed)	.000	.000	.	.000	.000	.000	.000	.000
	N	163	163	163	163	163	163	163	163
ASS	Correlation Coefficient	.347**	.379**	.367**	1.000	.129*	.565**	.390**	.406**
	Sig. (2-tailed)	.000	.000	.000	.	.035	.000	.000	.000
	N	163	163	163	163	163	163	163	163
DES	Correlation Coefficient	.239**	.279**	.262**	.129*	1.000	.233**	.276**	.270**
	Sig. (2-tailed)	.000	.000	.000	.035	.	.000	.000	.000
	N	163	163	163	163	163	163	163	163
CON	Correlation Coefficient	.250**	.339**	.302**	.565**	.233**	1.000	.385**	.373**
	Sig. (2-tailed)	.000	.000	.000	.000	.000	.	.000	.000
	N	163	163	163	163	163	163	163	163
CUSTOM	Correlation Coefficient	.399**	.408**	.406**	.390**	.276**	.385**	1.000	.482**
	Sig. (2-tailed)	.000	.000	.000	.000	.000	.000	.	.000
	N	163	163	163	163	163	163	163	163
CUSTSTS	Correlation Coefficient	.410**	.401**	.419**	.406**	.270**	.373**	.482**	1.000
	Sig. (2-tailed)	.000	.000	.000	.000	.000	.000	.000	.
	N	163	163	163	163	163	163	163	163

**. Correlation is significant at the 0.01 level (2-tailed).

*. Correlation is significant at the 0.05 level (2-tailed).

Additionally, collinearity analysis has also been done to make sure no multicollinear is happening in between the construct's structural paths. Table 17 has summarised the findings and the VIF value for the structural paths where the values are found to be less than 5. This further suggests that there are no multicollinear affects to significantly distorts the model.

Table 17. Collinearity statistics on the structural path towards dependent variables.

Structural Paths	VIF
Functionality → Customer Satisfaction	2.518
Enjoyment→ Customer Satisfaction	2.839
Security→ Customer Satisfaction	2.090
Assurance→ Customer Satisfaction	2.397
Design→ Customer Satisfaction	1.553
Convenience→ Customer Satisfaction	2.239
Customization→ Customer Satisfaction	1.976

4.3 Identifying and Analysing Key Dimensions (Considering Subgroup Differences)

This analysis investigates the impact of the most significant service quality dimensions perceived by the customers on their satisfaction, while also considering potential variation in perception across different customer groups.

4.3.1 Comparison of Customer Groups

Independent t-test and ANOVA (Analysis of Variance) is used for comparison analysis to assess the relationship between variables. However, they differ in the number of independent variables where independent t-tests compare the means of two groups such as males and females while ANOVA analyse the effect of one or more independent variables on a single dependent variable (Kim & Park, 2019; Mishra et al., 2019). For this research, independent t-test has been used to find any significance in genders. To know if there is a significance where the observed difference between groups is likely due to a real effect of the independent variables, rather than random chance, the p-value needs to be <0.05 (Mishra et al., 2019). For this research, it was found that there is no significance in differences between the genders across the constructs.

ANOVA test is then used to test the significance of marital status, employment status and monthly income range of the respondents. It is found that there is significance difference found in marital status for ASS and CON possibly due to have shared responsibility for shopping, being more risk-averse and having time constraints. Next, there is a significance difference found in employment status for FUN, ENJ, ASS and CON possibly due to the level of tech savviness, being more task-oriented due to time constrained, and different level of financial security concerns. Lastly, there is a significance difference found in monthly income range for ASS, DES and CON, possibly due to financial risk aversion, perceived value and familiarity with technology.

Table 18. Independent T-Test and ANOVA

	Gender	Marital Status	Employment Status	Income
FUN	0.307	0.205	0.021	0.138
ENJ	0.544	0.400	0.030	0.094
SEC	0.974	0.430	0.079	0.861
ASS	0.701	0.021	0.002	0.004
DES	0.505	0.368	0.552	0.019
CON	0.454	0.016	0.003	0.005
CUSTOM	0.921	0.496	0.411	0.254
CUSTSTS	0.631	0.983	0.206	0.250

4.3.2 Combined Analysis of Interaction Effects and Regression Results

The regression analysis from chapter 4.1.3 had identified that functionality and customization have a significant path coefficient. This may indicate that they are the most influential predictors of customer satisfaction in the context of Brunei SSK interactions. However, analysis from the survey answered by customers suggest that security is perceived to be the most crucial dimension of service quality, despite not being a statistically significant predictor in the regression analysis. Customers may possibly recognize security as a baseline expectation rather than a key differentiator for satisfaction. Other reason is the possibility of security having a moderating effect on the relationship between other dimensions and satisfaction.

4.3.3 Impact of The Most Important Dimension

From the analyses, while security is perceived to be the most important dimension through the survey questionnaire, statistically, functionality has been identified as the key dimension influencing customer satisfaction, as the dimension has been identified to have strong construct validity. Functionality has exhibited a strong loading (>0.5) and low VIF values (< 5), indicating a clear and distinct construct with low multicollinearity. The path coefficient 0.281 between functionality and customer satisfaction signifies a positive and statistically significant relationship. Additionally, functionality also has an effect size (f2) of 0.089, which suggests that it is affecting customer satisfaction with a small but notable influence. These may indicate that customer satisfaction can be enhanced when the effectiveness of SSKs in meeting customer needs is improved.

DISCUSSION

SSKs have evolved into an essential component in consumers' daily lives. SSKs have gained widespread acceptance globally, and this tendency is only growing. As SSKs become an increasingly important trend in service delivery, it is critical to research the effects of SSK service quality on customer happiness, as this can predict customer willingness to use the technology.

As this study is based in Brunei Darussalam, where technological acceptance is high, the country has experienced a significant penetration of smartphones, laptops, personal computers, and tablets. The national research institution situated in Bandar Seri Begawan, Universiti Teknologi Brunei (UTB), is a testament to the country's embracing of technology. They recently showcased cutting-edge holographic education technologies and prevailed at the Huawei ICT Competition 2023-2024 Global Final. (UTB'S SCHOOL OF COMPUTING AND INFORMATICS SHOWCASES CUTTING-EDGE HOLOGRAPHIC EDUCATION TECHNOLOGY | Universiti Teknologi Brunei, n.d.). Furthermore, a study conducted on the factors influencing students' acceptance of Learning Management Systems (LMS) in Brunei discovered that subjective norms and system design have a major influence on the students' LMS acceptance (Hamid et al., 2020). Additionally, according to a poll done by AITI, more than half of Bruneians' households have access to the internet, primarily via fixed broadband. (AITI - ICT Development Index 2023, 2023). Brunei Darussalam was also ranked 56th out of 142 countries on the Networked Readiness Index, which grades economies on their use, acceptance, and efficiency of ICT ("Brunei Darussalam in 2019: Issues Revisited," 2020). Due to the COVID-19 epidemic, Bruneians have recently begun to use and become accustomed to the online delivery platform (Almunawar & Anshari, 2022).

As a result, Bruneians are more likely to engage in technology-mediated interactions. They are quick to adapt to new situations, so they prefer to have services provided in novel and entertaining ways. In this context, individuals are continually in search of services that stimulate, challenge, and gratify them. If SSKs can deliver services according to these parameters, Bruneian customers will desire to use SSK to pay for their purchases. This study gives practical ideas for service providers to use technology-mediated interfaces to better serve their customers.

Data for this study were mostly acquired from SSK users, with target respondents chosen at random from across Brunei Darussalam. The current study uses the revised SSTQUAL scale to assess the service quality of SSK prepared by supermarkets and fast-food restaurants in Brunei (Lin & Hsieh, 2011; Radomir & Nistor, 2014). The results revealed that young and middle-aged people are more likely to use SSKs. Validation of the conceptual model among the constructs was achieved by reliability analysis, measurement model evaluation, and structural model assessment. The reliability study shows that all constructs have adequate reliability scores.

On the other hand, the measurement model assessment demonstrates that the model is accurate and valid based on concept loading, VIF, AVE, and discriminant validity, giving researchers confidence in their findings. Structural model assessment also assures that the study findings are credible and accurately depict the link between the underlying constructs under investigation. Model fit and explanatory power demonstrate acceptable levels of significance, although path coefficients and theoretical alignment simply indicate the importance of functionality and customization in affecting customer satisfaction with SSK.

5.1 Analyzing the Effect of SSK Service Quality Dimensions on Customer Satisfaction

The results demonstrate a positive and strong association between SSTQUAL dimensions and customer satisfaction, with only functionality and customization having a major impact on customer satisfaction. The findings reveal that superior quality in the dimensions of self-service kiosks (SSKs), especially in terms of functionality and customization, contributes to enhanced customer satisfaction. The linear regression analysis reveals that these two variables have a statistically significant effect on customer satisfaction (p-values < 0.05). This demonstrates that better functionality and flexibility will result in greater user satisfaction with SSKs. The remaining factors, while essential, did not have a statistically significant impact in this study.

5.1.1 Constructs With Direct Influence

The functionality of a kiosk is an important factor in customer satisfaction. When a kiosk performs its intended tasks flawlessly (high functionality), users can easily traverse the interface and complete their desired actions without experiencing mistakes or blockages. This results in a huge reduction in frustration, promoting a more favourable user experience. Furthermore, a functional SSK allows users to do their tasks efficiently, where task such as obtaining information needed, executing transactions and accessing available services are made simple. This efficiency immediately contributes to customer satisfaction because it saves time, resulting in a more enjoyable engagement experienced by the customers (Krishen, 2008). Beyond efficiency, a well-functioning kiosk gives consumers a sense of control. The ability to use the interface intuitively and successfully execute activities boosts confidence and enjoyment (Hattula et al., 2022; Pandita & Kiran, 2023). Finally, a highly functional kiosk reduces the cognitive load necessary for user interaction. Clear instructions, easy design, and smooth operation all help to reduce cognitive stress. This results in a less stressful and more fun experience, which leads to higher customer satisfaction(Lucia-Palacios et al., 2020). Therefore, the functionality of a kiosk can give a direct impact on user satisfaction because it meets expectations, eliminates aggravation, allows for efficient goal attainment, and encourages a sense of control and ease of use.

It has also been found in this study that employment status causes a significant difference in functionality. The first viable reason for this is that employed individuals might generally have a higher exposure to technology due to their workplace demands. This could translate to a greater appreciation for functionalities like intuitive interfaces, diverse payment options, and efficient product search features. On the other end of the spectrum, an unemployed individuals might prioritize ease of use and clear instructions for basic functionalities.

Additionally, employed individuals are more likely to be more time-constrained and prefer to complete shopping tasks quickly. Functionalities that facilitate an efficient and focused shopping experience are more valued. On the flip side, unemployed individuals might have more time to browse and explore the SSK, potentially placing less emphasis on task-specific functionalities.

The level of customization that is possible with SSKs is another key factor that affects customer satisfaction. Customers find satisfaction in being able to customize SSKs to their own preferences. For example, some fast-food establishments let customers build their own meals where they can add all the toppings they want and exclude the ones they do not want. Because there are more alternatives and everything is more accessible, the space feels cozier. This sense of customization extends beyond mere comfort by allowing consumers to connect with the kiosk familiarly, eliminating initial concerns and putting them at rest (Lin & Lin, 2023). Furthermore, customization can improve task efficiency since a kiosk may allow users to hide unimportant options or prioritize commonly used functions. The ability to customize the kiosk's functionality to meet their requirements minimizes the time and effort required to execute tasks.

Customization also improves the customer experience by streamlining interactions where, other than providing efficiency, customization gives users a sense of control (Kolasani, 2023). The flexibility to customize how they interact with SSKs, whether by changing the interface layout or selecting preferred input methods, develops a sense of agency. Customers are more satisfied since they can interact with SSKs in a way that seems obvious and natural thanks to this control over the encounter, where increased customization also raises customers' interest (Parasuraman et al., 2020). When they can customize an encounter, customers are more likely to participate. By becoming an extension of their preferences, these kiosks motivate people to engage more fully in the task at hand, due to the perception of more involvement experienced by customers. As a result, adequate customization provided by the shop's SSKs is essential to improve customer satisfaction since it promotes user engagement, comfort, efficiency, and a sense of control. After all, modern consumers need products and services to be as customized as possible rather than one-size-fits-all (Pech & Vrchota, 2022).

The primary goal of SSK interaction is to complete a specific task such as paying items, ordering foods or accessing information. The dimension of functionality and customization address this goal by allowing users to accomplish their tasks easily and efficiently. If the kiosk is complex or frustrating to use, it will not achieve user satisfaction. Both constructs are not just features of SSKs, they are the cornerstone of a positive user experience. By focusing on clarity, ease of use and user empowerment, SSKs can achieve their goal which is to provide convenient and efficient way for users to interact with businesses.

5.1.2 Constructs With Indirect Influence

As for the dimensions that are not significant, there may be several reasons why they might not be as significant as functionality and customization when analysing customer satisfaction with SSKs. Since customer satisfaction is affected directly by the completion of tasks when a customer is using SSK, it is possible that security andconveniencecan indirectly influence customer satisfaction.

In recent times, people have become increasingly cautious about sharing personal information and financial details electronically due to cases such as stolen money from bank accounts due to fraud activities (Borneo Bulletin, 2024; RBPF Issues Warning Against Phishing Scams, 2024; Xinhua, 2024). Using an SSK that appears insecure can trigger anxiety and reluctance for some customers to complete their transactions. This fear can lead to abandoning the use of SSK, ultimately hindering task completion and satisfaction. Therefore, businesses need to ensure it is safe to use their kiosk. Customers would want to feel confident that their personal information and transactions are safe when using SSK, and this can be seen in their responses, where security is the most important dimension from their perspective.

Things such as clear data protection policies, secure payment systems, and a well-maintained physical environment can contribute to enhancing the sense of security a customer can perceive. Security may not be directly involved in completing a specific task when using an SSK, but it plays a crucial role in creating a foundation of trust. Trust might act as a mediator between security and satisfaction, which this study is lacking in identifying. While direct links may not be strong enough, it is possible that security indirectly impacts user satisfaction by fostering trust and shaping perceptions (Martínez-Navalón et al., 2021). Customers might feel more satisfied if they perceive their transactions as secure.

Convenience, on the other hand, might not be directly involved in completing a specific task; however, the hallmark of a convenient SSK is its ease of use. SSK's clear instructions, intuitive navigation, and features like touch-screen accessibility make interacting with the kiosk effortless. When SSK is equipped with this leisure, it is easier for customers to quickly understand what needs to be done and complete tasks without unnecessary steps or confusion. This helps to reduce their frustration and time spent figuring out how the system works. By incorporating convenience factors such as location, flexible hours, and user-friendly features, SSKs become more user-friendly and accessible. This may translate to a smoother and less time-consuming experience, which leads to increased customer satisfaction. Customers will be more likely to use and appreciate an SSK that respects their time and makes interaction a breeze(Glushko & Nomorosa, 2012). However, convenience can also be subjective and context dependent. For instance, one customer might find the touch-screen interface to be convenient, while another customer with limited dexterity might find it difficult to use it.

Since these constructs are possible to indirectly affectcustomer satisfaction, they may not work in isolation and therefore rely on the direct constructs as the foundation. Since the primary purpose of an SSK is to enable users to complete their tasks, all other features from security protocols to user interfaces function as a support to the core objective. When the directly influencing construct are built as a solid foundation for customer experience, features such as security and convenience can then be layered on top to enhance the overall experience and lead to higher customer satisfaction (Chowdhury, 2023).

5.1.3 Constructs With Minimal to No Influence

Assurance has the possibility of indirectly affecting customer satisfaction. Assurance features focus on building trust and confidence when interacting with SSK. They do not directly complete tasks but empower customers, reduce anxiety, and foster a sense of control, which can be achieved when the SSK can give clear instructions, readily available help options, and progress indicators. SSK equipped with these functions can improve the user experience byreducing the potential barrier such as anxiety. Customers feel more confident and in control, knowing they can navigate the kiosk and find help if needed. Additionally, clear guidance and readily available support can give customers more empowerment to complete tasks independently, fostering a sense of accomplishment. However, this construct is unable to capture how the assurance of an SSK can affect customer satisfaction due to the questionnaire's design to know the reliability of the shop that is providing the SSK instead of trying to analyse the assurance the SSK itself is giving. Therefore, for future research, these need to be taken into consideration.

Furthermore, the way SSKs are used in Brunei might prioritise practicality and efficiency over entertainment. These SSKs are utilised for routine transactions or quick purchases, where getting the job done quickly is the main goal. Even though Brunei has a reputation for a slower pace of life, standard work hours can still be demanding. A 44-hour workweek is typical, and some professions might require longer hours, and this can possibly contribute to a limited time for errands and other tasks outside of work (BIPO Service, 2022).

Cultural values are also possible to affect how customers perceive SSK. Cultural values in Brunei might emphasize politeness and respect (Kristiawan et al., 2021; Sharbawi & Mabud, 2021). Therefore, an SSK with user-friendly interfaces, clear instructions in the preferred language, and a reliable transaction process would be more appreciated, as it would demonstrate respect for the user's time and needs. In this context, enjoyable features might be seen as less important. Customers would focus on well-designed functionalities and be less concerned with the enjoyment and design elements of SSK.

Additionally, the research fails to reveal a statistically significant direct effect of design on customer satisfaction with SSKs in Brunei. This is possibly due to how the construct is measured, which is unable to capture all its nuances. There is a possibility that design could still indirectly influence satisfaction through its interactions with other variables. For instance, functionality and design can be closely related to each other; however, the relationship is nuanced and may not be directly proportional and straightforward (Han et al., 2019). Therefore, future research exploring indirect effects and using context-specific design measurements could provide further insights.

5.2 The Interaction Effects Between the Construct of SSK Dimension Towards Customer Satisfaction

Since the dimensions of SSTQUAL are perceived to be the independent variables in this study, it is hypothesized that they are uncorrelated with each other in improving customer satisfaction (Ohira, 2018). However, from the correlation analysis, it has been found that two pairs of the constructs have a strong positive correlation due to having values ≥ 0.7 in the correlation matrix. The first pair of constructs that has been discovered to have a positive association are between functionality and enjoyment. The findings indicates that when functionality increases, the level of enjoyment also elevates. Convenience and assurance, on the other hand, also exhibit a substantial correlation, suggesting that greater convenience is associated with greater assurance, albeit neither significantly affects consumer satisfactiontowards SSKs and therefore will not be discuss further. This may imply that variables that are not correlated do not necessarily have to be independent, as they may still exhibit non-linear dependence (Franzese & Iuliano, 2019).

It is possible that when SSK operates properly, customers may do their activities quickly and effortlessly, which contributes to the favourable association between functionality and enjoyment. There are not many mistakes or problems that the customers may encounter when they are doing tasks such as obtaining the information they need, finishing transactions, and using the services provided by the stores' SSK when the functions are made easier for the customers to utilize. In addition, a properly operating SSKs can give customers a feeling of control over their interactions. When customers feel that they are controlling the technology instead of the opposite, this may enable them to successfully accomplish their desired task as they are able to navigate through the interface with ease without a sense of stress and discouragement (Flavián et al., 2019). This positive experience builds a sense of confidence and accomplishment towards them, which possibly decreases their own frustration(Vansteenkiste & Ryan, 2013). Moreover, alongside boosting efficiency, the positive outcomes can make the encounter more enjoyable.

Additionally, when customers can achieve their goals when using SSKs, they may be able to achieve the sensation of satisfaction and enjoyment. To ensure this, a SSK need to function well according to the needs of customers. Furthermore, a highly functional kiosk which only requires customers to use minimal mental effort required for the interaction can make the interaction less stressful. This can be achieved by providing a clear instruction and simple intuitive design together with smooth operation to reduce cognitive load. Together, this can provide positive reinforcement towards customers and increases their enjoyment of interaction with SSKs.

5.3 Optimising the Key Dimension of SSK in Supermarkets to Enhanced Customer Satisfaction

The analysed data provides information that functionality and customization play an important role in providing satisfaction when using SSKs. Even though, based on the responses from the customer, they are mostly prioritizing the security the SSKs can provide, especially in securing the transactions when making their payment.

Another aspect of security that is important is thief activities such as SWIPERS, in which the perpetrators trick the SSK system, where they only scan cheaper products to bring down the overall expenditure. This perception underscores the need for robust security measures to ensure customer trust and satisfaction and therefore increase the usage of SSKs both in supermarkets and fast-food restaurants. Therefore, for this part of the discussion, a robust SSK system with the implementation of AI is proposed to further enhance customer trust and satisfaction.

5.3.1 System Components of AI-Enhanced Smart Cart Checkout system

Since supermarkets are particularly susceptible to fraud activities, this proposed solution concentrates on SSKs within the supermarket. AI and machine learning technology can be used to improve the functionality, security, and efficiency of supermarket SSKs. The system will utilize an AI enhanced smart cart equipped with weight sensors and an AI powered camera for real-time recognition. Next, the SSK is equipped with an integrated scale, RFID reader, and an AI powered scanner. The last component that is needed is a central server. The purpose of this component is to manage inventory levels, pricing, transaction processes, and updating the AI model.

5.3.2 Operational Flow

The operational flow when utilizing this system will focus on two different aspects which are the shopping and checkout flow. First, from the shopping perspective, when the customer places the desired item onto the smart cart, the AI powered camera will identify the item. On the other hand, the weight sensors will record the weight of the item. Both components will send the information to the central server, where the server will verify the item. The server will do a cross-checking process where it will confirm the measured weight belongs to the image of the item it received. Additional function such as smart recommendations where the system may suggest complementary products based on the items in the cart and allergen alerts may also be added

During the checkout, customers will place the smart cart on the integrated scale of the SSK. The scanner, which will be powered by AI, will again do a cross-referencing. The weight of the item will be added up altogether and this will be matched up with the pictures taken. This is to ensure the consistency and validity of the items Customers can then make payments using their preferred method of transaction. When the products are scanned, nutritional information can be displayed on the SSKs screen to provide the nutritional value of the consumable items. To make SSK more interactive, recipes or meal ideas can be recommended based on the scanned items. After the payments have been made, the system will send information to remove the bought item from the list of inventories through the central server.

Figure 3. Process of AI enhanced Smart Cart and SSK

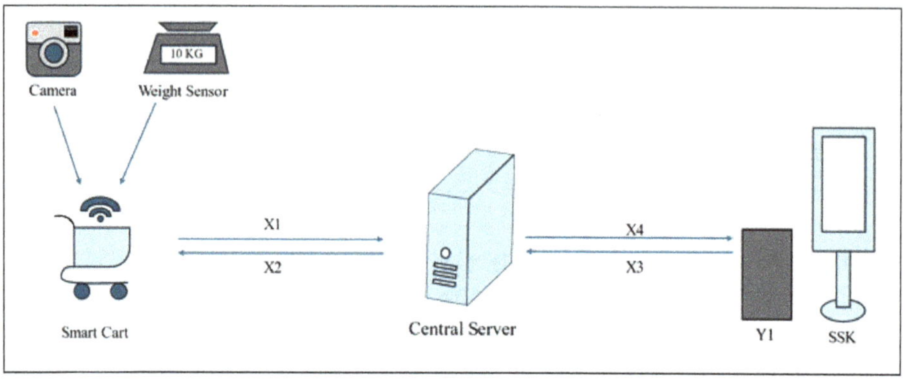

Note. X1 refers to information obtained by camera and weight sensor being sent towards the central server. X2 refers to information such as item recommendations sent toward the customer. X3 refers to information obtained by weight sensors and sent to the central server for cross-checking. X4 refers to information being sent back to SSK. This will include validity information, nutritional information and meal ideas.

Additionally, the SSK system will also offer language options for customers to choose their preferred language, especially for non-native speakers. The SSK will also be equipped with audio cues and visual indicators for differently abled customers. Lastly, after the transactions has been made, the SSKs may prompt customers to give feedback, where the data will be used to further enhance the system's functionality.

5.3.3 Security Measures

The SSK will also be equipped with respective security measures. One of the implanted securities will be based on correlating the weights with the picture taken. If the total weights and the pictures does not match with each other, the system will alert the supermarket staff to check the items and transactions manually. Another security measure is that the alarm will triggers if a customer attempts to leave the SSK area with an unpaid item. This will be based on the weighting and the visual discrepancies.

5.3.4 Benefits of the System

The system can significantly improve the shopping experience for a customer. Information such as complementary products, nutritional information, and allergen alerts can be passed on to customers when they are making item selections. Check-out processes are streamlined as customers only need to place the smart cart on the integrated scale, while security is also improved by the way items are verified. After customers make payment to obtain their desired item, the inventories of the shop will automatically be updated through the central server where the bought items are removed from the shops list of inventories. This ensures the stocks held by the supermarket are at an adequate level. Altogether, this ensures efficient and enjoyable experience towards both the supermarket and the customers.

CONCLUSIONS AND RECOMMENDATIONS

6.1 The Impact of Functionality and Customization Towards Customer Satisfaction

In line with research question, this study confirms that certain dimensions of service quality significantly influence customer satisfaction with SSK in Brunei. The construct functionality which focuses on giving clear instructions and easy navigation alongside the construct customization, which allows customers to personalize their language preferences and transaction options, emerged as the key driver of customer satisfaction. These findings suggest that when SSKs are designed to address user needs affectively and offer some level of control over the experience, customers are more likely satisfied with their experience. Therefore, retailers in Brunei can focus on improving the functionality and customization features of their SSKs as part of their strategic approach to enhance customer satisfaction which could potentially lead to increased loyalty and positive business outcomes.

SSKs play an important role in modern service delivery, especially in the retail and food service industries. The widespread adoption necessities for service organization to invest in understanding what factors influence customer satisfaction with these technologies. In this sense, it is critical for service organizations to invest significant resources in studying the elements that cause customer pleasure or discontent to improve their retention in using such technology. Retailers should prioritise maintaining high levels of functionality, and customization within their SSK interfaces. These efforts will enhance customers confidence in the technology and contribute to a more positive overall experience.

6.2 Exploring Interaction Effects in Future Research

The study has revealed a positive correlation between certain pairs of constructs, which are functionality and enjoyment as well as convenience and assurance. This suggests that improvements in one dimension can enhance perceptions in another, contributing to overall customer satisfaction. The findings have underscored the importance of considering the interplay between different dimensions of service quality. A holistic approach that optimizes the key dimensions in tandem rather than in isolation can lead to enhanced customer satisfaction with SSKs. Future research could explore these interaction effects through user experience studies or surveys that investigate how customers perceive the combined impact of various service quality features. Understanding these interactions can be crucial for designing SSKs that not only address functional needs but also create a more enjoyable, convenient, and secure user experience.

6.3 Tailoring Service Quality for Supermarket SSKs: Key Dimensions for Customer Satisfaction

The study has identified functionality as the key dimension influencing customer satisfaction with SSKs in supermarkets and fast-food restaurants, highlighting the importance of efficient and user-friendly interfaces. However, despite not being statistically significant in regression analysis, security is perceived by customers as crucial. This emphasizes that there is a need for robust security measures in SSKs. Therefore, the proposed AI-enhanced smart cart checkout system aims to optimize functionality, security, and customer satisfaction by integrating advanced technologies such as weight sensors and AI-powered cameras. Overall, retailers in Brunei should focus on improving SSK functionality and customization while ensuring strong security features to enhance customer satisfaction and align with Brunei's vision for 2035.

6.4 Further Recommendations

There are further limitations to this study that could affect how generalizable the findings are.The major limitation is concerned with the sample selections. Most of the respondents are young adults who are students while customers of different ages and professions may have different attitudes regarding the use of SSKs. Therefore, additional research should be initiated to closely observe the behavior of the customer within other ages and professions. Further research should be conducted using various methodological techniques, such as mixed methodology and qualitative approach, as this study was mostly quantitative in nature and has significant constraints surrounding the survey-based data collecting.

Another recommendation is based on the service quality.The higher the service quality supplied by SSKs, the greater the likelihood that consumers will adopt the technology. Supermarkets and fast-food restaurants must take the initiative to encourage customers' good attitudes regarding SSKs through constant monitoring and evaluation. These methods could assist supermarkets and fast-food restaurants in providing vital information about improving the service delivery process using SSKs. Business should focus on improving the quality of SSK to enhance customer satisfaction and retention. This aligns with the broader goals of technological advancement and economic sustainability in Brunei, contributing to the nation's vision 2035.

REFERENCES

Abdulhafedh, A. (2017). How to Detect and Remove Temporal Autocorrelation in Vehicular Crash Data. *Journal of Transportation Technologies*, *07*(02), 133–147. DOI: 10.4236/jtts.2017.72010

Acrelec United Kingdom. (2022, June 1). *Evolution of self-service technology*. LinkedIn. https://www.linkedin.com/pulse/evolution-self-service-technology-acrelec-united-kingdom/

Adil, M., Alghaswyneh, O. F. M., & Albkour, A. M. (2013, March 15). *SERVQUAL and SERVPERF: A Review of Measures in Services Marketing Research*. ResearchGate. https://www.researchgate.net/publication/304461403_SERVQUAL_and_SERVPERF_A_Review_of_Measures_in_Services_Marketing_Research

Ahn, J. A., & Seo, S. (2018). Consumer responses to interactive restaurant self-service technology (IRSST): The role of gadget-loving propensity. *International Journal of Hospitality Management*, *74*, 109–121. DOI: 10.1016/j.ijhm.2018.02.020

AITI - ICT Development Index 2023. (2023, December 29). Retrieved June 12, 2024, from https://aiti.gov.bn/news/2023/ict-development-index-2023/

AITI. (2019). Brunei Darussalam Information Communications Technology (ICT) Household Report 2019. In aiti.gov.bn

Alin, A. (2010). Multicollinearity. *Wiley Interdisciplinary Reviews: Computational Statistics*, *2*(3), 370–374. DOI: 10.1002/wics.84

Almunawar, M. N., & Anshari, M. (2022). Customer acceptance of online delivery platform during the COVID-19 pandemic: the case of Brunei Darussalam. *Journal of Science and Technology Policy Management/Journal of Science & Technology Policy Management, 15*(2), 288–310. DOI: 10.1108/JSTPM-04-2022-0073

Alolah, T., Stewart, R. A., Panuwatwanich, K., & Mohamed, S. (2014). Determining the causal relationships among balanced scorecard perspectives on school safety performance: Case of Saudi Arabia. *Accident; Analysis and Prevention*, *68*, 57–74. DOI: 10.1016/j.aap.2014.02.002 PMID: 24589246

Amu, A. (2021, September 16). Progresif debuts self-service kiosk at store front. *The Bruneian*. Retrieved August 28, 2023, from https://thebruneian.news/2021/09/16/progresif-debuts-self-service-kiosk-at-store-front/.

Anitha, M., Babel, A., Kumar, A., Rauniyar, A., & Zahid, K. (2019). Cloud-based secured QR code for self-service access control system at resort and hotels. Lecture Notes in Networks and Systems, 387–395. DOI: 10.1007/978-981-13-7150-9_41

Azahari, I. (2022, February 23). Riding the digitisation wave. *Borneo Bulletin*. Retrieved August 28, 2023, from https://borneobulletin.com.bn/riding-the-digitisation-wave/

Barnes, S., & Vidgen, R. (2001). Assessing the quality of auction Web sites. *Proceedings of the 34th Annual Hawaii International Conference on System Sciences*. DOI: 10.1109/HICSS.2001.927087

Bernoussi, R. E., & Rockinger, M. (2022). Rebalancing with transaction costs: Theory, simulations, and actual data. *Financial Markets and Portfolio Management*, *37*(2), 121–160. DOI: 10.1007/s11408-022-00419-6

Bhardwaj, P. (2019). Types of sampling in research. *Journal of the Practice of Cardiovascular Sciences*, *5*(3), 157. DOI: 10.4103/jpcs.jpcs_62_19

BIBD Debit Mastercard Contactless | Bank Islam Brunei Darussalam. (n.d.). https://www.bibd.com.bn/personal/cards/debit-card/debit-mastercard/#:~:text=Brunei%20Citizens%2C%20Permanent%20Residents%20and,BIBD%20Savings%20or%20Current%20Account

BIPO Service. (2022, November). Understanding Brunei's Employment Laws & Regulations. www.biposervice.com. Retrieved June 27, 2024, from https://www.biposervice.com/wp-content/uploads/2022/11/BIPO-Times-November-2022-en.pdf

Blut, M., Wang, C., & Schoefer, K. (2016). Factors influencing the acceptance of self-service technologies. *Journal of Service Research*, *19*(4), 396–416. DOI: 10.1177/1094670516662352

Bogicevic, V., Bujisic, M., Bilgihan, A., Yang, W., & Cobanoglu, C. (2017). The impact of Traveler-focused airport technology on traveler satisfaction. *Technological Forecasting and Social Change*, *123*, 351–361. DOI: 10.1016/j.techfore.2017.03.038

Bonett, D. G., & Wright, T. A. (2014). Cronbach's alpha reliability: Interval estimation, hypothesis testing, and sample size planning. *Journal of Organizational Behavior*, *36*(1), 3–15. DOI: 10.1002/job.1960

Borneo Bulletin. (2024, January 6). Brunei telco tells customers to be more vigilant after big surge in scams in the country. *The Star*. Retrieved June 21, 2024, from https://www.thestar.com.my/aseanplus/aseanplus-news/2024/01/06/brunei-telco-tells-customers-to-be-more-vigilant-after-big-surge-in-scams-in-the-country

Brunei Darussalam in 2019: Issues Revisited. (2020). In *ISEAS Publishing eBooks* (pp. 99–116). DOI: 10.1355/9789814881319-007

Bulmer, S., Elms, J., & Moore, S. (2018). Exploring the adoption of self-service checkouts and the associated social obligations of shopping practices. *Journal of Retailing and Consumer Services, 42*, 107–116. DOI: 10.1016/j.jretconser.2018.01.016

Campbell, C. S., Maglio, P. P., & Davis, M. M. (2010). From self-service to super-service: A resource mapping framework for co-creating value by shifting the boundary between provider and customer. *Information Systems and e-Business Management, 9*(2), 173–191. DOI: 10.1007/s10257-010-0142-4

Chan, C. Y. T., & Petrikat, D. (2022, November 30). Self-Service Technology: Benefits and Challenges. *Journal of Computer Science and Technology Studies, 4*(2), 118–127. DOI: 10.32996/jcsts.2022.4.2.14

Chanchaichujit, J., Balasubramanian, S., & Charmaine, N. S. M. (2020, January 1). A systematic literature review on the benefit-drivers of RFID implementation in supply chains and its impact on organizational competitive advantage. *Cogent Business & Management, 7*(1), 1818408. DOI: 10.1080/23311975.2020.1818408

Chang, H. J., Wu, C. H., Ho, J. F., & Chen, P. Y. (2008). On sample size in using central limit theorem for gamma distribution. *Information and Management Sciences, 19*(1), 153–174.

Chao, C. M. (2019, July 16). Factors Determining the Behavioral Intention to Use Mobile Learning: An Application and Extension of the UTAUT Model. *Frontiers in Psychology, 10*, 1652. Advance online publication. DOI: 10.3389/fpsyg.2019.01652 PMID: 31379679

Cheah, J. H., Magno, F., & Cassia, F. (2023). Reviewing the SmartPLS 4 software: The latest features and enhancements. *Journal of Marketing Analytics, 12*(1), 97–107. DOI: 10.1057/s41270-023-00266-y

Chen, W., Botchie, D., Braganza, A., & Han, H. (2022, January). A transaction cost perspective on blockchain governance in global value chains. *Strategic Change, 31*(1), 75–87. DOI: 10.1002/jsc.2487

Chih-Hung Wang, M. (2012, March 16). Determinants and consequences of consumer satisfaction with self-service technology in a retail setting. *Managing Service Quality, 22*(2), 128–144. DOI: 10.1108/09604521211218945

Chowdhury, R. (2023). Impact of perceived convenience, service quality and security on consumers' behavioural intention towards online food delivery services: The role of attitude as mediator. *SN Business & Economics*, *3*(1), 29. Advance online publication. DOI: 10.1007/s43546-023-00422-7 PMID: 36643187

Chun, R. (2019, January 31). The Dark Art of Stealing from Self-Checkouts. The Atlantic. https://www.theatlantic.com/magazine/archive/2018/03/stealing-from-self-checkout/550940/

Cohen, J. (2013). Statistical Power Analysis for the Behavioral Sciences. In *Routledge eBooks*. DOI: 10.4324/9780203771587

Collier, J. E., & Kimes, S. E. (2012). Only if it is convenient: Understanding How Convenience Influences Self-Service Technology Evaluation. *Journal of Service Research*, *16*(1), 39–51. DOI: 10.1177/1094670512458454

Collier, J. E., & Sherrell, D. L. (2010). Examining the influence of control and convenience in a self-service setting. *Journal of the Academy of Marketing Science*, *38*(4), 490–509. DOI: 10.1007/s11747-009-0179-4

Commisso, D. (2022, June 21). *Checking out in 2022, from self-checkout to Amazon's "Just walk out" tech*. CivicScience. https://civicscience.com/checking-out-in-2022-from-self-checkout-to-amazons-just-walk-out-tech/

Considine, E., & Cormican, K. (2016). Self-service Technology Adoption: An Analysis of Customer to Technology Interactions. *Procedia Computer Science*, *100*, 103–109. DOI: 10.1016/j.procs.2016.09.129

Cronin, J. J., & Taylor, S. A. (1994, January 1). *Servperf versus Servqual: Reconciling Performance-Based and Perceptions-Minus-Expectations Measurement of Service Quality*. Journal of Marketing; American Marketing Association. DOI: 10.1177/002224299405800110

Curran, J. M., & Meuter, M. L. (2005, March 1). Self-service technology adoption: Comparing three technologies. *Journal of Services Marketing*, *19*(2), 103–113. DOI: 10.1108/08876040510591411

Cuypers, I. R. P., Hennart, J. F., Silverman, B. S., & Ertug, G. (2021, January). Transaction Cost Theory: Past Progress, Current Challenges, and Suggestions for the Future. *The Academy of Management Annals*, *15*(1), 111–150. DOI: 10.5465/annals.2019.0051

Danjuma, I., Bawuro, F. A., Vassumu, M. A., & Habibu, S. A. (2018). The Service Quality Scale Debate: A Tri-Instrument Perspective for Higher Education Institutions. *Expert Journal of Business and Management, 6*(2), 127–133. https://business.expertjournals.com/ark:/16759/EJBM_612danjuma127-133.pdf

Davis, F. D. (1989). Perceived usefulness, perceived ease of use, and user acceptance of Information Technology. *Management Information Systems Quarterly, 13*(3), 319–340. DOI: 10.2307/249008

Definition of Customer Loyalty - Gartner Marketing Glossary. (n.d.). Gartner. https://www.gartner.com/en/marketing/glossary/customer-loyalty

Demirci Orel, F., & Kara, A. (2014, March). Supermarket self-checkout service quality, customer satisfaction, and loyalty: Empirical evidence from an emerging market. *Journal of Retailing and Consumer Services, 21*(2), 118–129. DOI: 10.1016/j.jretconser.2013.07.002

Demoulin, N. T. M., & Djelassi, S. (2016, May 9). An integrated model of self-service technology (SST) usage in a retail context. *International Journal of Retail & Distribution Management, 44*(5), 540–559. DOI: 10.1108/IJRDM-08-2015-0122

Department of Economic Planning and Statistics - Population. (n.d.). https://deps.mofe.gov.bn/SitePages/Population.aspx

Descriptive Research. (2021). Encyclopedia of Autism Spectrum Disorders, 1330–1330. DOI: 10.1007/978-3-319-91280-6_300489

Ding, D. X., Hu, P. J. H., & Sheng, O. R. L. (2011). e-SELFQUAL: A scale for measuring online self-service quality. *Journal of Business Research, 64*(5), 508–515. DOI: 10.1016/j.jbusres.2010.04.007

Dittig, J. (2023, January 13). *The rise of self-service - and how it's Reinventing Customer Service*. Samsung Business Insights. https://insights.samsung.com/2023/01/13/the-rise-of-self-service-and-how-its-reinventing-customer-service/

Djelassi, S., Diallo, M. F., & Zielke, S. (2018). How self-service technology experience evaluation affects waiting time and customer satisfaction? A moderated mediation model. *Decision Support Systems, 111*, 38–47. DOI: 10.1016/j.dss.2018.04.004

Dormann, C. F., Elith, J., Bacher, S., Buchmann, C., Carl, G., Carré, G., Marquéz, J. R. G., Gruber, B., Lafourcade, B., Leitão, P. J., Münkemüller, T., McClean, C., Osborne, P. E., Reineking, B., Schröder, B., Skidmore, A. K., Zurell, D., & Lautenbach, S. (2012). Collinearity: A review of methods to deal with it and a simulation study evaluating their performance. *Ecography, 36*(1), 27–46. DOI: 10.1111/j.1600-0587.2012.07348.x

Eisend, M., & Kuss, A. (2019). Testing Causal Relationships. *Research Methodology in Marketing*, 173–193. DOI: 10.1007/978-3-030-10794-9_8

El-Hashash, E. F., & Shiekh, R. H. A. (2022). A Comparison of the Pearson, Spearman Rank and Kendall Tau Correlation Coefficients Using Quantitative Variables. *Asian Journal of Probability and Statistics*, 36–48. DOI: 10.9734/ajpas/2022/v20i3425

Fernandes, T., & Pedroso, R. (2016). The effect of self-checkout quality on customer satisfaction and repatronage in a retail context. *Service Business*, *11*(1), 69–92. DOI: 10.1007/s11628-016-0302-9

Flavián, C., Ibáñez-Sánchez, S., & Orús, C. (2019). The impact of virtual, augmented and mixed reality technologies on the customer experience. *Journal of Business Research*, *100*, 547–560. DOI: 10.1016/j.jbusres.2018.10.050

Fornell, C., Johnson, M. D., Anderson, E. W., Cha, J., & Bryant, B. E. (1996). The American Customer Satisfaction Index: Nature, Purpose, and Findings. *Journal of Marketing*, *60*(4), 7–18. DOI: 10.1177/002224299606000403

Franzese, M., & Iuliano, A. (2019). Correlation Analysis. In *Elsevier eBooks* (pp. 706–721). DOI: 10.1016/B978-0-12-809633-8.20358-0

Geekturf. (2019, July 28). *Self-Ordering Kiosks. Coming soon to McDonald's Brunei Lambak*. Facebook. https://m.facebook.com/geekturf/photos/a.434705936971963/726222391153648/

Gelbrich, K. (2009). Beyond just being dissatisfied: How angry and helpless customers react to failures when using self-service technologies. *Schmalenbach Business Review*, *61*(1), 40–59. DOI: 10.1007/BF03396779

Ghasemaghaei, M. (2020). The role of positive and negative valence factors on the impact of bigness of data on big data analytics usage. *International Journal of Information Management*, *50*, 395–404. DOI: 10.1016/j.ijinfomgt.2018.12.011

Ghazali, F. B., Ramlee, S. N. S., Alwi, N., & Hizan, H. (2020). Content validity and test–retest reliability with principal component analysis of the translated Malay four-item version of Paffenbarger physical activity questionnaire. *Journal of Health Research. Warasan Wichai Witthayasat Kanphaet*, *35*(6), 493–505. DOI: 10.1108/JHR-11-2019-0269

Ghosh, M. (2020). Understanding the relationship of self-service technology quality with user adoption using SSTQUAL. *The TQM Journal*, *33*(2), 293–314. DOI: 10.1108/TQM-12-2019-0291

Glushko, R. J., & Nomorosa, K. J. (2012). Substituting Information for Interaction. *Journal of Service Research, 16*(1), 21–38. DOI: 10.1177/1094670512463967

Grand View Research. (2020, July 7). Self-service Technology Market Size Report, 2020-2027. https://www.grandviewresearch.com/industry-analysis/self-service-technology-market#

Gummerus, J., Lipkin, M., Dube, A., & Heinonen, K. (2019, February 11). Technology in use – characterizing customer self-service devices (SSDS). *Journal of Services Marketing, 33*(1), 44–56. DOI: 10.1108/JSM-10-2018-0292

Hajli, N. (2019). The impact of positive valence and negative valence on social commerce purchase intention. *Information Technology & People, 33*(2), 774–791. DOI: 10.1108/ITP-02-2018-0099

Halton, C. (2023, June 17). *Diffusion of Innovations Theory: Definition and Examples*. Investopedia. https://www.investopedia.com/terms/d/diffusion-of-innovations-theory.asp#:~:text=What%20Are%20the%20Steps%20In,later%20editions%20of%20his%20book

Hamid, M. A., Salleh, S., & Laxman, K. (2020). A Study on the Factors Influencing Students' Acceptance of Learning Management Systems (LMS): A Brunei Case Study. *International Journal of Technology in Education and Science, 4*(3), 203–217. DOI: 10.46328/ijtes.v4i3.101

Han, J., Forbes, H., & Schaefer, D. (2019). An Exploration of the Relations between Functionality, Aesthetics and Creativity in Design. Proceedings of the . . . *International Conference on Engineering Design*, 1(1), 259–268. DOI: 10.1017/dsi.2019.29

Haradhan, M. (2020). Quantitative Research: A Successful Investigation in Natural and Social Sciences. *Journal of Economic Development, 9*(4), 52–79. https://mpra.ub.uni-muenchen.de/105149/

Hattula, J. D., Herzog, W., & Dhar, R. (2022). The impact of touchscreen devices on consumers' choice confidence and purchase likelihood. *Marketing Letters, 34*(1), 35–53. DOI: 10.1007/s11002-022-09623-w

Hazzi, O. A., & Maldaon, I. S. (2015). A Pilot Study: Vital Methodological Issues. *Business: Theory and Practice, 16*(1), 53–62. DOI: 10.3846/btp.2015.437

Hennart, J. F., & Verbeke, A. (2022, September 27). Actionable and enduring implications of Oliver Williamson's transaction cost theory. *Journal of International Business Studies, 53*(8), 1557–1575. DOI: 10.1057/s41267-022-00558-y

Henseler, J., Hubona, G., & Ray, P. A. (2016). Using PLS path modeling in new technology research: Updated guidelines. *Industrial Management + Data Systems. Industrial Management & Data Systems, 116*(1), 2–20. DOI: 10.1108/IMDS-09-2015-0382

Henseler, J., Ringle, C. M., & Sinkovics, R. R. (2009). The use of partial least squares path modeling in international marketing. In *Advances in international marketing* (pp. 277–319). DOI: 10.1108/S1474-7979(2009)0000020014

Hilton, T., Hughes, T., Little, E., & Marandi, E. (2013). Adopting self-service technology to do more with less. *Journal of Services Marketing, 27*(1), 3–12. DOI: 10.1108/08876041311296338

Hoffman, K., & Bateson, J. (2016, January 1). *Services Marketing: Concepts, Strategies, & Cases*. Cengage Learning.

Independent Digital News and Media. (2011, March 15). *How Britain became a self-service nation*. The Independent. https://www.independent.co.uk/news/business/analysis-and-features/how-britain-became-a-selfservice-nation-2241830.html

Inman, J. J., & Nikolova, H. (2017). Shopper-Facing Retail Technology: A Retailer Adoption Decision Framework Incorporating Shopper Attitudes and Privacy Concerns. *Journal of Retailing, 93*(1), 7–28. DOI: 10.1016/j.jretai.2016.12.006

Jain, S. K., & Gupta, G. (2004, April). Measuring Service Quality: Servqual vs. Servperf Scales. *Vikalpa, 29*(2), 25–38. DOI: 10.1177/0256090920040203

Jie, N. X., & Kamsin, I. F. (2021). *Self- checkout service with RFID technology in supermarket*. Atlantis Highlights in Computer Sciences., DOI: 10.2991/ahis.k.210913.062

Joe, S., Kim, J. S., & Zemke, D. M. V. (2020, April 8). Effects of Social Influence and Perceived Enjoyment on Kiosk Acceptance: A Moderating Role of Gender. *International Journal of Hospitality & Tourism Administration, 23*(2), 289–316. DOI: 10.1080/15256480.2020.1746217

Johanson, G. A., & Brooks, G. P. (2009). Initial Scale Development: Sample Size for Pilot Studies. *Educational and Psychological Measurement, 70*(3), 394–400. DOI: 10.1177/0013164409355692

Johanson, G. A., & Brooks, G. P. (2009, December 18). Initial Scale Development: Sample Size for Pilot Studies. *Educational and Psychological Measurement, 70*(3), 394–400. DOI: 10.1177/0013164409355692

Johnson, E. (2021). Cronbach's Alpha. *Encyclopedia of Autism Spectrum Disorders*, 1248–1249. DOI: 10.1007/978-3-319-91280-6_307

Kaliyadan, F., & Kulkarni, V. (2019). Types of variables, descriptive statistics, and sample size. *Indian Dermatology Online Journal*, *10*(1), 82. DOI: 10.4103/idoj.IDOJ_468_18 PMID: 30775310

Kaminski, J. (2011). Diffusion of Innovation Theory. *Canadian Journal of Nursing Informatics*, *6*(2). https://cjni.net/journal/?p=1444

Kaushik, A. K., & Rahman, Z. (2017). An empirical investigation of Tourist's choice of Service Delivery Options. *International Journal of Contemporary Hospitality Management*, *29*(7), 1892–1913. DOI: 10.1108/IJCHM-08-2015-0438

Kiffle, K. (2019, November 29). Brunei Information & Communications Technology Sector Initiative: The next step on the path to a Smart Nation - Brunei NetCo. Brunei NetCo. Retrieved January 14, 2024, from https://www.bruneinetco.com/brunei-information-communications-technology-sector-initiative-the-next-step-on-the-path-to-a-smart-nation/

Kim, H.-J., & Lee, J.-M. (2020). Consumers' resistance and continued use intention of self-service kiosk. *Family and Environment Research*, *58*(3), 401–416. DOI: 10.6115/fer.2020.029

Kim, I., & Min, J. (2023). Working hours and the regulations in Korea. *Annals of Occupational and Environmental Medicine*, *35*(1), e18. Advance online publication. DOI: 10.35371/aoem.2023.35.e18 PMID: 37928375

Kim, T. K., & Park, J. H. (2019). More about the basic assumptions of t-test: Normality and sample size. *Korean Journal of Anesthesiology*, *72*(4), 331–335. DOI: 10.4097/kja.d.18.00292 PMID: 30929413

Kiosk Group. (n.d.). *Self Service kiosks 101*. https://www.kioskgroup.com/pages/self-service-kiosks-101

Kokkinou, A., & Cranage, D. A. (2013, June). Using self-service technology to reduce customer waiting times. *International Journal of Hospitality Management*, *33*, 435–445. DOI: 10.1016/j.ijhm.2012.11.003

Kolasani, S. (2023). Optimizing Natural Language Processing, Large Language Models (LLMs) for Efficient Customer Service, and hyper-personalization to enable sustainable growth and revenue. *Transactions on Latest Trends in Artificial Intelligence, 4*(4). Retrieved from https://ijsdcs.com/index.php/TLAI/article/view/476/196

Krishen, A. S. (2008). Perceived Versus Actual Complexity for Websites: Their Relationship to Consumer Satisfaction. *Journal of Consumer Satisfaction, Dissatisfaction & Complaining Behavior*, •••, 21.

Kristiawan, M., Suhono, S., Yussof, M. H. B., & Muslimah, M. (2021). The International School's Culture in Indonesia and Brunei Darussalam. *Jurnal Iqra*, *6*(1), 180–191. DOI: 10.25217/ji.v6i1.1263

Krithikadatta, J. (2014). Normal Distribution. *Journal of Conservative Dentistry. Journal of Conservative Dentistry : JCD*, *17*(1), 96. DOI: 10.4103/0972-0707.124171 PMID: 24554873

Ku, E. C., & Chen, C. D. (2013). Fitting facilities to self-service technology usage: Evidence from kiosks in Taiwan airport. *Journal of Air Transport Management*, *32*, 87–94. DOI: 10.1016/j.jairtraman.2013.07.001

Kwak, J. K. (2020). Empirical Analysis on Comparison between Self-checkout and Regular Staffed-checkout lanes in a Poland Retail Store, 8(1), pp. 56-61

Larivière, B., Bowen, D., Andreassen, T. W., Kunz, W., Sirianni, N. J., Voss, C., Wünderlich, N. V., & De Keyser, A. (2017). "Service encounter 2.0": An investigation into the roles of technology, employees and customers. *Journal of Business Research*, *79*, 238–246. DOI: 10.1016/j.jbusres.2017.03.008

Lawton, G. (2022, August 30). *9 disadvantages of self-service options: TechTarget*. Customer Experience. https://www.techtarget.com/searchcustomerexperience/tip/6-reasons-why-self-service-options-fail#:~:text=However%2C%20self%2Dservice%20initiatives%20can,before%20investing%20in%20this%20strategy

Lee, S. M., & Lee, D. (2019). "UNTACT": A new customer service strategy in the Digital age. *Service Business*, *14*(1), 1–22. DOI: 10.1007/s11628-019-00408-2

Leung, X. Y., Torres, B., & Fan, A. (2021). Do kiosks outperform cashiers? an S-O-R framework of restaurant ordering experiences. *Journal of Hospitality and Tourism Technology*, *12*(3), 580–592. DOI: 10.1108/JHTT-03-2020-0065

Lian, J. W. (2018, August 17). Why is self-service technology (SST) unpopular? Extending the IS success model. *Library Hi Tech*, *39*(4), 1154–1173. DOI: 10.1108/LHT-01-2018-0015

Liao, D., & Valliant, R. (2012). Variance inflation factors in the analysis of complex survey data. *Survey Methodology*, *38*, 53–62.

Lin, C. H., Shih, H. Y., & Sher, P. J. (2007). Integrating technology readiness into technology acceptance: The TRAM model. *Psychology and Marketing*, *24*(7), 641–657. DOI: 10.1002/mar.20177

Lin, J.-S. C., & Hsieh, P.-L. (2011). Assessing the self-service technology encounters: Development and validation of SSTQUAL scale. *Journal of Retailing*, *87*(2), 194–206. DOI: 10.1016/j.jretai.2011.02.006

Lin, S., & Lin, J. (2023). How organizations leverage digital technology to develop customization and enhance customer relationship performance: An empirical investigation. *Technological Forecasting & Social Change. Technological Forecasting and Social Change*, *188*, 122254. DOI: 10.1016/j.techfore.2022.122254

Loeb, S., Dynarski, S., McFarland, D., Morris, P., Reardon, S., & Reber, S. (2017). *Descriptive Analysis in Education: A Guide for Researchers. NCEE 2017-4023.* https://eric.ed.gov/?id=ED573325

Lucia-Palacios, L., Pérez-López, R., & Polo-Redondo, Y. (2020). Does stress matter in mall experience and customer satisfaction? *Journal of Services Marketing/the Journal of Services Marketing, 34*(2), 177–191. DOI: 10.1108/JSM-03-2019-0134

Khalufi, M. N. A., & Mohammad Shah, K. A. (2022). Factors Influencing Consumer's Intention to Use Self Service Technology in Retail. *Global Business and Management Research: An International Journal, 14*(3s), 1044–1052. http://www.gbmrjournal.com/pdf/v14n3s/V14N3s-71.pdf

Marcoulides, K. M., & Raykov, T. (2018). Evaluation of Variance Inflation Factors in Regression Models Using Latent Variable Modeling Methods. *Educational and Psychological Measurement*, *79*(5), 874–882. DOI: 10.1177/0013164418817803 PMID: 31488917

Martin, K. D., & Palmatier, R. W. (2020, December). Data Privacy in Retail: Navigating Tensions and Directing Future Research. *Journal of Retailing*, *96*(4), 449–457. DOI: 10.1016/j.jretai.2020.10.002

Martínez-Navalón, J. G., Gelashvili, V., & Gómez-Ortega, A. (2021). Evaluation of User Satisfaction and Trust of Review Platforms: Analysis of the Impact of Privacy and E-WOM in the Case of TripAdvisor. *Frontiers in Psychology*, *12*, 750527. Advance online publication. DOI: 10.3389/fpsyg.2021.750527 PMID: 34603166

McGrath, C., & Astell, A. (2016). The benefits and barriers to technology acquisition: Understanding the decision-making processes of older adults with age-related vision loss (ARVL). *British Journal of Occupational Therapy*, *80*(2), 123–131. DOI: 10.1177/0308022616667959

Meuter, M. L., Ostrom, A. L., Roundtree, R. I., & Bitner, M. J. (2000). Self-Service Technologies: Understanding Customer Satisfaction with Technology-Based Service Encounters. *Journal of Marketing*, *64*(3), 50–64. https://www.jstor.org/stable/3203487. DOI: 10.1509/jmkg.64.3.50.18024

Mishra, P., Singh, U., Pandey, C., Mishra, P., & Pandey, G. (2019). Application of student's t-test, analysis of variance, and covariance. *Annals of Cardiac Anaesthesia. Annals of Cardiac Anaesthesia*, *22*(4), 407–411. DOI: 10.4103/aca.ACA_94_19 PMID: 31621677

Mohamad, L. (2022, November 29). Cashless initiative empowers smart nation. *Borneo Bulletin*. Retrieved August 28, 2023.

Haming, M., Syaiful, A. Z., Putra, A. H. P. K., & Murdifin, I. (2019, February). The Application of SERVQUAL Distribution In Measuring Customer Satisfaction of Retails Company. *Journal of Distribution Science*, *17*(2), 25–31. DOI: 10.15722/jds.17.2.201902.25

Nasution, M. I., Fahmi, M., Jufrizen, N., Muslih, N., & Prayogi, M. A. (2020). The Quality of Small and Medium Enterprises Performance Using the Structural Equation Model-Part Least Square (SEM-PLS). *Journal of Physics: Conference Series*, *1477*(5), 052052. DOI: 10.1088/1742-6596/1477/5/052052

Noor, A. (2022, June 9). Digital payment, data storage, information key to vibrant digital ecosystem. *Borneo Bulletin*. Retrieved August 28, 2023, from https://borneobulletin.com.bn/digital-payment-data-storage-information-key-to-vibrant-digital-ecosystem/

Obermeier, G., Klingersberger, J., & Auinger, A. (2022, January 4). *Factors influencing usage intentions towards a self-service kiosk with biometric authentication*. Handle Proxy. http://hdl.handle.net/10125/79915

Ohira, T. (2018). On statistical independence and no-correlation for a pair of random variables taking two values: Classical and quantum. *Progress of Theoretical and Experimental Physics, 2018*(8). DOI: 10.1093/ptep/pty086

Othman, A. K., Hamzah, M. I., & Abu Hassan, L. F. (2020). Modeling the contingent role of technological optimism on customer satisfaction with self-service technologies. *Journal of Enterprise Information Management*, *33*(3), 559–578. DOI: 10.1108/JEIM-09-2019-0295

Otto, A. S., Szymanski, D. M., & Varadarajan, R. (2019, May 31). Customer satisfaction and firm performance: Insights from over a quarter century of empirical research. *Journal of the Academy of Marketing Science*, *48*(3), 543–564. DOI: 10.1007/s11747-019-00657-7

Pandita, A., & Kiran, R. (2023). The Technology Interface and Student Engagement Are Significant Stimuli in Sustainable Student Satisfaction. *Sustainability (Basel)*, *15*(10), 7923. DOI: 10.3390/su15107923

Parasuraman, A. (2000). Technology Readiness Index (Tri). *Journal of Service Research*, *2*(4), 307–320. DOI: 10.1177/109467050024001

Parasuraman, A. P., Zeithaml, V. A., & Berry, L. L. (1988, January 1). *SERVQUAL A Multiple-item Scale for Measuring Consumer Perceptions of Service Quality*. ResearchGate. https://www.researchgate.net/publication/200827786_SERVQUAL_A_Multiple-item_Scale_for_Measuring_Consumer_Perceptions_of_Service_Quality

Parasuraman, A., Ball, J., Aksoy, L., Keiningham, T. L., & Zaki, M. (2020). More than a feeling? Toward a theory of customer delight. *Journal of Service Management*, *32*(1), 1–26. DOI: 10.1108/JOSM-03-2019-0094

Parasuraman, A., Zeithaml, V. A., & Berry, L. L. (1985). A Conceptual Model of Service Quality and Its Implications for Future Research. *Journal of Marketing*, *49*(4), 41–50. DOI: 10.1177/002224298504900403

Parasuraman, A., Zeithaml, V. A., & Malhotra, A. (2005). E-S-QUAL. *Journal of Service Research*, *7*(3), 213–233. DOI: 10.1177/1094670504271156

Parise, S., Guinan, P. J., & Kafka, R. (2016). Solving the crisis of immediacy: How Digital Technology can transform the customer experience. *Business Horizons*, *59*(4), 411–420. DOI: 10.1016/j.bushor.2016.03.004

Park, J.-S., Ha, S., & Jeong, S. W. (2020). Consumer acceptance of self-service technologies in Fashion Retail Stores. *Journal of Fashion Marketing and Management*, *25*(2), 371–388. DOI: 10.1108/JFMM-09-2019-0221

Park, S., Lehto, X., & Lehto, M. (2021). Self-service technology kiosk design for restaurants: An QFD application. *International Journal of Hospitality Management*, *92*, 102757. DOI: 10.1016/j.ijhm.2020.102757

Pech, M., & Vrchota, J. (2022). The Product Customization Process in Relation to Industry 4.0 and Digitalization. *Processes (Basel, Switzerland)*, *10*(3), 539. DOI: 10.3390/pr10030539

Poister, T. H., & Thomas, J. C. (2011, April 12). The Effect of Expectations and Expectancy Confirmation/Disconfirmation on Motorists' Satisfaction with State Highways. *Journal of Public Administration: Research and Theory, 21*(4), 601–617. DOI: 10.1093/jopart/mur004

Poushneh, A., & Vasquez-Parraga, A. Z. (2018). The role of customer readiness and participation in non-technology-based Service Delivery. *Journal of Consumer Marketing, 35*(6), 588–600. DOI: 10.1108/JCM-11-2016-2006

Powering the new world of commerce. Ingenico. (n.d.). https://ingenico.com/en

Purba, R., Oesman, Y. M., & Komaladewi, R. (2022). The Effect of Self-Service Technology on Customers' Trust and Loyalty (Case Study at Terminal 3 Self-Service Technology Soekarno-Hatta International Airport). *Budapest International Research and Critics Institute-Journal (BIRCI-Journal), 5*(1), 6527–6542. https://doi.org/https://doi.org/10.33258/birci.v5i1.4380

Radomir, L., & Nistor, C. V. (2014). Comparing the Original and the Revised SSTQUAL Scale among High-educated Consumers in Romania. *Procedia Economics and Finance, 15*, 926–934. DOI: 10.1016/S2212-5671(14)00564-4

RBPF issues warning against phishing scams. (2024, January 30). RBPF Issues Warning Against Phishing Scams. https://borneobulletin.com.bn/rbpf-issues-warning-against-phishing-scams/

Rindfleisch, A. (2019, August 5). Transaction cost theory: past, present and future. DOI: 10.1007/s13162-019-00151-x

Robson, K., & Pitt, L. (2018). Tesco in South Korea: Strategic Localization. DOI: 10.4135/9781526439666

Rodrigues, L. L., Barkur, G., Varambally, K., & Golrooy Motlagh, F. (2011, October 4). Comparison of SERVQUAL and SERVPERF metrics: An empirical study. *The TQM Journal, 23*(6), 629–643. DOI: 10.1108/17542731111175248

Roopa, S., & Rani, M. (2012). Questionnaire Designing for a Survey. *The Journal of Indian Orthodontic Society, 46*(6), 273–277. DOI: 10.1177/0974909820120509S

Sakrabani, P., Teoh, A. P., & Amran, A. (2019). Strategic impact of retail 4.0 on retailers' performance in Malaysia. *Strategic Direction, 35*(11), 1–3. DOI: 10.1108/SD-05-2019-0099

Scherer, A., Wünderlich, N. V., & von Wangenheim, F. (2015). The value of self-service: Long-term effects of technology-based self-service usage on customer retention. *Management Information Systems Quarterly*, *39*(1), 177–200. DOI: 10.25300/MISQ/2015/39.1.08

Schweizer, A., Knoll, P., Urbach, N., von der Gracht, H. A., & Hardjono, T. (2020, November). To What Extent Will Blockchain Drive the Machine Economy? Perspectives From a Prospective Study. *IEEE Transactions on Engineering Management*, *67*(4), 1169–1183. DOI: 10.1109/TEM.2020.2979286

Shahid Iqbal, M., Ul Hassan, M., & Habibah, U. (2018, January 1). Impact of self-service technology (SST) service quality on customer loyalty and behavioral intention: The mediating role of customer satisfaction. *Cogent Business & Management*, *5*(1), 1. DOI: 10.1080/23311975.2018.1423770

Shahril, Z., Zulkafly, H. A., Ismail, N. S., & Sharif, N. U. N. M. (2021, January 28). Customer Satisfaction Towards Self-Service Kiosks for Quick Service Restaurants (QSRs) in Klang Valley. *International Journal of Academic Research in Business & Social Sciences*, *11*(13). Advance online publication. DOI: 10.6007/IJARBSS/v11-i13/8502

Shan, S., Wenger, E., Zhang, J., Li, H., Zheng, H., & Y. Zhao, B. (2020). Fawkes: Protecting Privacy against Unauthorized Deep Learning Models. USENIX, 1589–1604.

Sharbawi, S., & Mabud, S. A. (2021). Malay, Muslim and Monarchy: An Introduction to Brunei Darussalam and Its National Identity. In Springer eBooks (pp. 45–66). DOI: 10.1007/978-3-030-77119-5_3

Shim, H. S., Han, S. L., & Ha, J. (2020). The Effects of Consumer Readiness on the Adoption of Self-Service Technology: Moderating Effects of Consumer Traits and Situational Factors. *Sustainability (Basel)*, *13*(1), 95. DOI: 10.3390/su13010095

Singh, A. (2018). *Evaluating Passengers' Perceived Service Quality Towards Self-Service Luggage Check-In Technologies at Airports Using SSTQUAL Scale.* (Publication No. 10810972)[Master's thesis, Arizona State University]. ProQuest Dissertations and Theses Global.

Siva Durga Prasad Nayak, M., & Narayan, K. A. (2019). Strengths and weaknesses of online surveys. *IOSR Journal of Humanities and Social Sciences*, *24*(5), 31–38.

Skiera, B., Reiner, J., & Albers, S. (2021). Regression Analysis. In *Springer eBooks* (pp. 299–327). DOI: 10.1007/978-3-319-57413-4_17

Sohail, M. S., & Hasan, M. (2021, March 4). Students' perceptions of service quality in Saudi universities: The SERVPERF model. *Learning and Teaching in Higher Education: Gulf Perspectives*, *17*(1), 54–66. DOI: 10.1108/LTHE-08-2020-0016

Sreejesh, S., Mohapatra, S., & Anusree, M. R. (2013). Exploratory Factor and Principal Component Analysis. In *Springer eBooks* (pp. 207–228). DOI: 10.1007/978-3-319-00539-3_9

TabSquare. (2021, March 23). *Jollibee at Gadong, #Brunei is now successfully live with our Android Kiosk solution integrated with Aloha POS*. Facebook. https://www.facebook.com/TabSquare/posts/jollibee-at-gadong-brunei-is-now-successfully-live-with-our-android-kiosk-soluti/4072903862799467/

Tahta Kuswanto, D. R., & Rahmat Syah, T. Y. (2021). The influence of the self-service technology service quality, esatisfaction, and e-trust on e-loyalty. *American International Journal of Business Management*, *4*(8), 89–96. https://www.aijbm.com/wp-content/uploads/2021/08/J488996.pdf

Tawari, K. N. (2019, August 21). The impact of self-service technologies and other non-technology drivers on customer satisfaction in the Irish supermarket industry. NORMA@NCI Library. https://norma.ncirl.ie/4026/

Taylor, E. (2016, July 24). Supermarket self-checkouts and retail theft: The curious case of the SWIPERS. *Criminology & Criminal Justice*, *16*(5), 552–567. DOI: 10.1177/1748895816643353

Taylor, S. A. (1993, October). The Roles of Service Quality, Consumer Satisfaction, and Value in Quinn's (1992) Paradigm of Services. *Journal of Marketing Theory and Practice*, *2*(1), 14–26. DOI: 10.1080/10696679.1993.11501636

Thabane, L., Ma, J., Chu, R., Cheng, J., Ismaila, A., Rios, L. P., Robson, R., Thabane, M., Giangregorio, L., & Goldsmith, C. H. (2010). A tutorial on pilot studies: The what, why and how. *BMC Medical Research Methodology*, *10*(1), 1. Advance online publication. DOI: 10.1186/1471-2288-10-1 PMID: 20053272

Time In Brunei. (2021, November 10). *Self-Checkout your own grocery at Hua Ho Department Store*. Facebook. https://www.facebook.com/timeinbrunei/posts/self-checkout-your-own-grocery-at-hua-ho-department-store-the-first-in-brunei-da/2028381677311122/

Turkar, S. (2023, August 15). *At 7% cagr, self service technologies market size to hit USD 57.7 bn by 2028, says Intellectual Market Insights Research*. LinkedIn. https://www.linkedin.com/pulse/7-cagr-self-service-technologies-market-size-hit-usd-577-turkar/?trk=pulse-article_more-articles_related-content-card

Ul Hassan, M., Iqbal, M. S., & Habibah, U. (2020). Self-service technology service quality: Building Loyalty and intention through Technology Trust in pakistani service sector. *SAGE Open*, *10*(2), 215824402092441. DOI: 10.1177/2158244020924412

Ursavaş, M. F. (2022). Unified Theory of Acceptance and Use of Technology Model (UTAUT). *Springer Texts in Education*, 111–133. DOI: 10.1007/978-3-031-10846-4_6

UTB'S SCHOOL OF COMPUTING AND INFORMATICS SHOWCASES CUTTING-EDGE HOLOGRAPHIC EDUCATION TECHNOLOGY | Universiti Teknologi Brunei. (n.d.). https://www.utb.edu.bn/news/utb-s-school-of-computing-and-informatics-showcases-cutting-edge-holographic-education-technology/

Vakulenko, Y., Hellström, D., & Oghazi, P. (2018). Customer value in self-service kiosks: A systematic literature review. *International Journal of Retail & Distribution Management*, *46*(5), 507–527. DOI: 10.1108/IJRDM-04-2017-0084

Vakulenko, Y., Oghazi, P., & Hellström, D. (2019). Innovative Framework for self-service kiosks: Integrating customer value knowledge. *Journal of Innovation & Knowledge*, *4*(4), 262–268. DOI: 10.1016/j.jik.2019.06.001

Vansteenkiste, M., & Ryan, R. M. (2013). On psychological growth and vulnerability: Basic psychological need satisfaction and need frustration as a unifying principle. *Journal of Psychotherapy Integration*, *23*(3), 263–280. DOI: 10.1037/a0032359

Wang, Y., Wang, S., Wang, J., Wei, J., & Wang, C. (2018). An empirical study of consumers' intention to use ride-sharing services: Using an extended technology acceptance model. *Transportation*, *47*(1), 397–415. DOI: 10.1007/s11116-018-9893-4

WAWASAN BRUNEI 2035. (n.d.). Www.wawasanbrunei.gov.bn. https://www.wawasanbrunei.gov.bn/en/SitePages/home.aspx

Xinhua. (2024, June 16). War against the scammers - Brunei cybersecurity agency takes down 1,023 phishing websites. *The Star*. Retrieved June 21, 2024, from https://www.thestar.com.my/aseanplus/aseanplus-news/2024/06/16/war-against-the-scammers---brunei-cybersecurity-agency-takes-down-1023-phishing-websites

Yang, J., & Klassen, K. J. (2008). How financial markets reflect the benefits of self-Service Technologies. *Journal of Enterprise Information Management*, *21*(5), 448–467. DOI: 10.1108/17410390810904238

Yoo, B., & Donthu, N. (2014). Developing a Scale to Measure the Perceived Quality of an Internet Shopping Site (PQISS). *Proceedings of the 2000 Academy of Marketing Science (AMS) Annual Conference*, 471–471. DOI: 10.1007/978-3-319-11885-7_129

Yousuf, A. (2017). Transaction Costs: A Conceptual Framework. *International Journal of Engineering and Management Sciences*, *2*(3), 131–139. DOI: 10.21791/IJEMS.2017.3.13.

Zaveri, A. A., Faisal, N., Nazar, M., Shehmir, S., Parveen, M., Imtiaz, S., Matcharan, A., Akhter, S., & Sami, N. (2022, June 24). Easy-Mart: Self checkout System for Supermarkets. *Journal of Independent Studies and Research Computing*, *20*(1). Advance online publication. DOI: 10.31645/JISRC.22.20.1.2

Zhu, G., So, K. K., & Hudson, S. (2017). Inside the sharing economy. *International Journal of Contemporary Hospitality Management*, *29*(9), 2218–2239. DOI: 10.1108/IJCHM-09-2016-0496

APPENDIX A

SSTQUAL scale adopted from Lin & Hsieh (2011) and Radomir & Nistor (2014) which have been used to analyse the dimensions of service quality of SSK. For customer satisfaction, scale from three-item American Customer Satisfaction Index (Index) has been used. Questions mark ** are added question which were believe suitable to further enhance the information obtain.

Table 19.

Constructs	Items	References
Functionality	1. Icangetmyservicedonewiththefirm'sSSKinashorttime 2. The service process of the firm's SSK is clear 3. Using the firm's SSK requires little effort 4. I can get service done smoothly with the firm's SSK	(Lin & Hsieh, 2011)
Enjoyment	1. The operation of the firm's SSKs is interesting 2. I feel good being able to use the SSKs 3. The firm's SSKs have interesting additional functions 4. I lost track of what was going on around me when I am using Self-Service Kiosk** 5. I was proficient in using Self-Service Kiosk**	(Lin & Hsieh, 2011)
Security	1. I feel safe in my transaction with the firm's SSKs 2. Aclearprivacypolicyisstatedwhenlusethefirm'sSSKs 3. Each service item/function of the SSK is error-free 4. The shops SSK provide me with all relevant information 5. The shops SSK does not share my personal information with other website databases** 6. The shops SSK has adequate security features**	(Lin & Hsieh, 2011)
Assurance	1. The firm providing the SSK is well known 2. The firm providing the SSK has a good reputation	(Lin & Hsieh, 2011)
Design	1. The layout of the firm's SSK is aesthetically appealing 2. The firm SSK appears to use up-to-date technology 3. I am able to perform all the actions I want to on the Self-Service kiosk**	(Lin & Hsieh, 2011)
Convenience	1. The SSK has operating hours convenient to customers 2. It is easy and convenient to reach the firm's SSK	(Lin & Hsieh, 2011)
Customization	1. The firm's SSK understands my specific needs 2. The firm's SSK has my best interests at heart 3. The firm's SSK has features that are personalized for me	(Lin & Hsieh, 2011)
Customer Satisfaction	1. Overall, I am satisfied with the SSK offered by the firm 2. The SSK offered by the firm exceed my expectations 3. The SSK technologies offered by the firm are close to my idea	(Fornell et al., 1996).

Chapter 7

Navigating Ethical Frontiers in Automation and Human-Robot Synergy for Organizational Transformation

S. Pushpalatha
https://orcid.org/0009-0003-8730-0670
SIMATS, Saveetha University, India

R. Durai Pandian
https://orcid.org/0000-0003-2932-0270
Saveetha Institute of Medical and Technical Sciences, Saveetha University, India

ABSTRACT

This chapter examines the ethical boundaries in automation, concentrating on the incorporation of robots inside organizational structures and the ethical challenges arising from this transition. In an age where automation is transforming company operations, it is essential to assess the equilibrium between technological progress and the welfare of the human labor. The chapter explores human-robot collaboration, highlighting the necessity for ethical frameworks to govern the deployment of automated technology. The chapter provides insights into best practices and solutions for promoting ethical automation by examining conceptual and theoretical models, ethical difficulties, and organizational consequences. This document will examine

DOI: 10.4018/979-8-3373-3404-2.ch007

recommendations for addressing new trends, improving accessibility, and fostering collaboration between humans and robots, thereby offering a framework for ethical automation in corporate contexts.

INTRODUCTION

A revolutionary development that has evolved in the digital era is the inclusion of automation technology and robots into organizational frameworks. This is one of the most significant developments from the digital era. The discussion on ethical implementation and human-robot collaboration has become increasingly relevant as businesses rapidly migrate to artificial intelligence (AI), machine learning (ML), and robots in order to improve their productivity (Brynjolfsson & McAfee, 2022). The use of automation offers a number of distinct advantages, including scalability, cost reduction, and increased operational efficiency. In spite of this, its unchecked proliferation raises ethical concerns such as the displacement of employees, the openness of algorithms, discrimination, accountability, and the equitable distribution of chances (Cascio & Montealegre, 2016; Rahwan, 2018). The term "human-robot synergy" refers to the collaborative interaction that takes place between humans and intelligent robots, with the intention of increasing productivity while maintaining human-centric principles (Amodei et al., 2016). Not only does this synergy involve technological advancements, but it also calls for a careful examination of roles, responsibilities, and ethical concerns. The transformation of an organization that is driven by automation must incorporate ethical frameworks in order to guide the adoption of technology in a manner that respects human dignity, fosters trust, and encourages innovation that is inclusive at the same time (Calo, 2015; Dignum, 2019). When digital technologies begin to impact the future of work, businesses find themselves at a vital crossroads: they must either adopt automation without giving it considerable consideration or they must support transformation that is both ethical and sustainable. The purpose of this chapter is to provide a path by investigating the ethical boundaries of automation, the rules that govern human-robot collaboration, and the frameworks that are necessary for responsible execution. This research makes use of concepts from a variety of fields, including human resource management, organizational behavior, and technology ethics, in order to address the urgent need to bring technological innovation in line with social values (Coeckelbergh, 2020; Robert et al., 2018).

BACKGROUND

The history of automation is inextricably tied to the industrial revolutions that have had an impact on the world as we know it today. The nature of employment has been continually modified by technological advancements, beginning with mechanical looms in the 18th century and continuing into the present day with algorithmic intelligence (Frey & Osborne, 2017). The Fourth Industrial Revolution, which is characterized by artificial intelligence, robots, and data-centric systems, poses issues that are unprecedented and go beyond the concept of merely increasing efficiency and productivity (Almalki et al., 2016). These technologies have begun to replace tasks that formerly required human intelligence, and their impact can be seen in a variety of fields, including education, healthcare, finance, and logistics (Burrell, 2016). Despite the fact that the discussion of ethics in relation to technology has been going on for quite some time, the incorporation of artificial intelligence and robots into organizational settings poses new ethical challenges. Among the problems that need to be addressed are the lack of transparency in decision-making algorithms, surveillance and data privacy, unequal access to technology, and the psychological effects of working with computers (Zhang et al., 2019; Jaeger, 2012). In contrast to other industrial revolutions, the current digital transformation is not only reshaping tools but also reorganizing relationships. These relationships include those between employees and employers, between people and robots, and between businesses and society as a whole (Robins & Dautenhahn, 2014). To add insult to injury, the implementation of robotic systems frequently progresses at a faster pace than the adoption of acceptable ethical or regulatory safeguards. An ethical position that is reactive is insufficient in light of the growing prevalence of automation. It is vital to have frameworks that are proactive and motivated by principles in order to ensure that automation enhances human agency rather than diminishing it (Raji, 2020). In this context, human-robot synergy denotes a shift from replacement to cooperation, which calls for a more in-depth comprehension as well as a methodical modification to the organizational structure.

Importance and Significance of the Chapter

This chapter is significant as it examines a crucial turning point in the development of contemporary workplaces—the convergence of ethics, automation, and human capital. Automation has transcended mere repetitive manual labor and increasingly impacts intricate cognitive activities and decision-making processes (Mitchell, 2019; McBride et al., 2021). Consequently, corporations face mounting pressure to reevaluate their strategy for both operational efficiency and social and ethical responsibility. Ethical automation transcends mere damage avoidance; it involves the creation of systems that uphold human dignity, foster equity, and resonate with organizational principles (Floridi, 2018). This chapter's value is in its contribution to the ongoing discourse. It delivers an extensive examination of ethical principles, theoretical frameworks, and practical guidance, equipping organizations with the necessary capabilities to properly manage digital change. Moreover, it underscores the significance of human-robot collaboration not as a technological inevitability, but as a human-centric approach for sustainable innovation (Thomson & Alvarez, 2024). This chapter additionally addresses a significant research deficiency. Although extensive literature exists on the technical dimensions of AI and automation, there is very limited discourse on the ethical use of these technologies within businesses, particularly regarding their potential to bolster workforce morale, resilience, and inclusivity (Binns, 2018; Chock, 2020). The discourse seeks to reconcile this disparity by emphasizing strategic adaptation, ethical foresight, and the creation of synergistic work settings that integrate human intellect with technological capabilities.

Structure of the Chapter

This chapter is structured into a series of interconnected sections that lead the reader through the theoretical, ethical, and practical aspects of automation and human-robot collaboration in organizational transformation.

- **Conceptual Framework:** Defining and interlinking fundamental concepts: automation, ethics, human-robot collaboration, and organizational transformation.
- **Theoretical Framework:** Utilizing pertinent ideas such as socio-technical systems theory, utilitarian ethics, deontological ethics, and stakeholder theory.
- **Research Methodology:** Elucidating how literature-based qualitative investigation underpins the claims presented in this chapter.
- **Navigating Ethical Frontiers in Automation and Human – Robot Synergy for Organizational Transformation:** Examining significant ethical challenges in the deployment of AI and robotics.
- **Solutions and Recommendations:** Strategies and Proposals to Enhance Human-Robot Synergy.
- **Future Research:** This area aims to establish ethical automation ecosystems.
- **Conclusion:** This part culminates the chapter with integrated insights and anticipatory observations.

This framework offers a coherent progression that harmonizes theoretical depth with practical significance, guaranteeing that the subject is addressed with both academic precision and practical utility.

CONCEPTUAL FRAMEWORK

Definition of Key Concepts

This chapter employs a conceptual framework comprising four pillars automation, ethics, human-robot synergy, and organizational transformation to elucidate the ethical boundaries of automation and the evolution of human-robot collaboration.

Automation denotes the utilization of systems, software, or machinery to execute tasks that would typically necessitate human intervention. It encompasses robotic process automation (RPA), artificial intelligence decision systems, and physical robotics (Jobin et al., 2019).

Ethics of automation pertains to the values, norms, and principles that govern the design, implementation, and effects of these technologies (Rose, 2021). It encompasses inquiries of justice, accountability, bias, privacy, and the societal implications of technological systems.

Human-Robot Synergy denotes the deliberate cooperation between humans and machines aimed at improving joint performance while honoring human values and decision-making authority (Shi, 2016). This encompasses collaborative robots ("cobots") intended to aid rather than supplant workers (Wang et al., 2020).

Organizational Transformation is the process via which a firm realigns its structure, strategy, and operations to adapt to evolving circumstances. Upon the introduction of automation and artificial intelligence (Bryson, 2018),

Transformation encompasses not only technical advancements but also alterations in leadership, culture, and workforce strategy (Hevelke & Nida-Rümelin, 2023).

Interconnection of Concepts

Both of these ideas are intricately tied to one another. It is impossible for ethical automation to exist in isolation; rather, it must be in accordance with human values and their objectives for strategic transformation. The ethical framework ensures that automation enhances human capacity within the organization rather than diminishing (van Dijk, 2020). The implementation of AI-driven hiring systems without employing ethical oversight may lead to biased outcomes that have an effect on workforce diversity (Pereira & Liu, 2024). On the other hand, human-robot collaboration has the potential to improve decision-making, employee engagement, and sustainable innovation (Bommasani, 2021). This is because ethical principles can be incorporated into the design of artificial intelligence algorithms and organizational policies.

Visual Representation

Imagine a model that is concentric with four layers that are related to each other:

- The first core layer: Ethical values, including privacy, fairness, and accountability.
- The second Layer: Human-Robot Synergy, which includes the integration of workflow, task collaboration, and trust.
- The third layer consists of AI, machine learning, robotics, and robotic process automation (RPA) technologies.
- The fourth outer layer Organizational Transformation, which includes strategy, structure, culture, and reskilling of the workforce.

Finally, this model demonstrates how ethical underpinnings should be used to guide human-machine collaboration, which is then enabled by technology. This collaboration finally results in an organization that is transformed and prepared for the future.

THEORETICAL FRAMEWORK

The purpose of this chapter is to provide a critical analysis of ethical automation and human-robot synergy by drawing on four influential theories:

Socio-Technical Systems Theory

This idea proposes that the optimal functioning of organizations occurs when their social (people, culture) and technical (tools, technologies) systems are working together to achieve optimal performance (Story et al., 1998). The theory lends support to the process of building systems in which people and robots complement each other in ways that are consistent with ethical principles in the context of automation (Dwivedi, 2023).

Utilitarian and Deontological Ethics

Two approaches to ethics: utilitarianism and deontology

Utilitarianism, the philosophy of utilitarianism, which is credited to Bentham and Mill, is centered on the maximization of common good. Increasing overall productivity or decreasing expenses would be the result of this in the context of automation (Colgate, 1996).

Deontological ethics, which is connected with Kant, places an emphasis on obligations and rights regardless of the consequences. Even if automation appears to be more efficient, it places an emphasis on maintaining the dignity and privacy of workers (Braddock & Parish, 2001).

It is essential to implement both in a balanced manner in order to guarantee that automation will be beneficial to society without compromising individual liberties.

Stakeholders Theory

This contends that firms ought to take into account the interests of all stakeholders, not just shareholders (Breazeal, 2003). The employees, the consumers, the regulators, and society as a whole are all included in this category under ethical automation. Ignoring the concerns of stakeholders can result in mistrust, resistance, and pushback from the community (Broadbent et al., 2009).

Feminist ethics and the ethics of care providing

These theories place an emphasis on the importance of empathy, connections, and context in the process of making ethical decisions (Coeckelbergh, 2010). As they are applied to interactions between humans and robots, they indicate that relational ethics and emotional well-being should be taken into consideration during the design and implementation of the system.

Framework for Implementing Ethical Automation

Establishing ethical automation and human-robot collaboration in enterprises necessitates a comprehensive framework that incorporates ethical governance, stakeholder involvement, technological coherence, and ongoing assessment. The initial element of the framework is ethical foresight and governance, which involves the formation of a specialized ethics committee and the implementation of ethical auditing instruments to preemptively mitigate potential harms, including bias, job displacement, and surveillance threats (Floridi, 2018; Raji, 2020). Organizations must establish AI governance rules that conform to international standards, including the EU's AI Act and the OECD AI Principles, to guarantee accountability and transparency (European Commission, 2021). Secondly, stakeholder-inclusive co-design is crucial for aligning automation solutions with business culture and human values. Participatory design principles guarantee that employees, technicians, and customers engage in the decision-making process (Chock, 2020). This incorporation diminishes resistance to change and augments the reliability of systems (Rahwan, 2018). The third tier of the architecture emphasizes technical alignment and integration, guaranteeing the ethical and practical incorporation of automation and robotic systems into processes. Prioritize techniques such as human-in-the-loop architectures, explainable AI models, and context-aware robotics to maintain human agency and oversight (Bryson, 2018; Doshi-Velez & Kim, 2017). Moreover, technologies must be chosen and tailored according to contextual appropriateness solutions effective in a healthcare environment may not be applicable in a manufacturing facility (Broadbent et al., 2009). An essential fourth pillar encompasses accessibility, training, and reskilling to prevent the emergence of digital divisions. Facilitating digital accessibility via universal design principles and reskilling initiatives promotes equitable engagement among all employee demographics, including older adults and individuals with impairments (Lazar, 2017; Story et al., 1998). Human-robot collaboration should not exclude non-tech-savvy individuals but rather enhance their capabilities through focused educational programs. Ultimately, ongoing surveillance and flexible learning finalize the execution cycle. The ethical implementation is not a singular occurrence; it necessitates continuous audits, employee feedback mechanisms, and conformity with changing legal and social standards (Dignum, 2019). Utilizing AI lifecycle tools like model cards and algorithmic effect assessments enables firms to evaluate ethical performance and make necessary adjustments (Mitchell, 2019; Jobin et al., 2019). This approach guarantees that automation and human-robot collaboration are used not merely as efficiency tools, but as means of equity, responsibility, and sustainable transformation in contemporary businesses.

Steps for Integrating Ethical Automation and Human-Robot Collaboration Practices

Incorporating ethical automation and human-robot collaboration into organizational operations necessitates a systematic, staged methodology that harmonizes technology proficiency with human principles. The initial phase involves performing a corporate ethics audit to evaluate current rules, employee preparedness, and sectors susceptible to ethical issues. This diagnostic phase exposes deficiencies in openness, inclusivity, and accountability, establishing a foundation for subsequent treatments (Dwivedi, 2023). Subsequently, enterprises must establish or amend internal governance rules for AI and automation, ensuring alignment with international ethical standards, like the IEEE Ethically Aligned Design and the OECD AI Principles (Wang et al., 2020; Rose, 2021). The third step entails inclusive stakeholder involvement, incorporating multiple perspectives from CEOs and IT leaders to frontline workers and marginalized communities into the planning and design processes. This participatory design paradigm guarantees that systems fulfill the ethical, cultural, and operational requirements of all stakeholders (Müller & Bostrom, 2022). Subsequently, the choice of ethical technology is paramount. Organizations should prioritize explainable AI models, collaborative robots equipped with safety procedures, and privacy-by-design solutions that safeguard user data and foster trust (Doshi-Velez & Kim, 2017; Calo, 2015). The fifth phase involves investing in staff education, training, and reskilling initiatives. These initiatives guarantee that human workers are not supplanted but instead enabled to collaborate with robots and AI systems. This encompasses both technical training and an understanding of AI ethics, data literacy, and change management (Cascio & Montealegre, 2016; Brynjolfsson & McAfee, 2022). Concurrent with training, pilot testing and phased implementations must be executed to assess the ethical and operational efficacy of the integrated systems in real-world settings (Breazeal, 2003; Colgate, 1996). Upon deployment of technologies, it is imperative to implement ongoing feedback and monitoring measures to identify unintended repercussions, algorithmic drift, or adverse employee attitudes. Real-time feedback mechanisms, ethical assessments, and post-implementation evaluations can assist businesses in adapting swiftly (Raji, 2020). Ultimately, institutionalizing ethical contemplation and continuous enhancement—via frequent review boards, external assessments, and conformity with changing regulatory standards—guarantees the enduring viability of human-centered automation (Floridi, 2018; Thomson & Alvarez, 2024). This series of meticulously organized processes enables businesses to implement advanced technology responsibly, safeguarding human dignity, enhancing trust, and facilitating sustainable innovation through ethical human-robot collaboration.

RESEARCH METHODOLOGY

The purpose of this chapter is to investigate the ethical implications of automation and human-robot synergy in the context of organizational transformation. The research technique utilized in this chapter is qualitative, conceptual, and literature-based. Instead of depending on empirical data collecting, the study employs an interpretive research approach to assess and synthesize the current academic literature, as well as ethical norms documents pertaining to policy, as well as expert contributions from a variety of fields. Rather than attempting to generalize the findings statistically, the objective is to build a nuanced knowledge of evolving ethical dilemmas and organizational responses. This paradigm is suitable due to the fact that the ethical, theoretical, and strategic aspects of automation are highly contextual, value-driven, and always evolving (Braddock & Parish, 2001). In order to bring together a wide range of theoretical viewpoints and ethical principles from a variety of domains, such as artificial intelligence, robotics, corporate ethics, organizational studies, and socio-technical systems theory, an integrative literature review approach is utilized (Jaeger, 2012). It is possible to identify themes, patterns, and conceptual linkages within this body of literature with the assistance of a narrative synthesis technique, which provides interpretive depth rather than quantitative breadth (Breazeal, 2003). The IEEE Global Initiative on Ethics of Autonomous and Intelligent Systems are examples of global think tanks that have contributed to this collection of publications. These include white papers, conference proceedings, technical reports, and journal articles that have been subjected to peer review. For the purpose of ensuring that the sources were both recent and relevant, academic databases such as Scopus, Web of Science, and Google Scholar were utilized to extract materials that were published mostly between the years 2010 and 2024. Some of the keywords that were utilized were as follows: ethical governance of artificial intelligence, human-robot interaction, organizational change, explain ability, and ethics in automation. We chose the sources to use because of their scholarly reliability, the influence of their citations, the conceptual significance of their work, and the contribution they made to the ethical debate in automation. In order to represent the cross-sectoral relevance of the findings, studies that focused on healthcare, manufacturing, education, public policy, and human-computer interaction were given special consideration. Constructivist epistemology, which acknowledges that knowledge is socially created and contextually understood, is another source of inspiration for this approach. Because of this, the chapter analyzes ethical and strategic dynamics in automation through a situated lens, paying attention to specific organizational settings, stakeholder interests, and cultural expectations (Bryson, 2018). This is done rather than looking for answers that are universally applicable. The approach of the research helps to promote the building of a multidimensional conceptual model that blends ethical reasoning and human-centered transformation with automation technologies. This is accomplished by triangulating theoretical insights with normative ethical standards. Overall, this literature-driven qualitative technique offers a framework that is both adaptable and rigorous, making it suitable for conducting an in-depth analysis of the ethical frontiers of automation. The chapter is able to expose gaps in existing frameworks, highlight best practices, and provide solutions that are ethically grounded in order to achieve sustainable and inclusive organizational change through human-robot synergy as a result of this.

NAVIGATING ETHICAL FRONTIERS IN AUTOMATION

As automation technologies progressively infiltrate several areas, including manufacturing, logistics, healthcare, and finance, the ethical ramifications of their deployment have grown more intricate and consequential. These technologies offer unparalleled efficiency, precision, and scalability, but they also present issues about transparency and justice. Privacy, accountability, and human dignity (Amodei et al., 2016). A primary ethical issue in automation is the absence of explainability in decision-making processes, especially in intricate machine learning algorithms that operate as "black boxes" (Burrell, 2016). The incapacity to elucidate or rationalize algorithmic decisions might diminish trust among stakeholders and compromise corporate accountability (Doshi-Velez & Kim, 2017). Furthermore, automated systems frequently adopt and exacerbate societal prejudices present in historical data, resulting in discriminatory consequences, especially in recruitment, lending, and criminal justice sectors (Robert et al., 2018). These biases prompt inquiries on equity and the ethical application of AI in critical decision-making (Raji, 2020). For example, Amazon notably discontinued an AI-driven hiring tool upon realizing it discriminated against female applicants, demonstrating how algorithmic automation can unintentionally perpetuate systemic inequalities (Rahwan, 2018). Ethical automation necessitates bias auditing, inclusive design, and ongoing monitoring to avert harm (Binns, 2018). The influence of technology on employment and labor markets poses an ethical quandary. Although it enhances productivity, it may replace employees and exacerbate economic inequality (Frey & Osborne, 2017). The ethical response should encompass proactive workforce development, reskilling initiatives, and job redesign techniques that foster inclusivity and sustainability (Hevelke & Nida-Rümelin, 2023). The issue is not solely technological but fundamentally human, necessitating empathy and stakeholder involvement in transformation planning (McBride et al., 2021). Automation also presents concerns regarding privacy and data security, especially in workplaces and public environments where sensor-based monitoring and predictive analytics are utilized (Zhang et al., 2019). Ethical frameworks must provide informed consent, data reduction, and accountability systems that safeguard persons' rights (Floridi, 2018). Furthermore, issues of moral responsibility arise when autonomous systems are permitted to make judgments independently of human supervision, particularly in domains such as healthcare or autonomous cars (Jobin et al., 2019). This requires the integration of significant human oversight into system architecture (Brynjolfsson & McAfee, 2022). Ultimately, traversing ethical boundaries in automation involves not merely risk minimization but also the alignment of technical advancement with societal ideals. Organizations must integrate ethical concepts, including transparency, fairness, accountability, and human well-being, into the design, deployment, and governance of automated systems (van Dijk, 2020). The advancement of ethical AI must be iterative, inclusive, and attuned to changing norms and expectations across diverse cultural and institutional contexts.

HUMAN-ROBOT SYNERGY FOR ORGANIZATIONAL TRANSFORMATION

Human-robot synergy denotes the cooperative interaction between humans and robotic systems to accomplish common objectives, particularly in organizational contexts. Contrary to conventional perceptions of robots as simple instruments or substitutes, synergistic models highlight collaborative interactions, wherein robots enhance human capacities and cultivate novel organizational proficiencies (Mitchell, 2019). This paradigm change signifies a wider trend towards sociotechnical integration, where human-centered design is essential for the responsible and effective utilization of technology (Shi, 2016). The efficacy of human-robot collaboration depends on the development of systems that are adaptive, intuitive, and socially cognizant. In advanced manufacturing, collaborative robots (cobots) can now operate alongside humans, sharing tasks such as lifting, sorting, or assembling without jeopardizing safety (Zhang et al., 2019). These robots are designed to react to human gestures, speech, and proximity, thereby improving productivity while maintaining ergonomic health and psychological well-being. In this setting, ethical design emphasizes safety, inclusivity, and the respect for individual autonomy (Calo, 2015). Robotic assistants in healthcare aid nurses with patient lifting and vital sign monitoring, thereby alleviating physical strain and improving care quality (Broadbent et al., 2009). Robotic tutors are widely utilized in education to customize learning, particularly for children with specific needs (Dwivedi, 2023). These applications illustrate the capacity of robots to promote equity and accessibility when ethically aligned with user requirements and cultural sensitivities. Nonetheless, ethical issues persist—especially with emotional connection, manipulation, and role ambiguity, which may undermine interpersonal relationships or professional identities (Coeckelbergh, 2010). The human-robot synergy framework necessitates organizational reorganization, especially in leadership, communication, and decision-making processes. Organizations must promote interdisciplinary collaboration among engineers, ethicists, behavioral scientists, and frontline workers to develop solutions that are ethically sound and contextually relevant (Bryson, 2018). Leadership must advance to incorporate techno-ethical fluency, enabling managers to comprehend both the potential and the ramifications of robotic technologies (Wang et al., 2020). Transparent policies, inclusive training, and participatory execution are crucial for fostering employee trust and reducing resistance (Robins & Dautenhahn, 2014). Furthermore, attaining synergy is not merely a technical obstacle; it is also a cultural one. Trust, acceptance, and ethical alignment are fundamental to the relationship between workers and robotic systems (Rose, 2021). Effective synergy can only be achieved when businesses acknowledge the relational and emotional aspects of human-machine interaction. Honoring worker values, facilitating agency, and employing empathetic design are essential for constructing a sustainable future in which robots and people coexist harmoniously (Coeckelbergh, 2020). In conclusion, human-robot synergy pertains not to the substitution of humans by technology, but to the redefinition of work as a co-evolutionary process in which humans and machines together learn, adapt, and create. This synergy may empower businesses when directed by ethics, empathy, and equity to attain transformative results not alone regarding efficiency but also in resilience, adaptability, and human-centered creativity.

SOLUTIONS AND RECOMMENDATIONS

Exploring the ethical boundaries of automation and human-robot collaboration necessitates both recognizing the issues and executing pragmatic, morally sound solutions. Organizations must embrace a proactive approach that incorporates ethical issues at every phase of automation—from design and development to deployment and assessment (Floridi, 2018). A fundamental answer is the integration of ethics-by-design, a methodology that embeds ethical ideals like as justice, transparency, privacy, and human dignity into the architecture and operation of AI and robotic systems (Dignum, 2019). This necessitates cooperation among ethicists, technologists, and end-users to guarantee that technologies conform to human rights and cultural standards.

A further essential recommendation is the execution of algorithmic audits and ethical impact evaluations. These procedures facilitate the identification, documentation, and alleviation of potential biases, discriminatory results, or privacy infringements within automated decision-making systems (Binns, 2018; Raji, 2020). Through the implementation of ongoing supervision and impartial evaluation, organizations can maintain accountability and public confidence. Investing in digital literacy and upskilling initiatives is crucial for alleviating job displacement threats. Employees must be equipped with the skills required to collaborate effectively with intelligent systems, ensuring that automation serves as a means of augmentation rather than replacement (Thomson & Alvarez, 2024). Moreover, inclusive stakeholder involvement in technology adoption decisions guarantees that varied viewpoints particularly from marginalized or underrepresented groups are taken into account (Chock, 2020). The establishment of transparent governance systems is equally essential. These should delineate roles, supervisory frameworks, reporting protocols, and escalation procedures in the event of ethical infractions (Jobin et al., 2019). Organizations should contemplate the formation of AI Ethics Boards or analogous entities to direct policy, offer expert counsel, and address growing ethical challenges. These programs foster a culture of transparency, equity, and human responsibility. Finally, enterprises must to synchronize their automation strategies with worldwide standards, such the OECD AI Principles, UNESCO's AI Ethics Framework, and IEEE's Ethically Aligned Design guidelines, which offer extensive, internationally acknowledged ethical benchmarks. Adhering to these norms mitigates ethical risk and bolsters validity in the global marketplace.

Best Practices for Implementing Ethical Automation and Human- Robot Synergy

Formulating best practices is essential for directing ethical automation and promoting sustainable human-robot collaboration within enterprises. These principles function as versatile frameworks for various industries experiencing digital change. Initially, the implementation of a human-in-the-loop (HITL) model is a fundamental approach, including human oversight into the decision-making processes of automated systems. This method guarantees that essential evaluations—particularly those related to safety, equity, or ethical implications—are corroborated by human expertise (Rahwan, 2018). The notion of transparency by design must be implemented. This entails guaranteeing that algorithms are comprehensible, documentation is readily available, and system decisions are elucidated for non-technical stakeholders (Doshi-Velez & Kim, 2017). Organizations may utilize model cards, data sheets, and decision logs to elucidate the functionalities and constraints of AI systems (Mitchell, 2019). Third, efficient interdisciplinary collaboration is necessary. Ethics in automation necessitates contributions from computer scientists, engineers, social scientists, legal experts, and end users (Bryson, 2018). Interdisciplinary teams can identify ethical hazards early in the creative process and foster more comprehensive design results.

The fourth best practice is the implementation of employee-centric change management solutions. This include participatory design workshops, training initiatives, and feedback mechanisms that incorporate worker contributions into the deployment of automation technology (Cascio & Montealegre, 2016). This engagement fosters trust, diminishes resistance, and enables employees to collaboratively generate new value with emerging technology. Fifth, organizations must participate in ongoing ethical oversight and education. Ethics is not a mere checkbox for compliance but a dynamic and context-dependent activity. Post-deployment audits, ethical incident reporting, and feedback evaluations enable firms to adapt and address unforeseen repercussions over time (Rose., 2021). Ultimately, cultivating a culture of digital responsibility is arguably the most fundamental activity. This encompasses leadership dedication to ethical innovation, internal motivations for responsible technology utilization, and institutional principles that prioritize well-being over immediate efficiency improvements (Floridi, 2018). Organizations that integrate ethical integrity with their fundamental goal establish a robust foundation for sustainable automation.

Addressing Ethical Challenges

Notwithstanding the existence of ethical frameworks and technological norms, certain ethical dilemmas continue to arise in practical implementations of automation and human-robot interaction. A significant issue is algorithmic opacity, wherein users and even developers may not completely comprehend the methodology by which a system reaches its conclusions. This may result in reduced trust and impede accountability, particularly in sectors such as healthcare, finance, or public governance (Burrell, 2016). Addressing this necessitates both technical instruments for explain ability and policy measures that impose transparency standards for high-risk applications. A significant concern is value misalignment—the potential for automated systems to function according to objectives or rules that contradict human values or societal standards (Amodei et al., 2016). An AI-driven scheduling system may enhance productivity while unintentionally causing burnout or disregarding personal limits. Addressing this necessitates participatory design methodologies and ethical alignment assessments, engaging various stakeholders in establishing system objectives and permissible compromises (Dignum, 2019). Disruption of the workforce and inequality continue to be pressing issues. Automation may disproportionately impact lower-skilled workers and exacerbate existing gaps related to gender, ethnicity, and socioeconomic status (Frey & Osborne, 2017). Ethical remedies should extend beyond reskilling to encompass policy innovation, including trials of universal basic income, social safety nets, and equal access to digital resources (Almalki et al., 2016). Furthermore, emotional manipulation and excessive reliance on robotic systems may present psychological hazards, especially among susceptible groups. In domains like elder care or education, emotionally intelligent robots may inadvertently exploit connection or obscure the distinction between human and machine relationships (Coeckelbergh, 2010). Designers must establish ethical limits on emotional affordances to provide clarity and consent in human-robot interactions (Calo, 2015). Ultimately, global and cross-cultural disparities in ethical expectations hinder the universal application of AI ethics. What is deemed fair, courteous, or independent in one setting may be interpreted differently in another. Organizations ought to implement ethical pluralism, tailoring ethical frameworks to local cultural, legal, and moral contexts while maintaining fundamental human rights (Jobin et al., 2019). By directly addressing these ethical challenges with empathy, foresight, and interdisciplinary collaboration organizations may not only reduce risks but also harness the revolutionary potential of human-robot synergy for inclusive and ethical futures.

Enhancing Accessibility

Improving accessibility in automation and human-robot collaboration is not only a technological issue but also an ethical necessity that pertains to inclusivity, equity, and universal design. As firms include intelligent systems into work environments, it is essential to guarantee that individuals of various abilities, backgrounds, and socio-economic statuses may access and benefit from these technologies (Wang et al., 2020). In this setting, accessibility encompasses not just physical ergonomics but also cognitive, emotional, linguistic, and socio-cultural elements (Jaeger, 2012). Initially, automation and robotics should be developed in accordance with universal design principles guidelines intended to ensure that goods are accessible to all individuals, to the maximum extent feasible, without requiring modifications or specialist design (Story et al., 1998). Robotic interfaces ought to include voice commands, haptic feedback, multilingual options, and customizable visual interfaces to cater to users with diverse abilities and preferences (Broadbent et al., 2009). In professional environments, inclusive automation may encompass customized job processes that adjust to varying levels of physical ability, cognitive demand, or sensory processing. Robotic exoskeletons and collaborative robots can aid workers with mobility impairments or age-related decline in lifting activities, so enhancing job participation and ensuring safety (Lazar, 2017). Likewise, robots employed in educational or healthcare settings must be flexible to accommodate users with neuro divergent requirements, including those with autism or sensory sensitivities (Robins & Dautenhahn, 2014). Accessibility encompasses digital equity guaranteeing that everyone in both urban and rural settings, irrespective of income or infrastructure, possess equal access to the advantages of automation. This encompasses not only the technology but also the training, bandwidth, and cultural understanding requisite for its optimal utilization (van Dijk, 2020). Smart agricultural solutions utilizing drones and robotic sensors must be customized for low-connectivity settings and developed with input from local farmers to guarantee cultural compatibility and acceptance (Rose et al., 2021). Furthermore, linguistic and literacy obstacles may restrict access to intricate automated systems. Interfaces must employ clear language, culturally relevant metaphors, and iconographic assistance to facilitate intuitive engagement among diverse populations (Almalki et al., 2016). The incorporation of assistive technologies such as screen readers, closed captions, and real-time language translation effectively narrows communication disparities in global enterprises. Organizational policies must promote accessibility in procurement, onboarding, and performance assessment. Performance management systems utilizing AI must be adjusted to consider varying abilities and eliminate biases in the assessment of productivity or communication styles (Binns, 2018). Job redesign should similarly incorporate possibilities for those with impairments to utilize robotic assistance systems, enabling them to excel in previously inaccessible professions (Lazar et al., 2017). Improving accessibility necessitates community-driven innovation, involving people with disabilities or from marginalized communities in co-design initiatives. This method transforms accessibility from a compliance concern into a strategic opportunity for inclusion and creativity, enhancing usability for all individuals, not solely those with specific needs (Braddock & Parish, 2001). Ultimately, prioritizing accessibility guarantees that automation and human-robot collaboration do not exacerbate existing inequities but function as equalizers of opportunity in the digital economy. It demonstrates a dedication to fair organizational development, ensuring that no individual is marginalized owing to their physical, cognitive, or social circumstances (Floridi, 2018).

FUTURE RESEARCH DIRECTIONS

As automation and human-robot collaboration increasingly influence organizational transformation, future research must tackle essential ethical, social, and technological deficiencies. A primary focus is the creation of context-sensitive ethical frameworks that can adapt dynamically across various industries, geographies, and cultures. Although global standards from organizations like UNESCO or IEEE establish a foundation, further detailed research is essential to comprehend how ethical norms vary across diverse corporate cultures and socio-economic contexts (Jobin et al., 2019). A crucial domain for investigation is the enduring influence of human-robot collaboration on business culture and employee well-being. Contemporary research frequently emphasizes short-term usability or performance indicators; nevertheless, the psychological, emotional, and social ramifications of extended engagement with robots are still inadequately investigated (Robert et al., 2018). Investigating the evolution of trust, dependency, and work satisfaction over time may yield significant insights on sustainable human-robot collaborations. A third promising approach is the auditing of algorithmic justice and fairness. A critical necessity exists for effective, scalable approaches to identify and address biases in automated systems, particularly those influencing hiring, promotion, or workplace surveillance. Researchers ought to explore the implementation of such audits in real-time while ensuring the preservation of performance and privacy (Raji, 2020). Moreover, subsequent study ought to explore the development of emotionally intelligent robotic systems capable of navigating intricate human values while adhering to ethical constraints. This encompasses examining methods to avert emotional manipulation, guarantee informed consent in affective computing, and delineate distinct boundaries between tool and companion (Coeckelbergh, 2020). The significance of education and ethical upskilling remains little explored.

How can organizations provide workers, managers, and developers with the ethical literacy required to make responsible decisions regarding automation? Examining the efficacy of training programs, simulation-based learning, and interdisciplinary ethical teams could substantially improve organizational readiness (McBride et al., 2021). Subsequently, future research ought to embrace a systems-level view, investigating the interplay between automation and human-robot collaboration in relation to global challenges such as climate change, migration, and aging demographics. How may automation impact environmental deterioration, and in what ways can robots assist in care delivery for aging populations while preserving human empathy (Broadbent et al., 2009)? By investigating these new inquiries, can contribute to the development of a more inclusive, accountable, and future-oriented framework for automation within organizational settings.

Emerging Trends in Automation and Human- Robot Synergy

The domain of automation and human-robot collaboration is perpetually advancing, with numerous innovative trends shaping organizational change. A notable trend is the emergence of collaborative robots (cobots), engineered to operate with people in dynamic settings. In contrast to conventional industrial robots that function independently, cobots possess sophisticated safety mechanisms, adaptive artificial intelligence, and user-friendly interfaces, rendering them optimal collaborators in healthcare, logistics, and manufacturing (Colgate, 1996). A significant trend is the incorporation of generative AI and large language models (LLMs) in workplace automation. Instruments such as ChatGPT, Copilot, and Bard are revolutionizing the execution of knowledge work, including tasks from document composition to programming and decision-making. Nonetheless, these tools provoke inquiries regarding intellectual property, employment displacement, and ethical misapplication—compelling businesses to establish new governance rules (Bommasani, 2021). The regulation of ethical AI is increasingly becoming a primary concern. Governments and multilateral organizations are formulating legislation, including the EU AI Act and India's Digital Personal Data Protection Act, which will profoundly affect the design, deployment, and auditing of automation systems (European Commission, 2021). Organizations must proactively address compliance mandates by including ethical assessments into their research and development processes. The field of human-robot interaction is increasingly focused on affective computing and social robotics. Robots are being engineered not just to execute tasks but also to discern, understand, and react to human emotions, thereby enhancing their efficacy in caring, education, and mental health assistance (Breazeal, 2003). This tendency necessitates increased examination of psychological safety, emotional boundaries, and informed consent. Furthermore, decentralized and edge AI architectures are also gaining prominence. These models facilitate the operation of automated systems independent of continuous cloud connectivity, hence enhancing the speed and security of decision-making in real-time contexts such as autonomous vehicles or remote industrial processes (Shi, 2016). This improves performance but introduces new challenges with local accountability and data protection. Ultimately, inclusive design and ethical innovation ecosystems are developing as trends in which organizations collaboratively develop technologies with varied stakeholders, including individuals with disabilities, Indigenous communities, and workers in precarious industries. This transition indicates a Transition from top-down technical imposition to bottom-up ethical partnership (Chock, 2020). These tendencies suggest a future characterized by individualized, decentralized, emotionally intelligent, and ethically regulated automation.

Suggested Areas for Further Research

Notwithstanding considerable advancements in ethical automation and human-robot collaboration, numerous specialized and under examined domains persist as promising avenues for future scholarly and empirical investigation. A significant area of concern is the moral agency of autonomous systems, particularly in collaborative settings where robots independently make judgments in unanticipated circumstances. Despite the presence of theoretical discourse, empirical research on how companies and end-users perceive and react to robotic moral conduct is limited (van Dijk, 2020; Dignum, 2019). Interdisciplinary research on hybrid human-robot teams—focusing on role negotiation, trust calibration, and conflict resolution—is essential for comprehending team dynamics in mixed-intelligence work environments (Rose, 2021). This encompasses examining how human workers modify their communication techniques and expectations over time. A notable research vacuum exists on the influence of algorithmic and robotic technologies on leadership paradigms and workplace governance. As artificial intelligence and robotics increasingly take on supervisory or evaluative functions, it is imperative to investigate their impact on leadership accountability, employee autonomy, and ethical atmosphere (Brynjolfsson & McAfee, 2022; Calo, 2015). The legal frameworks regulating robot decision-making and culpability are currently developing and differ by country. Researchers may investigate harmonization methods, transnational ethical norms, and optimal regulatory approaches that reconcile innovation with the safeguarding of rights (Floridi, 2018; Jobin et al., 2019). An emerging domain is the ethical implementation of automation in public sector services, including policing, social welfare, and healthcare. Future research may investigate the influence of robotic systems on public trust in positions that conventionally necessitate empathy, discretion, or social judgment (Zhang et al., 2019; Dwivedi, 2023). Furthermore, longitudinal research is necessary to investigate the effects of ethical training and education programs in automation on worker preparedness, values alignment, and technological adaptability (Cascio & Montealegre, 2016). Ultimately, ethnographic and cross-sector case studies may provide valuable insights into how firms manage ethical challenges during automation transitions, uncovering both underlying tensions and effective coping strategies. These investigations can yield empirical, contextually relevant data that abstract frameworks frequently neglect. Collectively, these proposed topics necessitate a multidisciplinary study agenda encompassing ethics, robotics, sociology, law, management science, and organizational behavior to thoroughly understand the complexities of this developing field.

Integration of New Technologies

The incorporation of novel technology in the realm of ethical automation and human-robot collaboration is a pivotal catalyst for organizational change. Contemporary organizations are not simply implementing tools; they are fundamentally transforming the conceptualization and execution of work through technologies such as artificial intelligence (AI), collaborative robotics (cobots), and machine learning-based decision systems (Brynjolfsson & McAfee, 2022; Almalki et al., 2016). The ethical incorporation of these technologies necessitates a sensitive approach to socio-technical circumstances, human values, and long-term organizational sustainability (Dignum, 2019). Collaborative robots are employed in industry and healthcare to operate alongside human workers, executing repetitive or hazardous activities while improving safety and efficiency (Colgate, 1996; Broadbent, 2009). The implementation of these systems must incorporate ethical programming, including fail-safes, explainability, and adaptive learning, to prevent harm and preserve human trust (Calo, 20115; Doshi-Velez & Kim, 2017). Furthermore, AI-driven automation solutions that facilitate or inform decisions such as employee assessments, resource distribution, or consumer segmentation necessitate a stringent framework to guarantee fairness, accountability, and transparency (Lazer, 2017). Incorporating these technologies necessitates the application of explainable AI (XAI), algorithmic effect evaluations, and inclusive design methodologies to avert the perpetuation of societal biases (Raji, 2020; Mitchell, 2019). Technologies like Natural Language Processing (NLP), real-time emotion identification, and computer vision are revolutionizing human-robot interaction; yet, these advancements necessitate ethical regulations to safeguard privacy and autonomy (Floridi, 2018; Calo, 2015). Social robots in elder care might enhance emotional well-being; yet, in the absence of stringent ethical constraints, they may inadvertently infantilize users or unethically gather sensitive health information (Burrell, 2016). Ultimately, successful integration is not solely a technological task; it is also a human and organizational undertaking. It necessitates the alignment of technical capabilities with ethical considerations, the education of stakeholders, the assurance of equal access, and the ongoing assessment of impact (Müller & Bostrom, 2022; Chock, 2020). The transformation must be founded on a twin mandate: utilizing innovation for efficiency and growth, while maintaining human dignity, fairness, and organizational integrity.

CONCLUSION

The swift advancement of automation and human-robot collaboration has initiated a new phase of organizational change, characterized by significant opportunities and intricate ethical challenges. This chapter has examined the complex nature of these advances, highlighting that although automation can enhance efficiency, creativity, and global competitiveness, it must be reconciled with human values, inclusivity, and ethical responsibility. The ethical dimensions of automation have transitioned from theoretical issues to tangible challenges, encompassing algorithmic bias, workforce displacement, emotional manipulation, and inequitable access to technology. When tackled judiciously, human-robot synergy has the potential to enhance human talents, increase labor engagement, and stimulate innovation across several industries. This synergy must be established on the principles of trust, transparency, fairness, and inclusivity. Organizations must integrate ethical issues into all aspects of their automation strategy, encompassing system design, governance, deployment, and long-term impact assessment. This chapter delineated conceptual and theoretical frameworks, offered practical solutions, and examined best practices including human-in-the-loop designs, interdisciplinary collaboration, and digital accessibility. We emphasized new trends like as affective computing, collaborative robotics, and edge AI, while pinpointing future study domains including psychological effects, cross-cultural ethics, and sustainable automation. These discoveries collectively indicate a future in which technology and humanity collaborate rather than oppose one other in the development of egalitarian, ethical, and resilient organizations. To effectively address ethical challenges in automation and promote responsible human-robot collaboration, stakeholders including technologists, leaders, employees, policymakers, and researchers must collaborate. Only by collective engagement can we guarantee that technological advancement benefits all individuals, upholds human dignity, and fosters sustainable organizational transformation in the 21st century and beyond.

REFERENCES

Almalki, M., Gray, K., & Martin-Sanchez, F. (2016). The use of self-quantification systems for personal health information: Big data management activities and prospects. *Health Information Science and Systems*, *4*(1), 1–9. PMID: 26019809

Amodei, D., Olah, C., Steinhardt, J., Christiano, P., Schulman, J., & Mané, D. (2016). Concrete problems in AI safety. arXiv preprint arXiv:1606.06565.

Binns, R. (2018). Fairness in machine learning: Lessons from political philosophy. *Proceedings of the 2018 Conference on Fairness, Accountability and Transparency*, 149–159.

Bommasani, R. (2021). *On the Opportunities and Risks of Foundation Models*. Stanford University.

Braddock, D., & Parish, S. L. (2001). An institutional history of disability. Handbook of Disability Studies, 11(1), 11-68. DOI: 10.4135/9781412976251.n2

Breazeal, C. (2003). Emotion and sociable humanoid robots. *International Journal of Human-Computer Studies*, *59*(1–2), 119–155. DOI: 10.1016/S1071-5819(03)00018-1

Broadbent, E., Stafford, R., & MacDonald, B. (2009). Acceptance of healthcare robots for the older population: Review and future directions. *International Journal of Social Robotics*, *1*(4), 319–330. DOI: 10.1007/s12369-009-0030-6

Brynjolfsson, E., & McAfee, A. (2022). Human-compatible automation: Ethical and economic implications for the future workforce. *Journal of Artificial Intelligence Research*, *74*(2), 155–178. DOI: 10.1613/jair.1.13520

Bryson, J. J. (2018). Patiency is not a virtue: AI and the design of ethical systems. *Ethics and Information Technology*, *20*(1), 15–26. DOI: 10.1007/s10676-018-9448-6

Calo, R. (2015). Robotics and the lessons of cyberlaw. *California Law Review*, *103*, 513–563.

Cascio, W. F., & Montealegre, R. (2016). How technology is changing work and organizations. *Annual Review of Organizational Psychology and Organizational Behavior*, *3*(1), 349–375. DOI: 10.1146/annurev-orgpsych-041015-062352

Chock, S. (2020). *Design Justice: Community-Led Practices to Build the Worlds We Need*. MIT Press. DOI: 10.7551/mitpress/12255.001.0001

Coeckelbergh, M. (2010). Health care, capabilities, and AI assistive technologies. *Ethics and Information Technology*, *12*(3), 205–216.

Coeckelbergh, M. (2020). *AI Ethics*. MIT Press. DOI: 10.7551/mitpress/12549.001.0001

Colgate, J. E., Wannasuphoprasit, W., & Peshkin, M. A. (1996). Cobots: Robots for collaboration with human operators. *Proceedings of the ASME Dynamic Systems and Control Division, 58*(1), 433–439. DOI: 10.1115/IMECE1996-0367

Dignum, V. (2019). *Responsible Artificial Intelligence: How to Develop and Use AI in a Responsible Way*. Springer. DOI: 10.1007/978-3-030-30371-6

Doshi-Velez, F., & Kim, B. (2017). Towards a rigorous science of interpretable machine learning. arXiv preprint arXiv:1702.08608.

Dwivedi, Y. K. (2023). Responsible AI and robotics: A review of ethical frameworks, challenges, and governance. *International Journal of Information Management, 69*, 102605. DOI: 10.1016/j.ijinfomgt.2022.102605

European Commission. (2021). *Proposal for a Regulation laying down harmonized rules on artificial intelligence*. Artificial Intelligence Act.

Floridi, L. (2018). Soft ethics and the governance of the digital. *Philosophy & Technology, 31*(1), 1–8. DOI: 10.1007/s13347-018-0303-9 PMID: 30322997

Frey, C. B., & Osborne, M. A. (2017). The future of employment: How susceptible are job to computerisation? *Technological Forecasting and Social Change, 114*, 254–280. DOI: 10.1016/j.techfore.2016.08.019

Hevelke, A., & Nida-Rümelin, J. (2023). Ethical decision-making in autonomous robotic systems: Human oversight and accountability. *Robotics and Autonomous Systems, 162*, 104394. DOI: 10.1016/j.robot.2023.104394

Jaeger, P. T. (2012). *Disability and the Internet: Confronting a digital divide*. Lynne Rienner Publishers.

Jobin, A., Ienca, M., & Vayena, E. (2019). The global landscape of AI ethics guidelines. *Nature Machine Intelligence, 1*(9), 389–399. DOI: 10.1038/s42256-019-0088-2

Lazar, J. (2017). *Ensuring digital accessibility through process and policy*. Morgan Kaufmann.

McBride, N., Stahl, B. C., & Flick, C. (2021). Teaching responsible AI: Integrating ethics into technology education. *Communications of the ACM, 64*(4), 33–35.

Mitchell, M. (2019). Model cards for model reporting. *Proceedings of the Conference on Fairness, Accountability, and Transparency*, 220–229. DOI: 10.1145/3287560.3287596

Müller, V. C., & Bostrom, N. (2022). Future progress in artificial intelligence: Extent and limits of automation. *AI & Society*, *37*(4), 1239–1254. DOI: 10.1007/s00146-021-01191-0

Pereira, A. G., & Liu, S. (2024). Human-robot collaboration in digital workplaces: Ethical risks and organizational safeguards. *Journal of Technology Ethics*, *5*(1), 22–39.

Rahwan, I. (2018). Society-in-the-loop: Programming the algorithmic social contract. *Ethics and Information Technology*, *20*(1), 5–14. DOI: 10.1007/s10676-017-9430-8

Raji, I. D. (2020). Closing the AI accountability gap: Defining an end-to-end framework for internal algorithmic auditing. *Proceedings of the 2020 Conference on Fairness, Accountability, and Transparency*, 33–44. DOI: 10.1145/3351095.3372873

Robert, L. P., Dennis, A. R., & Ahuja, M. (2018). Social robots that tease: Understanding and leveraging humor in human–robot interaction. *Information Systems Research*, *29*(3), 525–545. DOI: 10.1287/isre.2018.0773

Robins, B., & Dautenhahn, K. (2014). Tactile interactions with a humanoid robot: Novel play scenario implementations with children with autism. *International Journal of Social Robotics*, *6*(3), 397–415. DOI: 10.1007/s12369-014-0228-0

Rose, D. C. (2021). Policy windows for the future of agriculture: Applying the multiple streams framework to the adoption of digital agriculture. *Journal of Rural Studies*, *82*, 226–236.

Shi, W., Cao, J., Zhang, Q., Li, Y., & Xu, L. (2016). Edge computing: Vision and challenges. *IEEE Internet of Things Journal*, *3*(5), 637–646. DOI: 10.1109/JIOT.2016.2579198

Story, M. F., Mueller, J. L., & Mace, R. L. (1998). *The universal design file: Designing for people of all ages and abilities*. Center for Universal Design.

Thomson, R., & Alvarez, P. (2024). AI-powered automation and the transformation of organizational structures: A socio-ethical analysis. *Journal of Organizational Transformation & AI*, *12*(2), 44–63.

van Dijk, J. (2020). *The Digital Divide*. Polity Press.

Wang, Y., Yu, C., & Fesenmaier, D. R. (2020). Accessibility and its ethical implications in digital tourism. *Tourism Management Perspectives*, *33*, 100606.

Zhang, B., Dafoe, A., & Dafoe, A. (2019). Artificial intelligence: American attitudes and trends. Center for the Governance of AI, Future of Humanity Institute, University of Oxford.

KEY TERMS AND DEFINITIONS

Affective Computing: Technology capable of identifying and reacting to human emotions.

Algorithmic Auditing: The evaluation of AI systems for equity and transparency.

Automation: Automation refers to the application of technology to execute tasks autonomously, without human involvement.

Bias in AI: Systematic and unjust prejudice in algorithmic determinations.

Collaborative Robots (Cobots): Robots engineered to operate securely in conjunction with humans.

Digital Accessibility: The creation of technologies that are usable by individuals with diverse abilities.

Digital Inequality: Discrepancies in access to digital technologies.

Edge AI: Edge AI refers to artificial intelligence systems that function locally on devices, independent of cloud infrastructure.

Emotional Manipulation: Utilizing technology to exploit emotional responses.

Ethical AI: Ethical AI refers to artificial intelligence created and implemented in accordance with human values and ethical standards.

Ethical Governance: Frameworks that guarantee principled decision-making.

Ethical Pluralism: A framework that recognizes the existence of various legitimate ethical viewpoints.

Human-in-the-Loop: Maintaining human participation in essential decision-making within automated systems.

Human-Robot Synergy: Cooperative engagement between humans and robots to enhance results.

Interdisciplinary Collaboration: The cooperation among diverse academic or professional disciplines.

Participatory Design: An inclusive design methodology that engages stakeholders in the process.

Sustainable Automation: Automation that is congruent with environmental and social sustainability.

Transparency by Design: System creation that guarantees comprehensible decisions and procedures.

Universal Design: A design methodology that guarantees products are accessible to all individuals.

Chapter 8
Building Trust in the Age of Intelligent Machines:
Rethinking Human–Robot Synergy

Satya Subrahmanyam
https://orcid.org/0000-0003-0441-2742
Holy Spirit University of Kaslik, Lebanon

ABSTRACT

In the age of AI and robotics, human-machine trust plays a pivotal role in the successful integration of technology into organizational workflows. This chapter explores the multidimensional nature of trust in human-machine collaboration, influenced by psychological, technical, and organizational factors. It delves into the significance of trust, key determinants like system reliability and transparency, and the challenges in fostering trust due to the absence of human-like qualities in machines. The chapter examines theoretical frameworks from psychology, human-computer interaction, and human-robot interaction, and offers insights into measuring and enhancing trust. It emphasizes the need for dynamic trust-building strategies, ethical considerations, and continuous calibration to support sustainable AI-driven collaboration, ensuring operational efficiency and reducing resistance to technological adoption.

DOI: 10.4018/979-8-3373-3404-2.ch008

1. INTRODUCTION: THE IMPERATIVE OF TRUST IN THE HUMAN-ROBOT ERA

1.1 The Growing Integration of Intelligent Machines and Robots in Organizational Ecosystems

In the contemporary digital landscape, intelligent machines and robots have evolved from peripheral tools to core elements within organizational ecosystems. From automated customer service agents to industrial cobots and algorithmic decision-making systems, artificial intelligence (AI) and robotics are transforming how work is structured, managed, and delivered (Brynjolfsson & McAfee, 2017). Organizations across manufacturing, healthcare, logistics, and even education increasingly rely on robotic systems to enhance productivity, precision, and scalability. This integration, often termed *Industry 5.0*, emphasizes a human-centric approach where technology augments rather than replaces human capabilities (Xu et al., 2021). As the boundaries between human and machine roles blur, the success of these collaborations hinges not only on technological sophistication but also on the relational dimension of *trust* between humans and intelligent systems.

1.2 From Cooperation to Cognitive Collaboration: Defining Human-Robot Synergy

Human-robot synergy represents a transformative paradigm in which humans and robots engage in dynamic, goal-oriented interaction that transcends simple cooperation. Traditional human-machine cooperation focused on functional task division, where robots performed repetitive physical tasks while humans handled cognitive ones. Today, advances in AI and machine learning have enabled *cognitive collaboration*, where robots can perceive, learn, and make autonomous decisions in tandem with human partners (Hancock et al., 2020). Autonomous vehicles collaborate with human drivers through adaptive assistance systems, while AI diagnostic tools work alongside clinicians in healthcare to interpret complex data (Topol, 2021). This new form of synergy requires mutual adaptation - humans must understand machine logic and limitations, while machines must interpret human intent and behavior. The success of this interdependence is contingent upon trust, which determines whether humans are willing to rely on, delegate to, or accept the guidance of intelligent systems.

1.3 The Centrality of Trust in Effective Human-Machine Interaction

Trust serves as the psychological and social glue that enables effective interaction between humans and intelligent machines. Without trust, even the most advanced robotic systems can face rejection, underutilization, or misuse. Trust in this context is defined as the willingness of a human to rely on a machine's actions under conditions of uncertainty (Subrahmanyam, 2025). Studies show that when humans perceive robots as competent, reliable, and aligned with their goals, they are more likely to engage positively and collaboratively (Glikson & Woolley, 2020).

On the other hand, mistrust - whether due to technical failures, opaque algorithms, or perceived threat - can undermine acceptance and performance. The "automation paradox" highlights this challenge: while automation promises efficiency, overtrust can lead to complacency and system vulnerability, while distrust can result in rejection and inefficiency (Chiou & Lee, 2021). Thus, achieving the right calibration of trust - neither excessive nor deficient - is fundamental to ensuring safe, ethical, and productive human-robot collaboration.

1.4 Purpose and Scope of the Chapter

This chapter explores the formation, calibration, and maintenance of trust in human-robot collaboration, emphasizing its role in driving organizational transformation. It argues that trust is not static but evolves through continuous interaction, transparency, and feedback. The discussion will unpack the stages of trust formation - initial trust based on design and reputation, dynamic trust shaped through experience, and sustained trust reinforced by consistent performance and adaptability. It will examine how organizations can leverage trust to optimize the balance between human oversight and machine autonomy, a balance essential for managing the autonomy–trust paradox (Hancock et al., 2011). Empirical cases from domains such as autonomous vehicle systems, healthcare robotics, and AI-driven decision-making will illustrate practical mechanisms for building and maintaining trust.

This chapter posits that trust is the cornerstone of human-robot synergy in the intelligent age. As robots become not just tools but cognitive collaborators, organizations must cultivate environments where trust is systematically designed, ethically grounded, and continuously nurtured. Doing so will not only enhance human-robot performance outcomes but also shape the moral and social fabric of the future workplace.

2. CONCEPTUAL FOUNDATIONS OF HUMAN-MACHINE TRUST

Understanding human-machine trust requires examining its conceptual and theoretical underpinnings within socio-technical systems. This section explores how trust between humans and intelligent machines differs from interpersonal trust, highlighting key theories such as Luhmann's trust theory, Mayer et al.'s organizational trust model, and frameworks from Human-Automation and Human-Robot Interaction that define trust's cognitive and affective dimensions.

2.1 Defining Trust in Socio-Technical Systems

In the context of socio-technical systems, *trust* serves as a critical construct that facilitates effective collaboration between humans and intelligent machines. While traditional notions of trust have focused primarily on interpersonal relationships, the rise of artificial intelligence (AI) and robotics has expanded this concept to encompass human-machine interactions. Human-human trust typically emerges from social, emotional, and moral factors - empathy, integrity, and shared experience - whereas *human-machine trust* is grounded in perceptions of technological reliability, predictability, and transparency (Glikson & Woolley, 2020). In socio-technical environments, machines act not only as tools but also as autonomous partners capable of decision-making and learning. Humans must develop a cognitive form of trust that is based less on emotional reciprocity and more on confidence in the machine's design, data integrity, and consistent performance (Hoff & Bashir, 2015). This shift signifies that trust in automation is not merely about believing in machine competence but also about understanding the logic and constraints that underpin its behavior.

2.2 Theoretical Underpinnings

Luhmann's Theory of Trust and Its Application to Automation

Niklas Luhmann's sociological theory of trust provides a foundational lens for understanding trust in socio-technical contexts. Luhmann (1979) conceptualized trust as a mechanism that reduces social complexity by allowing individuals to act in the presence of uncertainty. Applied to automation, this framework implies that humans place trust in machines to simplify complex decision-making environments where complete control or understanding is impossible (Schaefer et al., 2016). When an autonomous system recommends financial decisions or medical diagnoses, users rely on trust to bridge the cognitive gap between their limited understanding and the system's algorithmic reasoning. Luhmann's theory thus underscores the relational aspect of trust, humans must perceive that the machine operates within predictable boundaries and aligns with expected norms of reliability and safety. This theoretical lens helps explain why transparency, interpretability, and accountability are vital to fostering trust in automated systems.

Mayer, Davis, and Schoorman's Integrative Model of Organizational Trust

Mayer, Davis, and Schoorman's (1995) integrative model of organizational trust remains one of the most influential frameworks for analyzing trust dynamics, and it has been effectively adapted to human-machine relationships. The model posits that trust is influenced by three key antecedents: *ability*, *benevolence*, and *integrity*. In a human-machine context, *ability* translates to the perceived technical competence of the system - its accuracy, functionality, and effectiveness. *Benevolence* reflects the perception that the machine's actions are aligned with the user's best interests, which is increasingly relevant in ethical AI and safety-critical systems. *Integrity* refers to the degree to which the system adheres to transparent, fair, and consistent operational principles (Hancock et al., 2011). The model suggests that trust is dynamic and contingent on repeated interactions and feedback loops; trust grows when systems perform reliably and diminishes when they fail unpredictably. By integrating these dimensions, organizations can design human-robot interactions that cultivate sustained trust through iterative validation and ethical alignment.

Human-Automation Interaction (HAI) and Human-Robot Interaction (HRI) Frameworks

The fields of Human-Automation Interaction (HAI) and Human-Robot Interaction (HRI) offer applied perspectives for understanding trust in technological systems. HAI research emphasizes *appropriate reliance* - the optimal balance between overtrust (misuse) and distrust (disuse) of automation (Chiou & Lee, 2021). Within this framework, trust is conceptualized as a calibration process, where humans continuously adjust their reliance based on feedback and observed performance. HRI, on the other hand, focuses on social and cognitive factors in human-robot collaboration, recognizing that physical embodiment, anthropomorphic design, and communication cues shape trust perception (Hancock et al., 2011). Robots that communicate uncertainty provide explanations for their actions or exhibit adaptive learning behaviors are more likely to elicit calibrated and sustained trust (Bobko et al., 2022). Together, these frameworks highlight the multidimensional nature of trust as both a psychological construct and a system design challenge.

2.3 Dimensions of Trust in Machines: Competence, Predictability, Transparency, and Ethical Alignment

Trust in intelligent machines is not monolithic; it is composed of multiple interrelated dimensions. *Competence* refers to the perceived capability of a machine to perform its intended functions effectively and accurately. Systems that demonstrate consistent performance and error management foster higher trust (Hancock et al., 2011). *Predictability* involves consistency and stability of the machine's behavior over time. Humans are more likely to trust systems that act within expected parameters and produce reproducible outcomes (Daronnat et al., 2021). *Transparency* encompasses the extent to which users can understand how and why a machine makes specific decisions. Explainable AI (XAI) initiatives aim to enhance this dimension by making algorithmic reasoning accessible to non-experts (Samek & Müller, 2021). Finally, *ethical alignment* pertains to the degree to which machine behavior conforms to human values, fairness, and accountability standards. Ethical lapses - such as bias in AI algorithms or opaque decision-making - can severely erode trust (Rahwan et al., 2019). Thus, cultivating these four dimensions is essential for designing trustworthy socio-technical systems.

2.4 Cognitive vs. Affective Trust: Balancing Rational and Emotional Dimensions

Trust in machines can be broadly categorized into *cognitive* and *affective* dimensions. *Cognitive trust* arises from rational assessment - users trust a system because it demonstrates reliability, competence, and consistency. In contrast, *affective trust* is rooted in emotional attachment and perceived rapport, often influenced by the machine's anthropomorphic features or communication style (Glikson & Woolley, 2020). While human-machine interactions have historically prioritized cognitive trust, emerging research suggests that affective elements also play a crucial role, particularly in service robots and social AI contexts (Bobko et al., 2022). Emotional design features - such as empathetic responses, tone modulation, or expressive gestures - can foster user comfort and enhance trust resilience during system failures. Overreliance on affective cues without corresponding competence risks creating deceptive or misplaced trust. Therefore, effective trust calibration requires balancing rational evaluation with emotional engagement, ensuring that users both *understand* and *feel confident* in their technological counterparts.

The conceptual foundations of human-machine trust lie at the intersection of sociological, organizational, and cognitive theories. Trust functions as the bridge between human uncertainty and machine autonomy, determining whether humans will embrace or resist intelligent systems. By grounding trust in competence, predictability, transparency, and ethical integrity, while balancing cognitive and affective dimensions, organizations can cultivate resilient and adaptive forms of human-machine collaboration that are essential for the intelligent era.

3. DETERMINANTS OF HUMAN-MACHINE TRUST

The establishment and maintenance of trust in human-machine collaboration depend on a combination of technological, organizational, psychological, and ethical factors. As intelligent systems become more autonomous and embedded in decision-making processes, understanding the determinants of trust becomes essential for effective adoption and sustained use. These determinants interact dynamically - technical design shapes user perception, while organizational culture and ethical governance influence acceptance and reliance.

3.1 Technical Determinants: Reliability, Accuracy, and System Performance

Technical determinants form the foundation of human-machine trust. Reliability - defined as the consistency and dependability of a system's output - is central to whether users perceive technology as trustworthy. When a robot or AI consistently performs tasks with minimal errors, users develop confidence in their capabilities (Hancock et al., 2020). System failures or erratic performance can quickly erode trust, even if the system performs well most of the time.

Accuracy also plays a pivotal role; systems that deliver precise results aligned with user expectations enhance trustworthiness. In high-stakes domains such as healthcare and aviation, even minor inaccuracies can undermine confidence (Subrahmanyam, 2025). System performance - including speed, adaptability, and resilience - affects user satisfaction and reliance. Advanced AI systems that demonstrate learning capacity and self-correction foster greater trust by signaling intelligence and reliability (Daronnat et al., 2021). Thus, technical robustness and consistent performance form the core structural basis upon which trust is built.

3.2 Design and Interface Determinants: Explainability, Usability, and Human-Centered Design

The way users interact with machines significantly shapes their trust perceptions. *Explainability* - the degree to which a system can articulate its reasoning and decision-making process - is crucial for trust calibration. Explainable AI (XAI) enhances user understanding, reduces uncertainty, and mitigates the "black box" perception associated with opaque algorithms (Samek & Müller, 2021). When users comprehend why a system made a specific recommendation or decision, they can more accurately assess its reliability. *Usability* - encompassing intuitive interfaces, accessible feedback mechanisms, and adaptive interaction - fosters trust by making systems approachable and controllable.

Poorly designed interfaces that overwhelm or confuse users often generate frustration and distrust (Glikson & Woolley, 2020). Human-centered design further reinforces trust by aligning system functions with human cognitive patterns, emotional responses, and ergonomic needs. Robots designed with empathetic communication cues or familiar human-like gestures promote a sense of comfort and rapport. Hence, transparent, user-friendly, and empathetically designed systems are more likely to earn and sustain human trust.

3.3 Organizational Determinants: Culture, Leadership, Communication, and Training

Beyond technology design, the organizational environment strongly influences human-machine trust. An organizational culture that values innovation, openness, and learning encourage employees to experiment with and trust new technologies (Schaefer et al., 2016). Leadership plays an equally vital role: when leaders model confidence in technology adoption and communicate its purpose and benefits clearly, employees are more inclined to trust and embrace automation (Brynjolfsson & McAfee, 2017).

Effective communication about system capabilities, limitations, and safeguards also fosters realistic expectations and mitigates fear of technological displacement. *Training* is a key enabler, equipping users with knowledge about how AI or robotic systems function enhances their perceived control and competence, reducing anxiety and resistance (Chiou & Lee, 2021). Organizations that invest in continuous skill development and human-robot collaboration training thus create a psychological environment conducive to mutual trust between humans and machines.

3.4 Psychological and Social Determinants: Perception, Control, and Comfort with Autonomy

Human trust in machines is also shaped by cognitive and emotional factors. Perceived *control* - the extent to which individuals feel they can influence or override machine decisions - is a strong predictor of trust. Systems that allow users to intervene or adjust parameters tend to evoke higher trust levels than fully autonomous systems (Subrahmanyam, 2025). *Comfort with autonomy* varies across individuals and is influenced by factors such as prior experience with technology, personality traits, and cultural attitudes toward automation (Hoff & Bashir, 2015).

Users who view AI as augmentative rather than substitutive are more likely to trust and collaborate with machines. *Social perception* also matters; the anthropomorphic design of robots, such as expressive faces or natural speech patterns, can enhance trust by eliciting empathy and familiarity. Over-humanization risks creating unrealistic expectations, which can backfire when the machine fails to meet emotional or behavioral cues. Hence, balancing autonomy, user control, and realistic social interaction is essential for sustaining trust.

3.5 Ethical and Regulatory Determinants: Transparency, Accountability, and Data Integrity

In the era of AI-driven decision-making, ethical and regulatory considerations have become integral to human-machine trust. *Transparency* in algorithmic processes and data usage is critical - users must understand how information is collected, processed, and applied. Lack of transparency can lead to suspicion, particularly in sensitive areas like hiring or healthcare (Rahwan et al., 2019). *Accountability* mechanisms, including clear responsibility frameworks for system errors or biases, reinforce public confidence in AI governance. When users know that ethical oversight exists, they are more likely to engage with trust (Glikson & Woolley, 2020). Finally, *data integrity* - ensuring that information used by machines is accurate, secure, and free from manipulation - is vital for sustaining credibility. Regulatory frameworks such as the EU's AI Act and data protection regulations underscore the importance of ethical design and governance for maintaining trust at both individual and societal levels.

Figure 1. Determinants of Human–Machine Trust

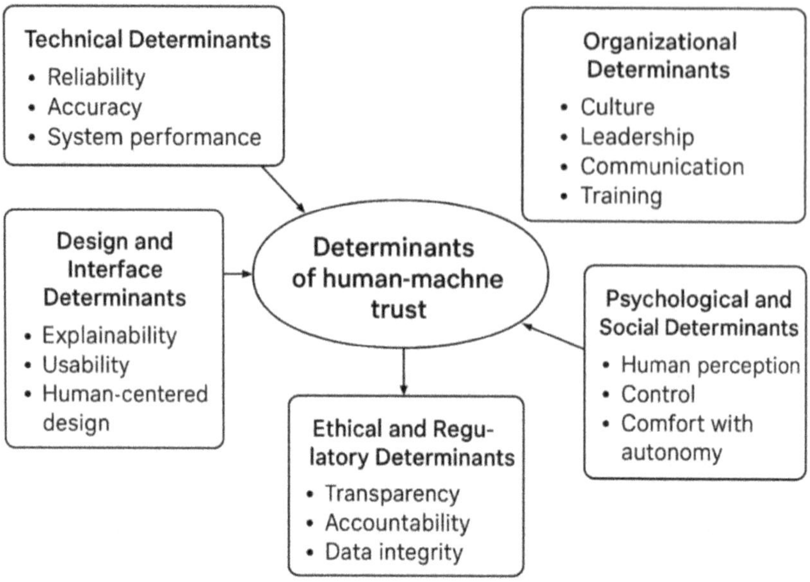

Author, (2025)

The determinants of human-machine trust span multiple dimensions—technical reliability, design transparency, organizational leadership, psychological comfort, and ethical governance. Trust cannot be engineered solely through advanced algorithms; it must be cultivated through holistic design, transparent communication, and responsible management. As machines become more autonomous and embedded in decision-making, fostering trust will remain central to achieving effective and ethical human-robot collaboration.

4. TRUST FORMATION AND CALIBRATION IN HUMAN-ROBOT COLLABORATION

Trust formation and calibration are central to achieving effective human-robot collaboration. As intelligent machines assume greater autonomy, understanding how trust develops, evolves, and stabilizes becomes essential. This section explores the progressive stages of trust, mechanisms for aligning human perception with machine capabilities, and strategies for sustaining balanced, adaptive trust in dynamic collaborative environments.

4.1 Stages of Trust Formation: Initial, Dynamic, and Sustained Trust

Trust formation in human-robot collaboration is a gradual process that evolves through distinct stages - initial, dynamic, and sustained trust. *Initial trust* is often based on user expectations, design cues, and prior experiences with technology. At this stage, visual design, interface aesthetics, and brand reputation play a significant role in shaping first impressions (Hancock et al., 2011). Robots with human-like features or intuitive interfaces tend to evoke higher initial trust by reducing perceived unfamiliarity. In contrast, systems that appear overly complex or opaque may trigger skepticism.

As collaboration progresses, *dynamic trust* emerges through direct experience and interaction. This phase involves continuous evaluation of the robot's behavior, reliability, and responsiveness to human input (Bobko et al., 2022). Trust can fluctuate depending on the robot's performance consistency and the user's ability to interpret its decision-making patterns. In industrial settings, if a robot consistently executes tasks safely and accurately, human operators are more likely to rely on it. Performance errors or unpredictable actions can erode confidence. Finally, *sustained trust* develops when the robot demonstrates long-term reliability, adaptability, and transparency. This stage is characterized by stable reliance, where the human no longer questions the robot's competence or motives (Glikson & Woolley, 2020). Sustained trust represents the culmination of effective design, consistent performance, and positive interaction history.

4.2 Trust Calibration: Aligning Perceived and Actual Machine Capabilities

Trust calibration refers to the alignment between a user's perceived trust in a system and the system's actual capabilities. Overtrust can lead to misuse, where humans rely excessively on automation, while undertrust results in disuse or rejection of valuable assistance (Chiou & Lee, 2021). Proper calibration ensures that trust levels correspond with system reliability and operational boundaries. In aviation, pilots using autopilot systems must understand both the strengths and limitations of automation to intervene appropriately.

Empirical evidence underscores the importance of transparent communication in achieving calibration. A study by Daronnat and others (2021) found that users who received real-time updates about a robot's confidence level in its decisions exhibited better-calibrated trust compared to those without such feedback. Likewise, when healthcare professionals interacted with diagnostic AI systems that explained their reasoning, their trust aligned more closely with the system's true accuracy (Topol,

2021). Thus, trust calibration is not merely a technical concern, it requires deliberate interface design and user education to synchronize perception with reality.

4.3 The Role of Feedback Loops: Continuous Learning Between Human and Machine

Feedback loops play a pivotal role in maintaining adaptive trust in human-robot systems. In an effective collaboration, both humans and machines engage in mutual learning, humans adjust their reliance based on the machine's performance, while robots refine their responses through human input. This bidirectional feedback fosters transparency and adaptability (Schaefer et al., 2016).

In robotic surgery, surgeons receive haptic and visual feedback from robotic instruments, which allows them to adjust movements and improve precision. The robotic system learns from human corrections to enhance its algorithms for future procedures. This process exemplifies "shared calibration," where trust evolves dynamically through reciprocal adaptation. Continuous feedback supports error detection and repair, critical factors in rebuilding trust after failure. As Subrahmanyam (2025) notes, timely acknowledgment and recovery from errors significantly improve trust resilience in automated systems.

4.4 Human Oversight and Control: Managing the Autonomy–Trust Paradox

A central challenge in human-robot collaboration is managing the *autonomy–trust paradox*, as robots become more autonomous, human oversight becomes both less frequent and more critical. Excessive autonomy can diminish human engagement, leading to complacency or overtrust, while too much oversight can hinder efficiency and innovation (Glikson & Woolley, 2020). The solution lies in designing systems that preserve *meaningful human control*.

Empirical studies in autonomous vehicle testing illustrate this balance. Drivers who received clear situational updates and had accessible manual override options maintained higher situational awareness and more appropriate trust levels (Walker et al., 2018). When automation concealed decision-making or restricted human intervention, drivers became disengaged and slow to react during emergencies. In healthcare, robotic assistants that allowed clinicians to confirm or adjust treatment recommendations encouraged collaboration rather than blind reliance. Thus, calibrated autonomy, supported by oversight mechanisms, ensures accountability without undermining trust.

Figure 2. Trust Formation and Calibration in Human-Robot Collaboration

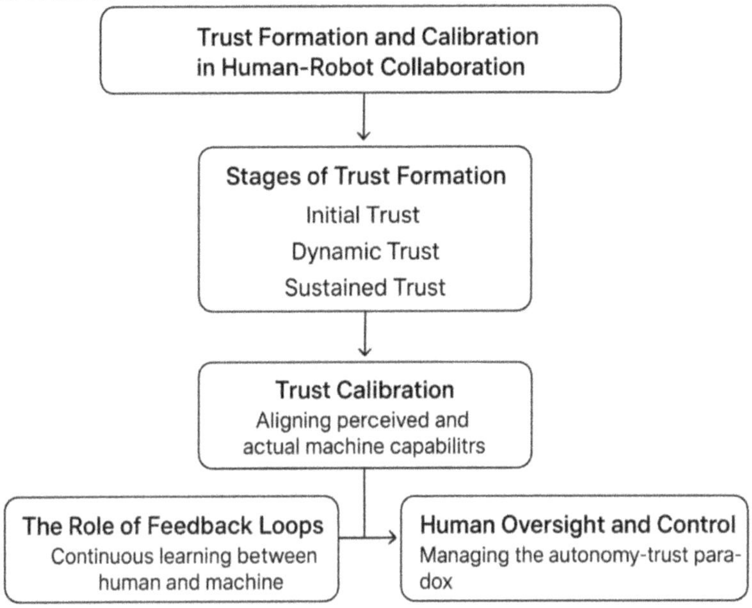

Author, (2025)

4.5 Case Insight: Trust-Building Mechanisms in Autonomous Vehicle Systems and Healthcare Robotics

Trust-building in human–robot collaboration becomes especially tangible when explored through real-world implementations. In domains such as autonomous vehicles and healthcare robotics, mechanisms like transparency, feedback, and shared control are critical for aligning human expectations with machine performance. These applications illustrate how users and organizations negotiate reliance, accountability, and shared autonomy with intelligent systems.

Trust Mechanisms in Autonomous Vehicle Systems

Autonomous vehicles (AVs) exemplify the complex challenge of calibrating human trust in adaptive, semi-autonomous systems. Research shows that users' trust depends not only on system reliability but also on perceived transparency, situational awareness, and predictability (Lee & See, 2004). When users understand the reasoning behind a vehicle's decisions—such as lane changes or braking, they are more likely to develop appropriately calibrated trust.

Empirical studies have demonstrated that providing drivers with feedback on system status and intent improves both situational awareness and trust (Petersen et al., 2019). Trust evolves dynamically over time, depending on driver experience and system performance consistency (Ayoub et al., 2021).

The autonomy–trust paradox is evident in real-world cases such as Tesla's Autopilot. Overtrust has occasionally led to driver complacency, contributing to safety incidents (National Transportation Safety Board [NTSB], 2020). To counteract this, driver monitoring systems are now being deployed to ensure sustained human engagement. This illustrates that both overreliance and underreliance can be harmful - trust must be continuously calibrated through transparency, feedback, and human oversight mechanisms.

Effective trust-building in AVs depends on visible decision transparency, system feedback loops, human-in-the-loop control, and consistent communication of system boundaries. These features collectively bridge the cognitive gap between human understanding and algorithmic behavior.

Trust Mechanisms in Healthcare Robotics

In healthcare, trust dynamics are even more critical due to patient safety and the ethical implications of clinical decision-making. The integration of robotic systems such as the *da Vinci Surgical System* has transformed surgical practice, where surgeons' trust develops progressively through training, feedback, and operational familiarity. Research highlights that trust in robotic surgery is associated with faith in technology, designer reputation, surgeon competence, and perceived reliability (Szabó et al., 2024). Fear of robotic error remains the most significant negative predictor of acceptance.

Simulation-based training has been shown to enhance surgeon confidence and reduce anxiety in robotic-assisted procedures by providing safe environments to understand the robot's capabilities and limitations (Lin et al., 2022). Hospitals increasingly employ team-based integration protocols that emphasize robots as collaborators rather than replacements, reinforcing collective accountability and shared control.

In eldercare and rehabilitation contexts, socially assistive robots (SARs) such as Pepper and Paro support emotional well-being, medication adherence, and social engagement. Studies show that emotional intelligence - such as empathic verbal and non-verbal behaviors - enhances affective trust and interaction quality among older adults (Abdollahi et al., 2022). Yet, trust must be safeguarded by maintaining transparency about data use and system limitations, preventing overpersonalization that could undermine credibility.

Comparative Insights

Both autonomous vehicle and healthcare robotics sectors reveal that trust is dynamic, context-dependent, and must be continuously calibrated. In AVs, trust is primarily centered on safety, control, and transparency, while in healthcare robotics, it focuses on competence, empathy, and accountability. Across both fields, mechanisms like real-time feedback, human oversight, and transparent communication are foundational to developing trust that is neither excessive nor deficient.

Importantly, trust-building is not achieved through technology alone - it must be institutionalized through governance, training, and ethical norms. The convergence of human and machine intelligence therefore represents not just technological progress, but a co-evolution of collaborative intelligence, where trust functions as both enabler and regulator of shared autonomy.

Trust formation and calibration in human-robot collaboration involve dynamic interplay among design, performance, feedback, and ethical oversight. Effective trust arises when users experience consistent reliability, understand system logic, and retain control over autonomous processes. As the boundary between human and machine cognition continues to blur, organizations must treat trust not as an incidental byproduct but as a strategic outcome of thoughtful design, communication, and shared learning.

5. ORGANIZATIONAL TRANSFORMATION THROUGH TRUSTWORTHY HUMAN-ROBOT SYNERGY

The integration of intelligent machines and robots into organizational systems has redefined the nature of work, leadership, and collaboration. Trust, as the core enabler of human-robot synergy, facilitates not only efficient task execution but also deeper organizational transformation. When organizations cultivate trust between humans and machines, they unlock higher levels of innovation, adaptability, and resilience - elements crucial in an era marked by rapid technological disruption.

5.1 Trust as an Enabler of Transformation

Trust in human-robot collaboration serves as a strategic asset that enhances an organization's capacity for transformation. In socio-technical systems, trust determines whether humans are willing to rely on autonomous systems to perform critical functions (Bobko et al., 2022). A trustworthy system allows human employees to shift from monitoring to strategic decision-making, enabling innovation and operational agility. In Amazon's fulfillment centers, the integration of trusted robotic systems such as Kiva robots has not only optimized logistics but also transformed workforce dynamics, leading to new models of human- robotic cooperation (Bhargava et al., 2021). When employees trust robots to perform repetitive or hazardous tasks reliably, they can redirect their efforts toward creative and analytical roles - fueling innovation and adaptability.

Empirical research supports this view. Hancock et al. (2020) found that trust mediates the relationship between automation reliability and human performance in collaborative contexts. Organizations that foster transparency, ethical governance, and continuous feedback mechanisms tend to exhibit higher levels of human-machine synergy, thereby facilitating sustainable transformation.

5.2 Collaborative Intelligence (CQ): Merging Human Intuition and Machine Precision

The emergence of Collaborative Intelligence (CQ) represents the next evolution in human-robot relationships. CQ emphasizes the complementarity between human intuition, creativity and emotional reasoning, and machine precision, speed, and data-driven analysis (Wilson & Daugherty, 2022). Trust acts as the bridge that enables humans to delegate analytical or predictive tasks to machines while maintaining agency over final decisions.

An empirical illustration can be seen in Siemens' deployment of AI-driven robotics in manufacturing plants. These systems collaborate with human technicians to monitor equipment performance and predict maintenance needs. The trusted relationship between human expertise and machine intelligence has improved operational efficiency by reducing downtime and enhancing predictive accuracy (Siemens et al., 2025). This synergy highlights that organizational transformation does not arise from replacing humans but from empowering them to work smarter alongside intelligent systems.

5.3 Redefining Roles and Responsibilities: From Operators to Collaborators

As trust in machines grows, employees' roles evolve from operators to collaborators and decision partners. Instead of viewing automation as a threat, trust-centered organizations position robots as teammates that enhance human capability. This transformation requires redesigning work processes and redefining accountability structures (Rahwan et al., 2019).

In Toyota's production systems, for instance, robots are trusted to execute repetitive assembly tasks while human workers oversee quality and customization. This trust-based division of labor improves both precision and adaptability (Toyota UK Magazine, 2021). By promoting psychological safety and technological confidence, organizations can mitigate resistance and build a culture where employees see machines as enablers rather than replacements.

5.4 Organizational Structures for Human-Robot Coexistence

Building a trust-based organizational structure entails developing new workflows, governance mechanisms, and decision-support systems. Trust-based workflows rely on transparent data sharing, explainable AI interfaces, and ethical oversight mechanisms (Subrahmanyam & Baz, 2025). Decision augmentation, rather than automation, becomes the primary model - where algorithms provide recommendations, but humans retain control over outcomes.

In logistics and service industries, DHL's "Smart Warehouse" initiative illustrates this transformation. Human workers collaborate with autonomous mobile robots that transport goods and update inventory in real-time. Transparent communication interfaces and safety assurance mechanisms foster mutual trust, allowing both humans and robots to work seamlessly (DHL Supply Chain. (2025). Alike, in healthcare, robotic systems such as da Vinci surgical robots have restructured surgical teams, emphasizing co-governance between human surgeons and robotic assistants - where trust in machine precision complements human judgment (Pelikan et al., 2018).

5.5 Case Insight: Trust-Centered Transformation Across Industries

Trustworthy human-robot synergy is reshaping industries by enhancing efficiency, safety, and adaptability while fostering collaborative intelligence between humans and machines. As organizations integrate robotics and artificial intelligence (AI) into their operations, trust becomes the central mechanism enabling smooth collaboration, innovation, and sustainable transformation. Real-world cases in manufacturing, logistics, and services illustrate how organizations build and maintain this trust to achieve competitive advantage.

Manufacturing: BMW Group's Human-Robot Collaboration

The manufacturing sector has been at the forefront of human-robot collaboration, particularly through collaborative robots (cobots) designed to work safely alongside human operators. BMW installed cobots in their production lines that operate in shared human workspaces without traditional fencing, allowing humans and machines to work side-by-side under enhanced safety frameworks (BMW Group, 2013).

In one prototype at BMW's motorcycle plant in Berlin, a collaborative robot with force control and vision processed bolting tasks while humans performed adjacent quality assurance steps - illustrating how human-robot work-sharing fosters trust, reduces physical strain, and increases productivity (Fraunhofer IPK, as reported by BMW) via human-robot collaboration (HRC) teams. These implementations suggest that workforce training, transparent robot behaviour (e.g., safe stops on human contact, clear work-scope boundaries), and participatory workflow design contribute to trust formation. The human operators' involvement in workflow design (e.g., co-design of tasks) and their experience working with cobots appears critical for trust calibration and collaborative performance.

Logistics: Amazon Robotics and Warehouse Optimization

In the logistics domain, the transformation driven by robotics is epitomized by Amazon's large-scale deployment of robotic systems in its fulfilment operations. Amazon has deployed more than one million robots across its facilities, working in tandem with human workers to pick, stow and move inventory (WSJ, 2025). Studies documenting the underlying algorithms show that Amazon improved human-robot collaboration by redesigning robotic picking algorithms, reducing robot travel distance by 62% and drive counts by 31%--which enhanced system predictability and thereby human trust in the robotic systems (Allgor et al., 2023).

From a trust-mechanism perspective, the logistics case reveals that: (1) predictable robot behaviour (structured patterns) reduces human uncertainty; (2) transparent robot-human interface or visual cues (robots signalling navigation paths, path planning) enhance situational awareness; (3) continuous performance monitoring and human oversight (humans remain in the loop) maintain appropriate trust levels and prevent over-reliance. This suggests that human-robot trust in logistics emerges through technical competence, transparency of operations, and balanced shared control rather than full autonomy.

Service Industry: Humanoid Service Robots (e.g., Pepper)

In the service industry, trust-based transformation is most visible in human-facing robotics applications such as hospitality and customer-service. For example, Pepper, a humanoid service robot developed by SoftBank Robotics, has been deployed across retail and hospitality environments with design emphasis on trust, interaction and human-robot social presence (Demaître, 2019).

Empirical research in service robotics shows: (a) trust in service robots significantly increases perceived usefulness and ease of use, and decreases perceived risk, which in turn improves user satisfaction and revisit intention (Seo & Lee, 2021). (b) Perceptions of competence and warmth (human-like attributes) influence trust and reduce social distance in human-robot interaction (Christoforakos et al., 2021). In organisational deployments, the human workforce remains engaged via oversight, escalation protocols (robot defers to human staff when uncertain), and clear visual cues of robot status, ensuring that the service robot is treated as an assistant rather than a replacement, which supports trust calibration.

Thus, in the service arena, trust mechanisms centre on social and relational dimensions (empathy, human-robot interaction quality), transparency of system limitations, and human-robot workflow co-governance.

Synthesis

Across manufacturing, logistics, and service industries, trust emerges as a strategic asset that enables effective human-robot collaboration. Whether through participatory co-design in manufacturing (BMW), transparency and oversight in logistics (Amazon), or relational interaction design in service robots (Pepper/SoftBank), organisations demonstrate that trust is not simply about machine performance - but about how humans perceive, understand and collaborate with machines.

Key lessons:

- Trust is dynamic and context-dependent, requiring continuous calibration rather than one-time acceptance.
- Trust in manufacturing emphasises safety and shared workspace, in logistics it emphasises predictability and transparency, in service it emphasises empathy and human-robot interaction quality.
- Technology alone is insufficient; trust must be institutionalised through training, governance frameworks, human-in-the-loop workflows, and cultural adaptation.

Trust-centred transformation means that humans and machines become reliable partners in decision-making, learning and innovation, blurring the boundary between automation and collaborative intelligence.

These cases collectively show that trust is not a peripheral factor but the foundation of transformation. It enables a co-evolutionary model in which humans and robots continuously adapt to one another, generating sustainable competitive advantage.

6. CHALLENGES AND BARRIERS TO BUILDING HUMAN-MACHINE TRUST

The successful integration of robots and intelligent systems into organizational ecosystems depends heavily on the establishment of trust. Despite technological progress, achieving trustworthy human-machine relationships remains a complex endeavor due to intertwined technological, human, organizational, ethical, and cultural challenges. Each of these dimensions presents unique barriers that must be addressed for sustainable collaboration and responsible AI deployment.

6.1 Technological Challenges: Bias, Opacity, and Algorithmic Unpredictability

One of the most persistent barriers to trust in machines lies in their technical design. Algorithmic bias - stemming from unrepresentative data, flawed modeling, or historical inequalities - undermines confidence in machine decisions. When algorithms replicate societal prejudices in hiring, credit scoring, or healthcare diagnostics, users question both fairness and accountability (Mehrabi et al., 2021). Opacity, often referred to as the "black-box problem," further erodes trust, as users are unable to understand how systems reach their conclusions (Hollanek, 2023).

Algorithmic unpredictability, particularly in machine learning models that evolve through autonomous learning, can lead to inconsistent outcomes. Autonomous vehicle systems have been criticized for erratic performance in edge-case scenarios, which limit user trust despite high average accuracy (Walker et al., 2018). Transparent and explainable AI mechanisms, therefore, are vital to mitigating these challenges and promoting informed reliance.

6.2 Human Challenges: Fear of Displacement, Cognitive Dissonance, and Overtrust/Undertrust

Human psychology significantly influences trust in machines. Employees often perceive automation as a threat to job security, leading to fear, resistance, and cognitive dissonance (Rayhan & Rayhan, 2023). This apprehension disrupts engagement and collaboration with robots. Overtrust can also pose risks when users become complacent, assuming machines are infallible. In contrast, undertrust results in the underutilization of automation, diminishing potential efficiency gains (Chiou & Lee, 2021).

Aviation pilots who over-rely on autopilot systems may experience "automation bias," failing to intervene during malfunctions. Radiologists may distrust AI-based diagnostic tools due to limited explainability, despite their proven accuracy. Balancing trust levels through proper training, human-centered interface design, and transparent performance feedback is therefore essential for psychological adaptation.

6.3 Organizational Challenges: Lack of Transparency, Inadequate Communication, and Resistance to Change

At the organizational level, barriers to trust often emerge from systemic mismanagement of technological integration. Many organizations fail to implement transparent communication strategies about AI systems' capabilities, limitations, and ethical use. Without clear communication, employees and stakeholders struggle to understand the rationale behind machine-driven decisions, fostering suspicion (Dennehy et al., 2023).

Resistance to change further compounds this challenge. Organizational inertia - rooted in established hierarchies and legacy systems - can delay adoption, while inadequate training undermines confidence in automation. Studies show that trust-based transformation succeeds when leaders model transparency, provide skill development programs, and establish feedback channels (Glikson & Woolley, 2020).

6.4 Ethical Challenges: Data Misuse, Accountability Gaps, and Loss of Human Agency

Ethical dilemmas represent a major impediment to trust formation. Data misuse - through unauthorized surveillance, biased decision-making, or breaches of privacy - damages public confidence in AI systems (Jobin et al., 2019). Accountability gaps arise when errors occur in semi-autonomous contexts, leaving uncertainty about whether responsibility lies with developers, users, or the organization.

A related concern is the erosion of human agency. As decision-making becomes increasingly automated, humans may feel marginalized or powerless, perceiving a loss of control over outcomes. Addressing these issues requires the development of governance frameworks emphasizing ethical AI principles such as transparency, fairness, and human oversight.

6.5 Cross-Cultural Challenges: Variations in Trust Perception Across Regions and Industries

Trust in technology is not universal - it is shaped by cultural, industrial, and societal factors. Research shows that users in high-trust cultures (e.g., Northern Europe) are more open to automation than those in low-trust societies (e.g., parts of Asia or Latin America), where skepticism toward machines is higher (Zhang & Dafoe, 2019). Industry-specific variations also exist; for instance, trust in health-care robots may be higher than in financial AI due to the perceived humanitarian purpose of the former.

Cross-cultural differences highlight the need for context-sensitive trust-building strategies. Multinational organizations must tailor communication, ethical standards, and regulatory compliance mechanisms to regional expectations to ensure equitable adoption.

Table 1. Key Challenges and Barriers to Human-Machine Trust Formation

Dimension	Key Issues	Empirical Examples	Implications for Trust
Technological	Algorithmic bias, system opacity, unpredictability	Autonomous vehicles' inconsistent performance (Walker et al., 2018)	Reduced perceived reliability and transparency
Human	Fear of job loss, cognitive dissonance, overtrust/undertrust	Workers' fear of displacement in logistics (Rayhan & Rayhan, 2023)	Decreased engagement and miscalibrated reliance
Organizational	Lack of communication, training, and leadership support	Poor communication in AI-driven HR systems (Dennehy et al., 2023)	Resistance to adoption and low system credibility
Ethical	Data misuse, accountability ambiguity, loss of human agency	Biased healthcare algorithms (Jobin et al., 2019)	Ethical skepticism and legal uncertainty
Cross-Cultural	Regional and industrial variation in trust perceptions	Variability in AI acceptance across Asia vs. Europe (Zhang & Dafoe, 2019)	Uneven adoption and need for localized trust frameworks

Building trust in human-machine relationships demands a multidimensional approach that addresses technical transparency, human adaptation, ethical integ-rity, and cultural sensitivity. Organizations must design inclusive and explainable systems, foster psychological readiness, and institutionalize ethical governance to overcome the identified barriers. Only through proactive engagement across these domains can organizations create a foundation for trustworthy and transformative human-robot collaboration.

7. FRAMEWORK FOR CULTIVATING TRUST IN HUMAN-ROBOT SYNERGY

Trust is essential for human-machine cooperation as companies incorporate robots and intelligent technologies. Organizational aim, ethics, and human-centered engagement create, nurture, and calibrate trust in robots. Transparency by design, ethical governance, empowerment and training, feedback integration, and human-centric design are backed by quantitative trust metrics to evaluate and maintain trustworthy human-robot synergy.

7.1 Transparency by Design: Explainable AI and Open Communication

Transparency serves as the foundation for building human-machine trust. "Transparency by design" ensures that users understand how intelligent systems make decisions, what data they use, and what limitations they possess. Explainable AI (XAI) plays a vital role by translating complex algorithmic processes into human-understandable reasoning (Hollanek, 2023).

In practice, companies such as IBM have adopted explainability frameworks like "AI FactSheets," which provide end-users with documentation about model performance, data provenance, and fairness indicators (Rai et al., 2019). This level of openness enhances perceived reliability and reduces user anxiety. Additionally, open communication - through regular reporting, visual dashboards, and AI literacy sessions - enables stakeholders to make informed judgments about machine trustworthiness.

7.2 Ethical Governance: Establishing Norms for Fairness and Accountability

Ethical governance creates a moral and procedural infrastructure to ensure that human-robot collaboration aligns with societal and organizational values. Trust flourishes when systems operate within transparent ethical boundaries, emphasizing fairness, accountability, and respect for privacy (Jobin et al., 2019).

Organizations are increasingly forming AI ethics boards and adopting international guidelines such as the EU's "Ethics Guidelines for Trustworthy AI," which advocate for human oversight, transparency, and data integrity (European Commission, 2019). Microsoft's Responsible AI framework embeds ethical audits throughout product development, ensuring that AI systems are designed and deployed responsibly (Raji et al., 2021). Establishing these governance mechanisms signals organizational commitment to trustworthy innovation, reinforcing confidence among users and regulators alike.

7.3 Empowerment and Training: Enhancing Digital Trust Literacy

Human trust in machines depends not only on the technology's performance but also on users' understanding of it. Empowering employees through digital trust literacy - the ability to comprehend, evaluate, and interact confidently with intelligent systems - is essential (Dennehy et al., 2023).

Organizations can promote trust literacy through structured training, experiential workshops, and cross-disciplinary learning initiatives. For example, Siemens' "AI Learning Campus" provides interactive modules that teach employees how to interpret algorithmic recommendations and recognize biases (Siemens et al., 2025). Such initiatives bridge the psychological gap between human intuition and machine logic, encouraging appropriate trust calibration and informed reliance.

7.4 Feedback Integration: Continuous Trust Assessment and Recalibration

Trust in human-robot systems is dynamic rather than static. Therefore, organizations must establish feedback loops that enable continuous monitoring, assessment, and recalibration of trust levels. This process involves collecting user feedback, analyzing performance discrepancies, and adapting system behavior to maintain appropriate trust alignment (Bobko et al., 2022).

In industrial robotics, feedback systems that integrate user evaluations into machine learning updates have been shown to increase reliability and operator satisfaction. Toyota's Human Support Robot (HSR) program collects operator feedback to refine task responsiveness and predictability (Toyota UK Magazine, 2021). Such participatory design strengthens user confidence by demonstrating that the organization values human judgment and experience in system evolution.

7.5 Human-Centric Design Principles: Emphasizing Empathy and Inclusivity

Trustworthy human-robot collaboration depends on empathetic, inclusive, and context-aware design. Human-centric design ensures that robots and AI systems align with human cognitive, emotional, and social needs (Glikson & Woolley, 2020).

Designers must consider factors such as nonverbal communication cues, adaptive interfaces, and cultural sensitivity to foster emotional comfort and usability.

Empirical studies in healthcare robotics show that robots with empathetic design features - such as expressive gestures and contextually appropriate speech - enhance patient trust and acceptance (Pelikan et al., 2018). Service robots that adapt to users' emotional states and preferences contribute to sustained relational trust.

Figure 3. Framework for Cultivating Trust in Human-Robot Synergy

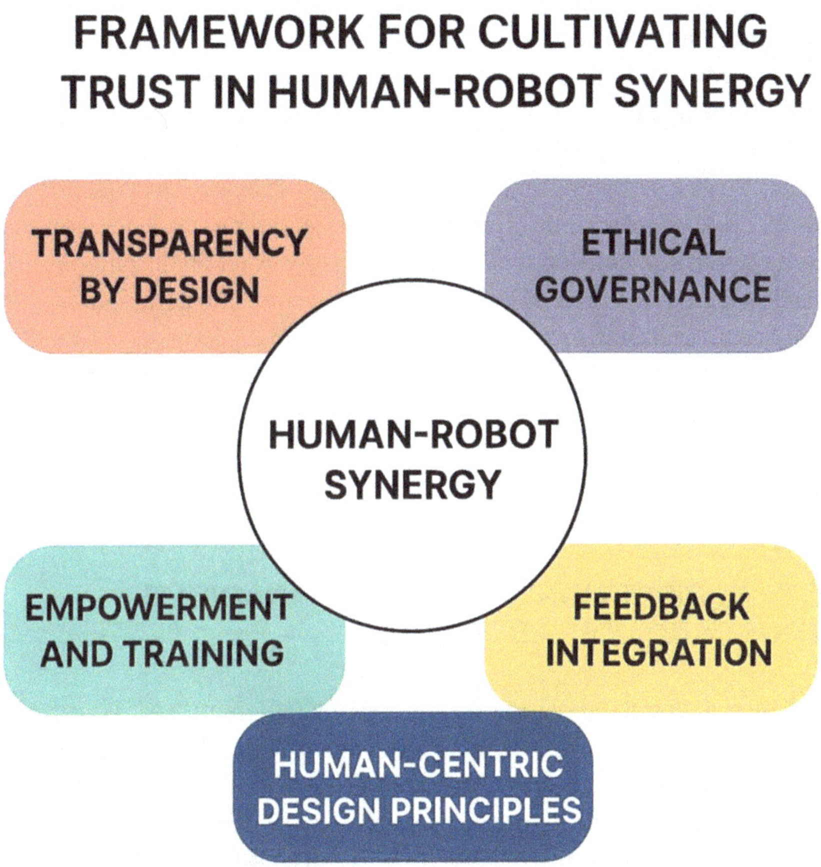

Author, (2025)

7.6 Metrics for Trust Measurement: Behavioral Indicators, Performance Outcomes, and Sentiment Analytics

To effectively cultivate and sustain trust, organizations must establish measurable indicators that capture their multidimensional nature. Metrics may include:

- **Behavioral Indicators:** Frequency of human reliance on automated recommendations, willingness to delegate tasks, and system utilization rates.
- **Performance Outcomes:** Improvements in collaboration efficiency, error reduction, and productivity gains following human-robot integration.
- **Sentiment Analytics:** Real-time monitoring of user attitudes through surveys, feedback forms, and natural language processing of communication data (Subrahmanyam, 2025).

A data-driven trust measurement system enables organizations to identify emerging issues, evaluate interventions, and refine trust-building strategies continuously.

The proposed framework provides a holistic pathway for organizations seeking to build and sustain trust in human-robot synergy. By embedding transparency, ethics, empowerment, feedback, and empathy into design and governance, organizations can move beyond transactional collaboration toward transformative partnership. Trust becomes both the enabler and the outcome of responsible human-robot interaction, ensuring that technological innovation strengthens, rather than replaces, human capability and judgment.

8. FUTURE DIRECTIONS: EVOLVING PARADIGMS OF TRUST AND COLLABORATION

As human-robot collaboration matures, the paradigm of trust is shifting from static, rule-based interactions to adaptive, co-evolving relationships where both humans and machines learn, adapt, and calibrate mutual expectations. Future directions in trust and collaboration are increasingly characterized by dynamic feedback mechanisms, emotional intelligence in machines, decentralized trust infrastructures, and the emergence of symbiotic organizational systems that reimagine the human-machine relationship as a partnership of equals.

8.1 Adaptive Trust Models

Adaptive trust models represent the next frontier of human-machine collaboration, emphasizing the dynamic evolution of trust based on contextual learning and user feedback. These models leverage machine learning algorithms to personalize system responses according to user trust signals - such as hesitation, acceptance, or error correction patterns (Lee & See, 2004). By continuously monitoring user behavior, adaptive systems can recalibrate their transparency levels, communication frequency, and autonomy accordingly. In aviation and autonomous driving, adaptive automation adjusts control handovers based on the operator's cognitive workload and trust state, reducing overreliance or distrust (Subrahmanyam, 2025). This adaptability transforms trust from a static precondition into a continuously optimized relational variable.

8.2 Emotional AI and Empathetic Machines

Another emerging dimension in trust development is the rise of *emotional AI* - machines capable of perceiving, interpreting, and responding to human emotions. Emotional AI fosters *affective trust*, a form of trust grounded in perceived empathy and relational warmth rather than mere technical reliability (Nguyen et al., 2024). In healthcare, empathetic robots such as the humanoid assistant *Pepper* have demonstrated increased patient engagement and compliance through emotion-aware interactions (Bittencourt et al., 2025). Social robots used in eldercare or customer service contexts that express empathy or mirror human affective cues tend to be perceived as more trustworthy. Emotional AI thereby deepens the psychological bond between humans and machines, enhancing collaboration outcomes in contexts requiring sensitivity and emotional labor.

8.3 Blockchain for Trust Assurance

Blockchain technology offers a promising infrastructure for *trust assurance* by enabling transparency, traceability, and immutability in data transactions (Zhou et al., 2023). In human-robot ecosystems, blockchain can validate decision provenance, ensuring that machine actions are auditable and ethically compliant. In supply chain robotics, blockchain-based ledgers can log autonomous decisions about inventory movement or resource allocation, thereby ensuring accountability and data integrity (Fernández-Caramés et al., 2024). By decentralizing trust, blockchain minimizes reliance on centralized authorities, supporting a more transparent and equitable model of human-machine collaboration.

8.4 Human Digital Twins

The concept of *Human Digital Twins (HDTs)* - virtual replicas that simulate human cognitive, behavioral, and trust dynamics, represents a radical step toward predictive trust modeling (Nguyen et al., 2024). HDTs can anticipate user reactions and calibrate robotic responses, accordingly, creating *co-adaptive systems* that evolve with individual users. In manufacturing and healthcare, digital twins are already being deployed to forecast human performance and risk factors; integrating trust analytics into these models can further refine collaborative safety and efficiency (Bittencourt et al., 2025). HDTs thus transform trust into a quantifiable and trainable element of human-machine ecosystems.

8.5 Toward Symbiotic Organizations

The culmination of these innovations points toward *symbiotic organizations*—entities where human and artificial agents share cognitive workloads and decision authority through mutual adaptation. In such environments, trust becomes a two-way construct, where machines also "trust" human input and intentions (Hancock et al., 2011). This mutual cognitive alignment requires continuous negotiation of autonomy, accountability, and ethical norms. For instance, in logistics and manufacturing, companies like Siemens and Amazon deploy co-bots (collaborative robots) that learn from human operators while optimizing workflows, symbolizing a movement from substitution to *synergy*.

Table 2. Emerging Paradigms in Human-Machine Trust and Collaboration

Paradigm	Core Concept	Key Technologies	Trust Mechanism
Adaptive Trust Models	Dynamic recalibration of trust through continuous learning	Reinforcement learning, user feedback loops	Behavioral monitoring and algorithmic adjustment
Emotional AI	Machines capable of perceiving and responding to emotions	Affect recognition, sentiment analysis	Empathy-driven interaction and affective bonding
Blockchain for Trust Assurance	Decentralized validation of machine actions	Distributed ledger, smart contracts	Transparency, accountability, and immutability
Human Digital Twins	Simulation of human behavior and trust patterns	Digital modeling, predictive analytics	Anticipatory trust calibration
Symbiotic Organizations	Cognitive co-dependence between humans and robots	IoT, cognitive computing	Mutual trust and adaptive collaboration

The evolution of human-machine trust is increasingly driven by adaptive, empathic, and decentralized paradigms. As emotional AI, blockchain, and human digital twins mature, trust will no longer be a fragile prerequisite but a dynamically maintained equilibrium. Future symbiotic organizations will depend on shared ethical norms and mutual cognitive understanding, marking a profound transformation in how humans and machines collaborate for innovation, resilience, and shared intelligence.

9. CONCLUSION: TOWARD A TRUST-CENTRIC FUTURE OF ORGANIZATIONAL TRANSFORMATION

Trust stands as the cornerstone of sustainable human-robot collaboration, shaping how organizations innovate, adapt, and thrive in the age of intelligent machines. Across the preceding discussions, it is clear that trust is not a static attribute but a dynamic, evolving construct rooted in transparency, ethical governance, and mutual adaptation. Whether through explainable AI, empathetic design, or co-evolving feedback loops, trust defines the quality and depth of human-machine synergy.

For leaders and policymakers, the imperative lies in building trust governance structures that institutionalize ethical standards, accountability mechanisms, and inclusivity across all levels of automation. Trust cannot be legislated - it must be designed into systems, nurtured through training, and reinforced through transparent communication and responsible leadership.

As organizations transition toward hybrid ecosystems where humans and robots co-create value, the ultimate challenge and opportunity lie in designing workplaces that uphold human dignity while amplifying machine intelligence. Trust-centered transformation will enable organizations to move beyond efficiency-driven automation toward purposeful, ethical, and resilient collaboration - ushering in an era where human creativity and artificial intelligence operate in true cognitive partnership. This future, grounded in trust, represents the most profound evolution of organizational life in the 21st century.

REFERENCES

Abdollahi, H., Mahoor, M. H., Zandie, R., Siewierski, J., & Qualls, S. H. (2022). Artificial emotional intelligence in socially assistive robots for older adults: A pilot study. *IEEE Transactions on Affective Computing, 13*(3), 1123–1137. DOI: 10.1109/TAFFC.2022.3143803 PMID: 37840968

Allgor, R., Cezik, T., & Chen, D. (2023). Algorithm for robotic picking in Amazon fulfillment centers enables humans and robots to work together effectively. *INFORMS Journal on Applied Analytics, 53*(4), 266–282. DOI: 10.1287/inte.2022.1143

Ayoub, J., Avetisyan, L., Makki, M., & Zhou, F. (2021). An investigation of drivers' dynamic situational trust in conditionally automated driving. *arXiv preprint arXiv:2112.04095.* https://doi.org//arXiv.2112.04095DOI: 10.48550

Bhargava, A., Bester, M., & Bolton, L. (2021). Employees' perceptions of the implementation of robotics, artificial intelligence, and automation (RAIA) on job satisfaction, job security, and employability. *Journal of Technology in Behavioral Science, 6*(2), 106–113. DOI: 10.1007/s41347-020-00153-8

Bittencourt, V., Saakes, D., & Thiede, S. (2025). Surrogate modelling for continuous ergonomic assessment and adaptive configuration of industrial human-centered workplaces. *Journal of Manufacturing Systems, 79*, 383–397. DOI: 10.1016/j.jmsy.2025.02.001

Bobko, P., Hirshfield, L., Eloy, L., Spencer, C., Doherty, E., Driscoll, J., & Obolsky, H. (2022). Human-agent teaming and trust calibration: A theoretical framework, configurable testbed, empirical illustration, and implications for the development of adaptive systems. *Theoretical Issues in Ergonomics Science, 24*(3), 310–334. DOI: 10.1080/1463922X.2022.2086644

Brynjolfsson, E., & McAfee, A. (2017). The business of artificial intelligence: What it can—and cannot—do for your organization. *Harvard Business Review.*

Chiou, E. K., & Lee, J. D. (2021). Trusting automation: Designing for responsivity and resilience. *Human Factors, 63*(1), 5–28. DOI: 10.1177/00187208211009995 PMID: 33906505

Christoforakos, L., Gallucci, A., Surmava-Große, T., Ullrich, D., & Diefenbach, S. (2021). Can robots earn our trust the same way humans do? A systematic exploration of competence, warmth, and anthropomorphism as determinants of trust development in HRI. *Frontiers in Robotics and AI, 8*, 640444. Advance online publication. DOI: 10.3389/frobt.2021.640444 PMID: 33898531

Daronnat, S., Azzopardi, L., Halvey, M., & Dubiel, M. (2021). Inferring trust from users' behaviours; agents' predictability positively affects trust, task performance and cognitive load in human-agent real-time collaboration. *Frontiers in Robotics and AI, 8*, 642201. Advance online publication. DOI: 10.3389/frobt.2021.642201 PMID: 34307467

Demaître, E. (2019). Trust is key to service robot design, says SoftBank Robotics. *The Robot Report*. Retrieved from https://www.therobotreport.com/trust-key-service-robot-design-softbank-pepper/

Dennehy, D., Griva, A., Pouloudi, N., Dwivedi, Y. K., Mäntymäki, M., & Pappas, I. O. (2023). Artificial intelligence (AI) and information systems: Perspectives to responsible AI. *Information Systems Frontiers, 25*(1), 1–7. DOI: 10.1007/s10796-022-10365-3

European Commission. (2019). *Ethics guidelines for trustworthy AI*. Publications Office of the European Union. https://digital-strategy.ec.europa.eu/en/library/ethics-guidelines-trustworthy-ai

Fernández-Caramés, T. M., Blanco-Novoa, O., Fraga-Lamas, P., & Froiz-Míguez, I. (2024). Towards an autonomous industry 4.0 warehouse: A UAV and blockchain-based system for inventory and traceability applications in big data-driven supply chain management. *arXiv*. https://doi.org//arXiv.2402.00709DOI: 10.48550

Fraunhofer, I. P. K. (2022). Bolting Cobots – Human and Machine in the Same Workspace. *FUTUR* (Fraunhofer IPK publication). Retrieved from https://www.ipk.fraunhofer.de/en/media/futur/futur-2022-2/bolting-cobots.html

Glikson, E., & Woolley, A. W. (2020). Human trust in artificial intelligence: Review of empirical research. *The Academy of Management Annals, 14*(2), 627–660. DOI: 10.5465/annals.2018.0057

Group, B. M. W. (2013, September 10). *Innovative human-robot cooperation in BMW Group production*. Press release. https://www.press.bmwgroup.com/global/article/attachment/T0209722EN/301973

Hancock, P. A., Billings, D. R., & Schaefer, K. E. (2011). Can You Trust Your Robot? *Ergonomics in Design, 19*(3), 24–29. DOI: 10.1177/1064804611415045

Hancock, P. A., Kessler, T., Kaplan, A. D., Brill, J. C., & Szalma, J. L. (2020). Evolving trust in robots: Specification through sequential and comparative meta-analyses. *Human Factors, 63*(7), 1196–1229. DOI: 10.1177/0018720820922080 PMID: 32519902

Hoff, K. A., & Bashir, M. (2015). Trust in automation: Integrating empirical evidence on factors that influence trust. *Human Factors*, *57*(3), 407–434. DOI: 10.1177/0018720814547570 PMID: 25875432

Hollanek, T. (2023). AI transparency: A matter of reconciling design with critique. *AI & Society*, *38*(5), 2071–2079. DOI: 10.1007/s00146-020-01110-y

Jobin, A., Ienca, M., & Vayena, E. (2019). The global landscape of AI ethics guidelines. *Nature Machine Intelligence*, *1*(9), 389–399. DOI: 10.1038/s42256-019-0088-2

Lee, J. D., & See, K. A. (2004). Trust in automation: Designing for appropriate reliance. *Human Factors*, *46*(1), 50–80. DOI: 10.1518/hfes.46.1.50.30392 PMID: 15151155

Lin, C.-Y., Liu, Y.-C., Chen, M.-C., Chiang, F.-F., & Lee, T.-Y. (2022). Learning curve and surgical outcome of robotic assisted colorectal surgery with ERAS program. *Scientific Reports*, *12*(1), 20566. Advance online publication. DOI: 10.1038/s41598-022-24665-w PMID: 36446802

Luhmann, N. (1979). *Trust and power*. Wiley.

Mayer, R. C., Davis, J. H., & Schoorman, F. D. (1995). An integrative model of organizational trust. *Academy of Management Review*, *20*(3), 709–734. DOI: 10.2307/258792

Mehrabi, N., Morstatter, F., Saxena, N., Lerman, K., & Galstyan, A. (2021). A survey on bias and fairness in machine learning. *ACM Computing Surveys, 54*(6). *ACM Computing Surveys*, *115*(6), 1–35. DOI: 10.1145/3457607

National Transportation Safety Board. (2020). *Collision between a sport utility vehicle operating with partial driving automation and a crash attenuator—Mountain View, California.* Washington, DC: U.S. Government Printing Office. https://www.ntsb.gov/investigations/Pages/HWY18FH011.aspx

Nguyen, D., Cohen, M. C., Kao, H.-T., Engberson, G., Penafiel, L., Lynch, S., & Volkova, S. (2024). Exploratory models of human-AI teams: Leveraging human digital twins to investigate trust development. *arXiv*. https://doi.org//arXiv.2411.01049DOI: 10.48550

Pelikan, H. R. M., Cheatle, A., Jung, M. F., & Jackson, S. J. (2018). Operating at a distance – How a teleoperated surgical robot reconfigures teamwork in the operating room. *Proceedings of the ACM on Human-Computer Interaction, 2*(CSCW), Article 54. https://doi.org/DOI: 10.1145/3274407

Petersen, L., Robert, L. P., Yang, X. J., & Tilbury, D. (2019). Situational awareness, driver's trust in automated driving systems, and secondary task performance. *SAE International Journal of Connected and Automated Vehicles*, *2*(2), 65–76. DOI: 10.4271/12-02-02-0009

Rahwan, I., Cebrian, M., Obradovich, N., Bongard, J., Bonnefon, J. F., Breazeal, C., Crandall, J. W., Christakis, N. A., Couzin, I. D., Jackson, M. O., Jennings, N. R., Kamar, E., Kloumann, I. M., Larochelle, H., Lazer, D., McElreath, R., Mislove, A., Parkes, D. C., Pentland, A. S., & Wellman, M. (2019). Machine behaviour. *Nature*, *568*(7753), 477–486. DOI: 10.1038/s41586-019-1138-y PMID: 31019318

Rai, A. (2019). Explainable AI: From black box to glass box. *Journal of the Academy of Marketing Science*, *48*(1), 137–141. DOI: 10.1007/s11747-019-00710-5

Raji, I. D., Scheuerman, M. K., & Amironesei, R. (2021). You can't sit with us: Exclusionary pedagogy in AI ethics education. In *Proceedings of the 2021 ACM Conference on Fairness, Accountability, and Transparency* (FAccT '21), March 3–10, 2021, Virtual Event, Canada. ACM, 515–525. https://doi.org/DOI: 10.1145/3442188.3445914

Rayhan, S., & Rayhan, A. (2023). The psychological impact of AI: Adapting to a world of smart machines. *ResearchGate*. https://www.researchgate.net/publication/372724209_The_Psychological_Impact_of_AI_Adapting_to_a_World_of_Smart_Machines

Samek, W., & Müller, K. R. (2021). Towards explainable artificial intelligence. In *Explainable AI: Interpreting, Explaining and Visualizing Deep Learning* (pp. 5–22). Springer. https://doi.org/DOI: 10.1007/978-3-030-28954-6_1

Schaefer, K. E., Billings, D. R., & Hancock, P. A. (2016). A meta-analysis of factors influencing the development of trust in automation: Implications for understanding autonomy in future systems. *Human Factors*, *58*(3), 377–400. DOI: 10.1177/0018720816634228 PMID: 27005902

Seo, K. H., & Lee, J. H. (2021). The emergence of service robots at restaurants: Integrating trust, perceived risk, and satisfaction. *Sustainability (Basel)*, *13*(8), 4431. DOI: 10.3390/su13084431

Siemens. (2025). *AI-based predictive maintenance*. Retrieved from https://www.siemens.com/global/en/products/automation/topic-areas/industrial-ai/usecases/ai-based-predictive-maintenance.html

Subrahmanyam, S. (2025). Development of Research Design. In Segura, L., Vasconcelos, A., & Slomski, V. (Eds.), *Evolving Designs, Applications, Technological Advances, and the Future of Qualitative Research* (pp. 31–58). IGI Global Scientific Publishing., DOI: 10.4018/979-8-3373-4432-4.ch002

Subrahmanyam, S. (2025). Emerging Technologies and Shift in Employment in the Transportation Sector. In Olga Titovna, E., Elango, D., Kumar, P., Mehra, P., & Verma, R. (Eds.), *Integrating Sustainability and Autonomous Transportation for Enhanced Urban Mobility* (pp. 149–178). IGI Global Scientific Publishing., DOI: 10.4018/979-8-3373-0882-1.ch006

Subrahmanyam, S. (2025). Human Factors and Workforce Adaptation in Digital Transition. In Martínez-Falcó, J., Sánchez-García, E., & Marco-Lajara, B. (Eds.), *Navigating Digital Transformation Through Business Process Management* (pp. 95–126). IGI Global Scientific Publishing., DOI: 10.4018/979-8-3373-3336-6.ch005

Subrahmanyam, S. (2025). Leveraging AI and ML for Enhanced Efficiency and Innovation in Manufacturing. In Vyas, V., Bi, X., Pandey, D., Xu, Z., & Pandey, B. (Eds.), *AI's Role in Enhanced Automotive Safety* (pp. 319–344). IGI Global Scientific Publishing., DOI: 10.4018/979-8-3373-0442-7.ch021

Subrahmanyam, S. (2025). Sustainability and AI. In Younas, A., & Kassim, A. (Eds.), *Transforming Business Through Digital Sustainability Models* (pp. 363–392). IGI Global Scientific Publishing., DOI: 10.4018/979-8-3373-0608-7.ch016

Subrahmanyam, S., & Baz, B. E. (2025). AI and Diversity: Ensuring Fairness in Automated Decision-Making. In El Imrani, O. (Ed.), *Navigating Responsible Business Practices Through Ethical AI* (pp. 19–50). IGI Global Scientific Publishing., DOI: 10.4018/979-8-3373-3104-1.ch002

Supply Chain, D. H. L. (2025). DHL's Smart Warehouse initiative: Human-robot collaboration in logistics. https://www.dhl.com/global-en/delivered/innovation/warehouse-robotics-and-automation.html

Szabó, B., Őrsi, B., & Csukonyi, C. (2024). Robots for surgeons? Surgeons for robots? Exploring the acceptance of robotic surgery in the light of attitudes and trust in robots. *BMC Psychology*, *12*(1), 45. DOI: 10.1186/s40359-024-01529-8 PMID: 38267998

Topol, E. (2021). *Deep medicine: How artificial intelligence can make healthcare human again*. Basic Books.

Toyota, U. K. Magazine. (2021). *Toyota Human Support Robot: What is it and how can it be used?* https://mag.toyota.co.uk/toyota-human-support-robot/#:~:text=The%20original%20prototype%20was%20developed,practical%20adoption%20of%20the%20HSR

Walker, F., Boelhouwer, A., Alkim, T., Verwey, W. B., & Martens, M. H. (2018). Changes in trust after driving level 2 automated cars. *Journal of Advanced Transportation, 2018*, 1045186. Advance online publication. DOI: 10.1155/2018/1045186

Wilson, H. J., & Daugherty, P. R. (2022). *Radically human: How new technology is transforming business and shaping our future*. Harvard Business Review Press.

WSJ. (2025). Amazon is on the cusp of using more robots than humans in its warehouses. *The Robot Report*. Retrieved from https://www.livemint.com/companies/news/amazon-is-on-the-cusp-of-using-more-robots-than-humans-in-its-warehouses-11751343089546.html

Xu, X., Lu, Y., Vogel-Heuser, B., & Wang, L. (2021). Industry 4.0 and Industry 5.0—Inception, conception and perception. *Journal of Manufacturing Systems, 61*, 530–535. DOI: 10.1016/j.jmsy.2021.10.006

Zhang, B., & Dafoe, A. (2019). *Artificial intelligence: American attitudes and trends*. Center for the Governance of AI, University of Oxford.

Zhou, Q., Li, B., Han, L., & Jou, M. (2023). Talking to a bot or a wall? How chatbots vs. human agents affect anticipated communication quality. *Computers in Human Behavior, 143*, 107674. DOI: 10.1016/j.chb.2023.107674

KEY TERMS AND DEFINITIONS

Adaptive Trust Models:AI and robotic systems capable of adjusting their behavior and recommendations based on human feedback: experience, and environmental context to maintain optimal trust levels.

Affective Trust:An emotional and relational form of trust where humans feel comfort: safety, and confidence in interacting with machines, influenced by perceptions, empathy, and social cues.

Cognitive Trust:A rational: knowledge-based trust in a machine's abilities, performance, and reliability, grounded in evidence and experience rather than emotion.

Collaborative Intelligence (CQ):The integration of human judgment: creativity, and intuition with machine computational power, data processing, and precision, producing superior decision-making and problem-solving outcomes.

Ethical Governance in AI:Establishing organizational norms: policies, and accountability structures to ensure AI systems operate fairly, responsibly, and in alignment with human values.

Human Oversight and Control:Mechanisms allowing humans to monitor: intervene, or override automated systems, balancing autonomy with accountability to maintain trust and safety.

Human-Robot Synergy:The cooperative and cognitive interaction between humans and robots: combining human intuition with machine precision to achieve shared goals and enhance organizational performance.

Transparency by Design:A principle in system design that ensures explainability: visibility, and clarity in machine operations and decision-making processes to foster trust and informed human oversight.

Trust Calibration: The process of aligning human perception of a machine's capabilities with its actual performance to prevent overtrust or undertrust and optimize collaboration.

Trust in Socio-Technical Systems:The belief that a system or machine will act competently: predictably, and ethically, allowing humans to rely on it while maintaining oversight. It differs from interpersonal trust as it includes technological, ethical, and cognitive dimensions.

Chapter 9

Logistics Robotics in Action: Enhancing Speed, Accuracy, and Operational Efficiency

Sucheta Sachin Yambal
https://orcid.org/0000-0002-3113-9609
Dr. Babasaheb Ambedkar Marathwada University, India

Yashwant Arjunrao Waykar
https://orcid.org/0000-0002-9693-9738
Dr. Babasaheb Ambedkar Marathwada University, India

ABSTRACT

The integration of robotics is rapidly transforming global logistics by enhancing storage, processing, and distribution. It speeds up operations, improves accuracy, and makes them more efficient. Robotics is changing the way conventional logistics work by adding automated guided vehicles (AGVs), robotic arms, drone-based delivery, and smart sorting systems. Driven by demands for speed, accuracy, and cost-efficiency, logistics robots—like AGVs, robotic arms, drones, and smart sorting systems—boost operational efficiency and reduce human error. These technologies optimize warehouse use, streamline inventory management, and improve last-mile delivery (Zonta et al., 2020). With advancements in AI, machine learning, and sensors, robotics now offers scalable, customizable solutions for various industries. This chapter explores their real-world applications, innovations, and strategic impact on the future of supply chains.

DOI: 10.4018/979-8-3373-3404-2.ch009

1. INTRODUCTION

In today's fast-paced, highly competitive global economy, the logistics industry is under immense pressure to deliver products to customers faster, precisely, and of course for lesser money. To meet these growing needs of the markets, almost all businesses are employing robots in one or the other department as a revolutionary force in modern logistics. Logistics robotics uses smart robots and automated systems to accomplish complex tasks like picking, packing, sorting, moving, and delivering goods. According to (Boone, 2023) logistics operations have become efficient, accurate, and scalable than ever before, thanks to the robotic solutions that use technologies like AI, machine learning, and the Internet of Things (IoT).

Robotics is changing the whole logistics ecosystem, from autonomous mobile robots (AMRs) moving around complicated warehouse settings to robotic arms quickly completing orders with greater accuracy and drones making speedy deliveries. The changes brought forth make it human workout to stay away from tedious tasks, reduced errors, increase storage space, and ensure operations to run smoothly; even when demand is much more than the store could stomach. Real-time data integration also makes supply chains smarter and stronger by enabling better demand forecasting, inventory monitoring, and decision-making.

(Bondar, 2023) shows that using robots wisely not only helps companies run more smoothly, but it also provides them an advantage over their rivals by making them faster and happier with their consumers. (FreightAmigo, 2023) says that logistics robots will probably have a greater part to play in making supply chains more flexible, environmentally friendly, and innovative in the future. This is because technology is growing better.

The chapter discusses the challenges relevant to Logistics robotics along with its implementation in Industry; it also details about the issues and the possible future of this industry with respect to its rising usage in various sectors at large.

1.1. Overview of Logistics Challenges

Logistics is very important to the global economy today because it makes sure that goods move quickly and efficiently from producers to consumers. There are a number of problems that the shipping company has that make it hard for it to do its best work. Concerns include rising business costs, rising customer demands, a lack of workers, complicated supply networks, and the possibility of global calamities like pandemics or political problems. As e-commerce grows and people want faster delivery, transportation companies are under more pressure to make their operations more efficient. Because more and more people want their goods delivered the same day or the next day, businesses need to rethink how they store and move things around.

DHL, a well-known company, noted that transportation used to depend a lot on human labor, but it doesn't work for today's business needs for speed and scale. A lack of workers, high unemployment rates, and widespread safety concerns all make it harder for businesses to get things done. People regularly make mistakes when cleaning, packing, and arranging that cause delays and big costs. If transportation operations don't handle the large amount of data they acquire appropriately, things become a lot more complicated.

Transportation has to use more environmentally friendly methods because of government rules and worries about the environment. This shows how important it is to come up with new ways to cut down on carbon emissions. This means that there is a greater demand for choices that are quick, accurate, scalable, long-lasting, and affordable (Supply Chain Digital, 2024).

1.2. Types of Robotics Used in Logistics

Even if managing the supply chain and transportation is becoming harder, automation is making a big difference. In the transportation business, robots use smart systems and self-driving tools to undertake jobs that people used to do, such choosing, packing, sorting, moving, and delivering things (Berkers et al., 2023) Robots could be able to fix some of the problems that come up when people do things by hand. Also, activities may be done faster, more accurately, and for less money.

Over the last decade, an increasing number of robots have been developed to move objects. FedEx, DHL, and Amazon have spent a lot of money on automating package delivery and making their businesses easier to run in order to stay ahead of the competition (Park et al., 2021). The COVID-19 epidemic made this tendency go even faster by showing how weak systems that rely on humans are and how important it is to have solutions that are simple to use, work well, and are available to a lot of people.

Machine learning, artificial intelligence (AI), and the Internet of Things (IoT) are some of the technologies that have made this robotic change possible. Thanks to these technologies, robots can now move around large, complex buildings, make decisions quickly, and work with people without any problems (Banur et al., 2024). Robots may also help humans collect and analyze data better, which helps them make better decisions.

Not only are robots being made to take over jobs, but they are also being designed to help people do their jobs better. (Soteriades & Psarommatis, 2021) says that robots can handle boring, repetitive, and hard work, which gives people more time to concentrate on important tasks like planning, monitoring, and quality control. Using connections this way makes the supply chain more trustworthy and useful.

1.3. Chapter Outline

The chapter outlines the contents that specify how robots are used in contemporary logistical scenarios thus explain the supply chain is becoming more efficient, accurate, and faster as a result of these technological advancements. This chapter gives you a full picture of logistics robots by first explaining what they are and what their main parts are. After that, it speaks about the many kinds of robots that are used for various kinds of logistical tasks.

This content also talks about the technology that makes these robots operate, such AI, the Internet of Things (IoT), sensors, and automation software. It talks about how big shipping companies have successfully used robots in their work and what real-world benefits they have found via examples and case studies. It also talks about the problems that come up when robots are used, such how to connect them to older systems, how to keep costs low, and how to get employees to change. Finally, it speaks about what will happen in the future and how robots will change their roles in a supply chain that has gone digital.

At the end of this chapter, readers will know how robotics is changing logistics and why it is so important for the 21st century supply chain to work.

1.4. Understanding Logistics Robots

Logistics robots are a combination of mechanical + electronic devices thus could be called mechatronic machines that evolve the entire mechanism of supply chain and logistics activities making them easier and more efficient. They can do the chores of directing warehouse operations, delivering items, satisfying client requests, and keeping an eye on inventory levels, according to (Deng et al., 2023). These robots are trained to do their work regularly and precisely with minimal help from people, which improves their overall performance and reliability.

Most likely, transportation robots have all or either of the following sub-parts:

1. Robotic arms- It could be an AI-powered mobile robots, drones, automated guided vehicles (AGVs), or a conveyor system.
2. Control a system that instructs robotic devices and are a part of software tools and programs. They utilize software like ERP and WMS to conduct their jobs. ERP stands for "enterprise resource planning" while WMS stands for "warehouse management systems."

3. Using sensors and cameras, robots can find their way around, figure out where they are, and understand what they need to do. Few examples of the same are cameras, proximity sensors, and gadgets that can detect and measure light.
4. Robots can talk to one other and run system in real time thanks to a communication infrastructure. This network usually works well with the Internet of Things.
5. People may utilize AI engines and data analytics to make better decisions, work more efficiently, and improve operations by using data collected by robots.

There are lots of components that collaborate to lets transportation robots to operate alone or with little assistance. (Wurman et al., 2022) concluded that they can respond to environmental changes if the need arises and still facilitates smoother operations. Numerous types of robots are used in logistics, where each one designed to perform a specific function in the supply chain. Some of the major ones include Autonomous Mobile Robots (AMRs); they are the robots that are capable of navigating buildings independently. They use maps and sensors to navigate and transfer items from one spot to another. In contrast to AGVs, AMRs do not need adherence to predetermined paths. They may alter their trajectories in response to their surroundings.

Automated Guided Vehicles (AGVs) transport goods following designated routes, often defined by lines, magnetic strips, or lasers. They are extensively used in commercial enterprises and industries for the repetitive transportation of things. Robotic arms appendages may be stationary or mobile, enabling them to grasp, release, and categorize items. They are often used in pallet manufacturing sites, packing factories, and high-velocity sorting lines. Drones are becoming used for inspecting products at big merchants and for delivering stuff to residences in certain regions. They can do tasks swiftly and via many methods that would need people some time to complete.

2. TECHNOLOGIES ENABLING ROBOTICS (AI, IOT, SENSORS, ETC.)

As technology becomes better, logistics robots are becoming smarter and virtually necessary; with these improvements, robots may now work on their own, make their own decisions, and communicate with nearby systems and the real world if need be. AI lets robots learn from data, adapt to new situations, and train themselves over time. Machine learning algorithms now enable robots find patterns, figure out the best ways to do things, and make decisions on their own. For example, robots with AI could be able to figure out the best way to arrange things to please customers.

(Yang et al., 2023) has made it evident that the Internet of Things (IoT) allows computers, robots, and other devices speak to one other in real time. IoT sensors embedded into computers and transportation systems let us gather and analyze data all the time. This makes it possible for everyone in the supply chain to observe, monitor, and respond faster. Robots can't see or hear anything without sensors. LiDAR, vision systems, barcode and RFID readers, and proximity sensors all let robots detect objects, figure out what they are, and navigate about securely. Robots can accomplish customizable duties extremely well using these technologies, even in places that are busy or changing.

Computer vision is a technology that, when paired with AI, allows robots see what's going on around them mentioned (Belter & Skrzypczyński, 2019). This is important for things like spotting defects in merchandise, reading labels, and getting things ready for shipment or packaging just right. Cloud platforms let you store and work with massive volumes of data that robots produce when it comes to cloud computing and big data. This makes it possible to have unified control, measure performance, and use predictive analytics to assist people make wise decisions.

Edge computing allows data be processed directly on the robot itself when real-time processing is required, such when a robot needs to find its way or avoid obstacles. This makes the robot more responsive and cuts down on delays. Automation Software: Robotics process automation (RPA) software connects robot tasks to existing IT systems to assist to keep track of work flows. WMS, ERP, and client order systems work together, in order to evolve current working style of many. When used in tandem, these technologies let transportation bots act as smart, connected systems that improve the performance of the whole supply chain by making it more productive and making fewer errors.

2.1. Uses of robotics in logistics: automating warehouses (storage, packing, and picking)

(Gunasekaran et al., 2018) has made a mark when it rightly revealed that operations in and around warehouse take the huge time and are the most likely to go haywire. Since these operations form the backbone of the logistics system, robotics has fundamentally revolutionized how warehouses are run by automating tasks that are the same every time, such picking, packing, sorting, and storing. Robotic arms, mobile robots, and automated storage and retrieval systems (AS/RS) are working together or on their own more and more in large fulfillment centers to make warehouse procedures more efficient.

Picking and Packing: Robotic picking systems use machine learning and vision-guided technology to find, grasp, and remove objects/items from shelves. These systems can handle thousands of SKUs (Stock Keeping Units) precisely and their working can remain unaffected to different weights and sizes. When items are chosen, robotic packaging devices, which typically has tasks like printing labels and sealing and eventually putting them into boxes or shipping containers for further journey. This cuts down on the need for people to have contact with them.

Storage Automation: AS/RS systems use robotic cranes, shuttles, and shelves to automatically store and retrieve items/ products. These systems make the most of available space, which allows for high-density vertical storage and faster access times while allowing things to be stored at multiple levels. They generally reduce the need for people to intervene to handle things and the need for large warehouse spaces as well.

Also, robotic warehouse automation makes things safer by doing rid of the requirement for individuals to carry big things or drive forklifts through small places. With the additional augmentation of Internet of Things (IoT) sensors and Warehouse Management Systems (WMS) to the mix, the entire warehouse becomes a more controlled and an error-free place.

2.2 Transportation and delivery (AGVs, drones, AMRs)

Robotics has changed the way goods/items/ products are handled both within and outside of warehouses. Drones, autonomous mobile robots (AMRs), and automated guided vehicles (AGVs) are the three key technologies that provide power to internal logistics robotics and transportation.

Automated guided vehicles, or AGVs, in large warehouses and factories are a usual scene. They go along the set paths marked by wires, magnetic strips, or other means to transport pallets, raw materials, or finished goods. AGVs are reliable for tasks that need to be done over and over again in places where the layout doesn't change much.

Autonomous mobile robots (AMRs), on the other hand, use LiDAR, SLAM (Simultaneous Localization and Mapping), and sensors are dynamic in nature thus incorporating changing ecosystem. They quickly get adapted to the course in order to avoid obstructions and don't need to follow predetermined paths. AMRs are great for modern warehouses where the layout is dynamic since they are used to move products/ items between picking stations, restock stock, and sort packages zone by zone.

Also, Drones are beginning to be utilized in logistics for deliveries and inspecting inventories outdoors nowadays. Drones using RFID or barcode scanners can be seen flying around the aisles of big warehouses to quickly and accurately count supplies, which cuts down on downtime and is almost humanely impossible in such restricted timeline. Drones are being tested and deployed to move small products/ objects to places that are difficult to reach or are far away, avoiding traffic and the problems that come with human delivery declared (Majeed & Rupasinghe, 2023). These robotic vehicles and logistics management systems work in tandem to make sure that products/ items are moved smoothly, that throughput goes up, that traffic goes down, and that people don't have to handle things in an inefficient way thus eventually boiling everything down to maximized efficiency.

2.3. Managing orders and inventory

Efficient inventory management is needed to keep supply and demand in check, cut down on waste, and make sure that orders are fulfilled on time. Robotics has shown sky-rocket pace of automation and visibility in this industry to levels never seen before.

In the past, counting inventory required physical audits done by the staff where human err was not to be denied; which were time-consuming and repetitive. Robotics makes it possible for flying drones and mobile scanning robots to do regular inventory checks using RFID or visual recognition that does not as much time is accurate. These robots collect information in real time and send it to computerized inventory systems, which can tell them exactly how many items there are and where they are; this further can lead to better management of purchase order release, which further leads to cuts down on stockouts, overstocking, and loss of product.

Smart shelf systems: That work with robots make sure that goods are constantly in the optimum place based on real-time demand estimates thus robots transfer fresh stock to its allotted spot in the warehouse as per the specific instructions. This speeds up the process of fulfilling orders much quicker and makes selecting routes more efficient.

Order fulfillment: Robots work with order management systems to acquire/ select the items needed as soon as an order is made. This automated procedure, which includes selecting, packing, and labeling, is faster and more accurate than doing it by hand. This is particularly significant in e-commerce because of the short delivery times and large order sizes which is a regular scenario as such.

Robotics makes managing inventories and orders proactive, smart, and flexible, thus enables the supply chain to react rapidly as indicated previously.

2.4. Last-Mile Delivery Solutions

The "last mile," or the last part of delivery to the intended customer is one of the most expensive and logistically difficult parts of logistics. Robots are making a big impact on last-mile delivery, which is a resourceful way to get packages to people who live far away in rural areas or are in cities.

Delivery Drones: In areas with bad roads or a lot of traffic, unmanned aerial vehicles (UAVs), sometimes known as drones, can transport things in a much better way. Companies like Zipline, UPS, and Amazon have set up drone delivery services that bring lighter products right to people's homes or safe drop-off spots; these drones are quite helpful for medical supplies and deliveries that need to serve the customer real fast.

Autonomous Delivery Robots: These robots are usually seen on college campuses and in cities more often than not. They can roam freely over streets and sidewalks without clogging. These portable robots have GPS, cameras, and sensors that let them move food, groceries, and other packages to their destination. They are getting increasingly popular during hectic times and emergencies since they provide frictionless, efficient delivery options.

Logistics businesses are also looking at robotic locker systems that might serve as local delivery hubs today and in near future too. Robots bring things to community lockers, where customers may utilize smartphone apps or QR codes to receive their packages anytime they want almost eliminating the human force who would process the same order but by being on land a bit late. (Zonta et al., 2020) cited that by deploying robots for last-mile delivery, businesses may make customers happier, provide more flexible service, and cut their costs, delivery times, and effect on the environment.

2.5. Benefits of Using Robotics

(Cherubini et al., 2016) remarked that one of the best things about using robots in logistics is that they cut down on the time it takes to complete a cycle of production. Robots can work all day and night without getting weary or needing breaks, which speed up the flow of goods and the processing of goods whichever is the case.

Robots make it a lot faster to choose and pack things in warehouses without erring. A robot can execute a job that would take a human picker several minutes in a fraction of that time even when asked to do it repeatedly. AMRs and AGVs also have cut down on the time it was needed earlier to move things/products around by making internal transit more efficient.

Drones and self-driving delivery robots speed up delivery times by avoiding traffic congestion and other manual delays in transit and last-mile delivery notified

(Krüger et al., 2019) in general that is fantastic. Because of this level of efficiency, businesses may provide high-end services like delivery on the same day or in one hour becoming more of a trend now a days.

Shorter cycle times (production, distribution etc.) contribute to a supply chain that is more flexible, can quickly respond to changes in demand, cut lead times, and make customers happier.

Figure 1. Benefits of Robotics Logistics over Traditional Logistics

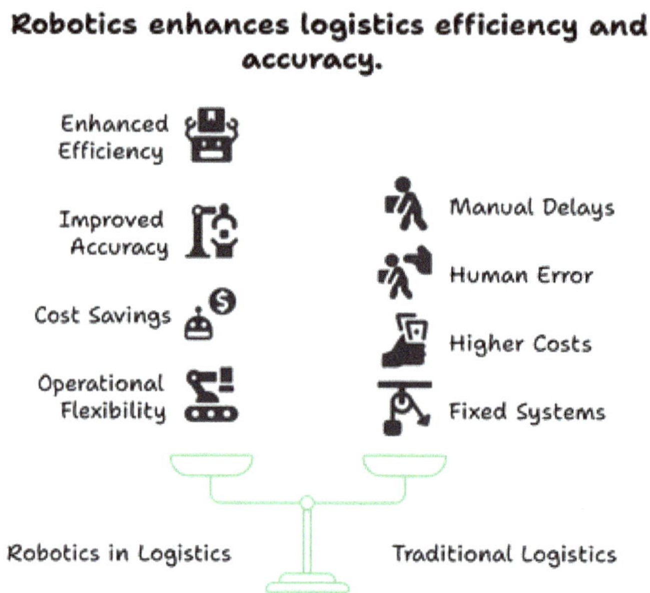

2.6. Better accuracy and fewer mistakes

Human error is one of the most common causes of logistical problems, especially when it comes to things like choosing an order, packing it, changing the inventory, and figuring out the exact path through which to get it to the customer. Robotic systems are designed to work with great accuracy as they use sensors, software, and AI-based decision-making to do this.

Robotic picking systems can find and choose the proper item by utilizing barcodes, weights, shapes, or even visual clues to the destined location (Zhang & Chen, 2020). This reduces the percentage of errors that happen when people pick the wrong item. Automated inventory solutions assist to reduce mistakes and stop stockouts or too much inventory by making sure that every stock movement is recorded and updated right away without procastinating. Appropriate Delivery Routing; Drones and self-driving delivery robots utilize GPS and AI to find the best delivery routes reducing the hassles. This lowers the chance of deliveries being missed or delayed on regular basis. Robotics makes sure that logistical operations are accurate even when there is a lot of demand, and it cuts down on the number of returns from wrong orders by a huge amount eliminating redundancy to a greater extent.

2.7. Cost Savings and Workforce Optimization

- o *Robots may cost a lot at first, but they save a lot of money over time with better ROI in longer runs. Robots lower expenses for labor, mistakes, and losses from broken things or inventory that makes it better alternative to human labor.*
- o *Labor Efficiency: Robots execute physically demanding, repetitive, or hazardous tasks so that people may focus on tasks that need supervision, analysis, or talking to customers leading to saving human from labor-intensive tasks. This makes workers/labors happier and minimizes the costs of accidents and burnout.*
- o *Robotic systems are designed to utilize space and energy as effectively as possible through the use of visualization tool. Vertical AS/RS systems make warehouses compact and mobility robots use lesser power when they move about properly doing the intended tasks.*
- o *Operational Scaling: Robotic systems may be gigantic or quite miniscule with minimal trouble during busy times or peak seasons. This flexibility keeps the costs of overtime and temporary workers from being too high in an organization.*

Robotics gives logistics companies a good return on investment (ROI) since it makes their processes leaner, with fewer errors, faster processing, and less need for human labor as per the info passed on by (Boysen et al., 2019).

2.8. Enhanced Operational Efficiency and Flexibility

Robotics systems are a perfect alternative for changing logistical scenarios since they can be changed, are based on data input, and are adaptable. Unlike older fixed systems, modern robots don't require as much reconfiguration to do new tasks, move around, or meet new needs and are almost self-reliant.

∇ Operational Flexibility: AMRs can adapt to new procedures or deal with changes in the layout of the warehouse if need arises. Robotic picking systems may be modified to work with new SKUs without needing to change the real infrastructure in general.
∇ Real-Time Monitoring and Optimization: IoT-connected robots provide manager's real-time data on operations so they can keep an eye on their performance, find problems in real-time, and make speedy improvements. This visibility cuts down on downtime thus the profits and makes things more responsive.
∇ Multi-Channel Fulfillment: Robotics lets organizations fulfill orders from many channels (including retail, B2B, and e-commerce) from the same location by changing procedures and resource allocation on the go without adding fuss.
∇ Eco-Efficiency & Sustainability: Many robot systems are built to use lesser energy. They may be utilized with green logistics initiatives to cut down on carbon emissions/ carbon footprints and improve sustainability measures.

Robotics helps firms stay competitive in a logistics world that is growing less predictable and more focused on customers by making operations quicker, adapting to be flexible as per the scenario. Using robots in logistics is no longer just a sci-fi fantasy; it's becoming a real need divulged turning dream into reality (Wurman et al., 2022). Robotic systems are changing the way things are moved and handled in big organizations with ease. For example, they are changing how items are delivered and moved, how warehouses are automated, and how inventory is managed in much better way than earlier mechanism. In a competitive global market, these technologies provide firms a big edge over each other by making things rapid, more accurate, cheaper, and agile. As robotics continues to grow with improvements in AI, machine learning, and networking, its role in logistics will become irreplaceable in coming future times. This will lead to a future where operations are smart, independent, and environment- friendly.

3. CASE STUDIES AND EXAMPLES FROM THE FIELD

Using robots in logistics has made things a lot faster, more accurate, and more efficient making space to incorporate changing scenario with no difficulty. To stay ahead of the competition and meet expanding market demands, workforce shortages, and changing customer expectations, industry leaders have turned to robotic automation and are happy about it. This part looks at case studies from well-known logistics organizations throughout the world, focusing on how they put their plans into action,measure results, and the most important -lessons learned.

3.1. Real-World Implementations in Leading Logistics Firms

I. Amazon Robotics, which used to be called Kiva Systems

In 2012, Amazon bought Kiva Systems, which was a big deal for warehouse robots then. Amazon Robotics, which is what Kiva's robotic technology is currently called, was employed by Amazon to automate picking and moving of merchandise in its fulfillment centers with perfection.

Implementation: Amazon sent out a number of orange pod-carrying autonomous mobile robots to pick up shelving units and bring them to human workers at picking stations reducing the involvement of human soul to half. This idea of goods-to-person cuts down on the time workers spend walking around the warehouse plummeting labor-intensive, tiring jobs .

Key outcomes:

There are more than 750,000 robots working in over 300 fulfillment centers across the globe.

The time it took to fulfill orders went down from 60 to 75 minutes to less than 30 minutes.

Increased the amount of storage space in the warehouse by 40% by finding associated space adjustments.

Lowered operating costs by making better use of space and people.

II. DHL and Locus Robotics

DHL has tested a lot of Locus Robotics' collaborative robots (cobots) says (Sivakumar & Li, 2020) at various locations in the US and Europe. These AMRs make it easier for people to work by moving about the warehouse on their own and carrying certain items/products (Caputo et al., 2021).

Implementation: LocusBots were added to the order picking process, which made workers less weary and increased efficiency in e-commerce fulfillment centers dropping the time-lag caused earlier.

Key outcomes:

Productivity has gone up by two times than that of the past.
Cutting down travel time for people by up to 50%.
Checking on the progress of an order in real time reflecting current status.
Less physical strain leads to happier/ satiated employees.

III. Roxo and SameDay Bot from FedEx

FedEx introduced "Roxo," a robot that delivers packages on sidewalks for the people who walk. Roxo, which was created with DEKA Development & Research, uses technology that keeps pedestrians safe to deliver packages in cities without much ado. LiDAR technology and a number of cameras make sure that Roxo can snake through safely (Guerra et al., 2023). It is designed to go over curbs, over rough terrain, and interact with traffic signals, in short as per the circumstances.

Key Outcomes (pilot study results):

Successful Delivery of items without contact during the COVID-19 epidemic.
Faster delivery for short-distance packages that calls for delivery on the same day.
Better brand presence/penetration and customer participation.

IV. JD.com in China

JD.com, a well-known Chinese e-commerce site, uses robots, drones, and self-driving delivery trucks to automate its warehouses and other parts of its supply chain thus curbing the overall timeframe.

Twenty robots run JD's fully automated warehouse in Shanghai, and just five folks are in charge of it (Li et al., 2022). The business uses self-driving delivery vehicles and drones to transport goods to remote regions.

Key outcomes:

∇ Over 99.9% of orders are fulfilled correctly.

∇ Over 90% of Tier 1 and Tier 2 cities are able to provide same-day delivery with ease.

∇ Automated warehouses are able to cut-down operational costs almost by 70%.

3.2. Impact Assessment: Important Performance Indicators Before and After Automation

The usage of robots has a magnanimous effect on logistics key performance indicators -KPIs.

These numbers show that robots not only make entire procedure run smoothly, but they also lower risks and make customers more-contented.

Table 1. KPI Indicators to show changes before and after automation

KPI	Before Automation	After Automation
Order Picking Time	60–90 minutes per batch	15–30 minutes
Order Accuracy	93–96%	99.5–99.9%
Labor Dependency	High manual involvement	Up to 50–70% tasks automated
Operational Downtime	Frequent (due to shifts, fatigue)	Minimal (24/7 operation)
Fulfillment Throughput	1000–1500 orders/day	3000–5000+ orders/day
Cost per Order	\$3.00–\$5.00	\$1.00–\$2.00
Inventory Audit Cycle	Monthly/Quarterly	Weekly or real-time

3.3. Lessons Learned from Successful Adoption

I. Some common lessons can be drawn from the case studies above that highlight the key needs for successful robotics integration for better future use.

II. Initializing with the pilot projects: Big companies perform small-scale pilots to see whether the suggested plot is possible, measure its effects, and improve processes before fully implementing it; This in turn lowers the associated risk and builds confidence among all parties that are involved.

III. Focusing on Hybrid Human-Robot Models: The best way to use automation is to combine it with human skills as augmentation. Cobots and AMRs are meant to make human labor better, not to replace it totally; this encourages teamwork and productivity.

IV. Invest in Training and Change Management: It's important to teach frontline workers to work with robots and to change management roles to fit new procedures to incorporate robots in regular work process. Companies that employed structured training programs had more people using their services with fewer problems as compared to those without trainings.

Figure 2. Steps to Achieve Robotics Integration

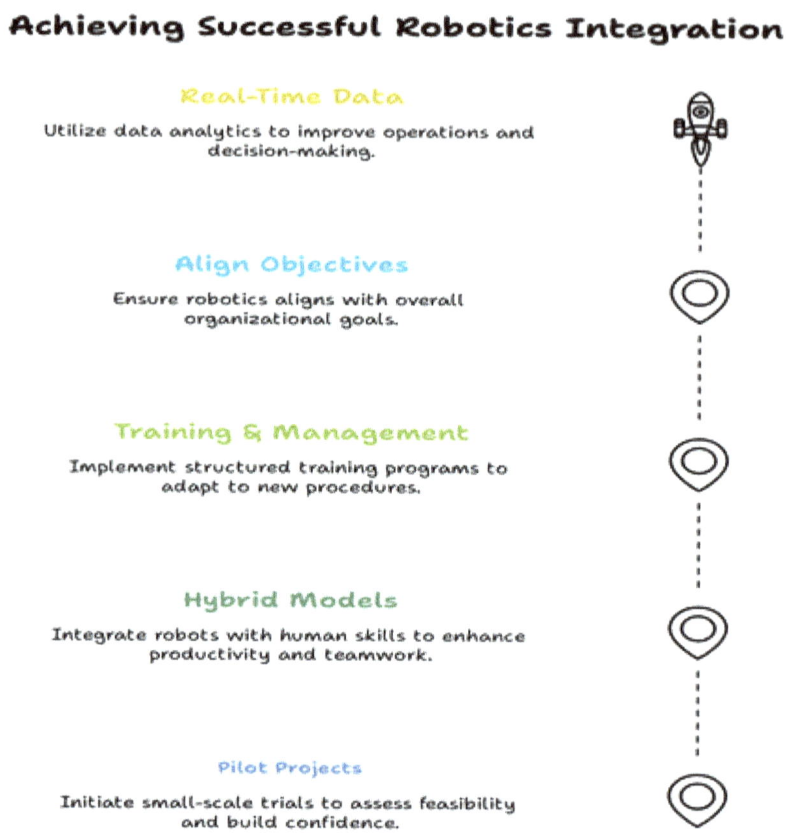

3.4. Tribulations and Limitations

Despite all the benefits robots may bring to logistics, there are still major problems that make it hard to use them more widely and thus impacts their popularity at large. These include problems with capital- investment, technology, culture, and security that needs to be addressed before deployment can be called successful.

High Initial Investment Costs

Robotic systems, especially large-scale warehouse automation, require a lot of money up front to buy hardware, build software, and change infrastructure to suit the new architecture plan. Payment for buying, integrating, and keeping up with modular systems like AMRs or drones is huge.

a) Examples of cost parts:

∇ Basic robotic arm: $25,000 to $100,000 approximately;
∇ AMR fleet (per robot): $20,000 to $35,000 approx;
∇ For a medium-sized warehouse, an Automated Storage and Retrieval System (AS/RS) costs between $1 and $5 million as per some published reports;
∇ Integration with WMS/ERP: More than $100,000 approx;
∇ Training employees and downtime for operations are hidden costs and are variable.

Even though the ROI is usually good in the long term, smaller businesses with limited resources may find it hard to justify the cost at least in their initial years. This problem could be easier to deal with via public-private partnerships, leasing deals (Robotics-as-a-Service), and tax breaks that would give a breather to the small company.

b) Integration with Legacy Systems

Many logistics organizations still utilize old IT systems that may not work with modern, high-end robotic platforms. Generally, problems with integration include:

∇ Data Silos: The contemporary systems may not be able to talk to robots in real time, which makes it harder to see and coordinate impacting the functioning altogether.
∇ Middleware Requirements: To connect robots to ERP or WMS systems, customized APIs or middleware is needed.
∇ Hardware Conflicts: Robotic activities may not work with old conveyor belts or forklifts that are too outdated.
∇ Delays in integration might raise costs, cause problems with operations, and slow down the move to automated processes resulting in greater time-lag. Companies need to provide money to improve system design and make sure that modules work together so that migrations from legacy to advanced systems are smooth enough.

c) Workforce Adaptation and Training Needs

Workers who are worried about losing their jobs or are scared of new technology may be against the deployment of robots resulting in ineffective communication thus this may lower morale and productivity.

Challenges faced by the workforce:

- ∇ Job Stability: Machines might take over jobs that people do leading to insecure mindset.
- ∇ Lack of Skills: Older workers don't have enough digital skills and mostly are unskilled labors.
- ∇ Cultural Resistance: Not trusting AI-generated decisions or not believing in automation because of lack of everyday use.

Some of the suggested remedies are:

- ∇ A clear description of what automation does and what it doesn't.
- ∇ Training and retraining programs (for example, learning how to operate robots or read data).

Some of the new jobs that will be created include robot maintenance technicians, data analysts, and automation supervisors among many others.

Figure 3. Challenges in Robotics Integration

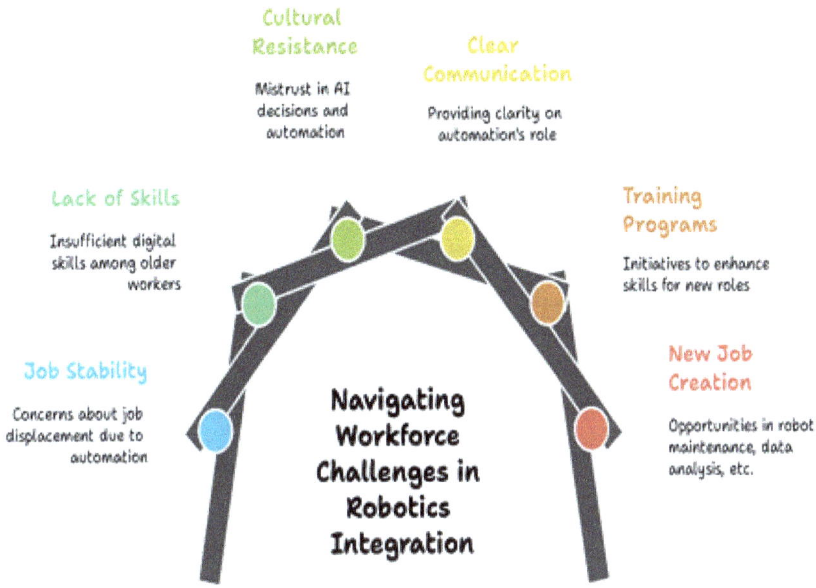

It is also reported that companies that regard robots as helpers instead of threats create a culture of collaboration and get more employees involved preparing for better environment in long run.

d) Cybersecurity and Data Concerns

As logistics robots grow better at working with the Internet of Things (IoT), cloud platforms, and AI, they also become more susceptible to be attacked by hackers; security and integrity of supply chain data are at risk because of the following:

∇ Unauthorized Access: Cybercriminals might get a revelation of real-time supply information or working of robots.
∇ Ransomware attacks: When hackers stop autonomous systems in order to extract money to restart the entire system.

∇ Data Leakage: Getting to use sensitive customer or inventory information without authorized permission.

Here are some of the ways to lessen/reduce the damage:

∇ Try to set up secure communication protocols and end-to-end encryption without fail.
∇ Do random/regular audits for vulnerabilities and penetration testing.
∇ Use AI-based technology to identify unusual pattern in network.
∇ Set up role-based access controls and multi-factor authentication for the users.
∇ Regulatory compliance, including GDPR and ISO 27001, as well as industry requirements for robotics security, is continually changing and should be implemented.

Robots have been used in logistics beyond pilot studies, and they have been shown to be a reliable way to speed up, improve accuracy, and make operations more efficient; For example, Amazon's robot-enabled fulfillment facilities, JD.com's drone deliveries, and DHL's collaborative AMRs implementation have shown that robots can make the supply chain better in measurable ways says (Carvalho et al., 2019).

Key performance indicators including order cycle time, picking accuracy, and fulfillment throughput thatshowcase significant improvement with automation. But these betterments come with problems as well; strategic planning and proactive management are required for big capital investments, technology integration, workforce readiness, and cybersecurity in general.

Companies that have successfully adopted robotics provide important lessons likewise- start small, focus on how people and robots can work together, utilize data/info wisely, and give workers the tools along with the training they need to learn new skills. There is no one way to get to robotic transformation, but it is increasingly necessary for businesses to stay competitive and strong in a rapidly changing global logistics environment normally.

As robots grow cheaper, more adaptable, and smarter, they shall be employed by smaller and mid-sized businesses instead of large ones only. This might be the next mammoth step in supply chain innovation.

4. FUTURE TRENDS AND INNOVATIONS

As technology improves, global competition rises, and environmental concerns grow, robots will play enormous - innovative role in the logistics industry. The initial steps in integrating robots were to automate tedious chores and make internal processes better. Innovations in robotics are pushing the boundaries by creating smart, collaborative, sustainable, and service-oriented robotic ecosystems that are changing the future of supply chains as a rule.

4.1. Predictive robotics that use artificial intelligence

Customarily, the use of artificial intelligence (AI) in designing logistics robots better at predicting things is a big step forward. Predictive robotics employs AI and machine learning to predict operational needs, customer behavior, and system performance as foreseen by (Carvalho et al., 2019); robots take preventive steps instead of just reacting to problems routinely, for example:

∇ Smart Decision-Making: Robots are going from being programmable machines to self-driving agents that can make complex decisions as and when the need arises. AI-powered robots at warehouses may look at order history, real-time inventory data, and demand forecasts to move shelves around, figure out which items to select first, and improve traffic flow at much better pace. Typically, this kind of predictive intelligence shrink downtime, speeds up throughput, and makes the system more responsive.

∇ Predicting Maintenance: AI systems can almost predict when equipment will break down. Robots can find early signs of failure and schedule repairs before a breakdown by looking at how things are used, measuring temperature, and collecting vibration data leading to reliable system on the whole. This kind of predictive maintenance cuts down on costly downtime and makes robotic equipment last longer giving superior ROI generally.

∇ Route and Load Optimization: In transportation and last-mile logistics, predictive robots may improve delivery routes in real-time by looking at traffic patterns, weather conditions, and how easy it is for customers to land to their destinations; AI-powered robots routinely dynamically reschedule delivery to avoid delays and increase fuel efficiency that guarantees on-time deliveries and eventually happier customers.

AI-driven robotics goes beyond just automating tasks- it also includes making adaptive learning systems that change for the better. As AI models become improved, it would lead to next generation of logistics robots that can manage whole processes without manual intervention.

4.2. Working together with robots

The future of robotics in logistics is about working together instead of replacing humans with robots for better functionality. This hybrid environments lets people and robots work together to produce the best possible results. This strategy, which is also called Human-Robot Collaboration (HRC), takes use of the best characteristics of both robots and human as per (Villani et al., 2018).

"Collaborative robots," or "cobots," are designed to provide work safely to humans specifically to the tasks that are hazardous in nature. Collaborative robots make it easier to choose, sort, label, and package things in logistics; they are easy train thus are particularly useful in places where dexterity, flexibility, and a lot of human engagement are expected.

Cobots don't take the position of warehouse workers; instead, they make things run more smoothly by doing tasks that are repetitive and are labor-intensive. For instance, a human operator may focus on checking the quality of the goods while the collaborative robot takes care of moving them to the right place.

Future logistics operations surely will feature more flexible workplaces where tasks are planned out between humans and robots. Robots will handle activities that are predictable and need a lot of data, while people will take on strategic, managerial, and exception-handling roles in general. This relationship makes work safer, more efficient, and more satiating for workers.

Intuitive Interfaces and Training- Voice commands, gesture recognition, and augmented reality (AR) are all examples of user-friendly interfaces that will make it easier even for non-technical people to work with robots. Virtual reality (VR) training programs will help people quickly get used to new ways of working together, which makes it easier for robots to integrate their functionality with the workforce seamlessly.

Working together with robots is more than just a change in technology- it means a change in culture and organization that changes the way people work or were working in the logistics sector. As robots become more like people and less like tools, a new era of co-evolution between humans and machines has already begin.

4.3. Sustainable and Green Robotics

∇ Logistics now places sustainability high on its priority list- Because of growing worries about the environment, government rules, and corporate social responsibility goals, it is very important to run businesses in a way that is more sustainable and lesser energy. (Bag et al., 2021) has rightly pointed out that robots are a good way to resolve this problem.

∇ Energy Efficiency and Optimization- Modern robotic systems are designed to use less energy by planning routes intelligently, reducing idle time, and using mechanism that calls for less energy. AMRs may be set up to find the fastest way to go there while avoiding obstacles; this saves energy and time. Some warehouse systems include robots with regenerative braking that converts kinetic energy into electrical energy-thus it's clear that advanced robots can go into low-energy modes when they aren't doing anything and charge themselves when they are idle.

Figure 4. Pathways to Green Logistics

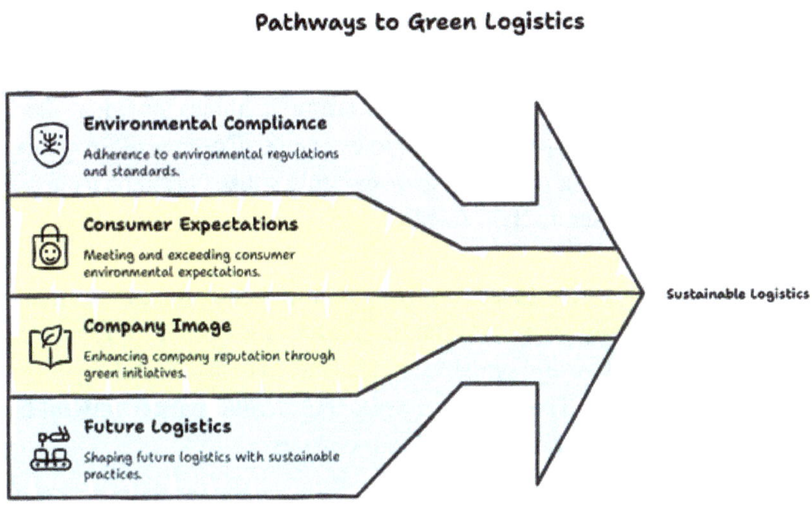

∇ Reducing Waste and Overproduction- AI-powered robots improve inventory management by keeping track of stock levels, which cuts down on overproduction and unnecessary storage at large. Automated packaging robots make the best use of materials by making packages that best-fit the size of the product. This cuts down on plastic and cardboard waste naturally.
∇ Better Last-Mile Delivery- Electric self-driving delivery cars and drones are environment-friendly than other traditional delivery options. They easily cut

down on traffic jams, carbon emissions, and noise pollution, especially when combined with algorithms that help find the best routes possible.
∇ Recycling and lifespan -things to think about; The longevity of the equipment is another something that sustainable robotics looks at. Modular designs make it easier to upgrade or recycle pre-used parts. More and more, robotics companies are adopting recyclable materials and technical approaches to take apart robots when they are obsolete.
∇ Green robotics not only meets environmental rules, but it also meets consumer expectations and improves the company's image. In future, the main thing that will be looked at when judging logistics systems will be how they affect the environment. Robots will be a big part of making logistics more sustainable and predictable.

4.4. RaaS, or Robotics as a Service

∇ Robotics-as-a-Service (RaaS) is becoming a game-changing idea that makes advanced automation available to everyone. Companies may now pay for robotic services on a pay-per-use or monthly subscription basis instead of having to buy and operate their own robotic systems- similar to Software-as-a-Service (SaaS).
∇ Lowered Entry Barriers; historically, high initial costs have made it hard for companies, especially small and medium-sized ones (SMEs), to use robots. RaaS is a cost-effective and scalable choice since it turns capital expenses into operational costs only. This lets businesses try out automation without having to pay for it preventing initial big investments.

RaaS lets businesses change the number of robots they use based on changing 'seasonal demand', which makes it easy to scale up or down. (Wamba & Queiroz, 2020) narrated that throughout the holiday season, an online store may require more robots than usual, but during the rest of the year, it may need fewer. RaaS businesses provide flexible contracts to deal with these changes effectively.

4.5. Analytics and Insights:

The service provider is in charge of maintenance, software updates, and performance improvements especially in case of RaaS. This takes some of the stress off of the IT and engineering team within the company while making sure that businesses always employ the most up-to-date technology as disclosed by 'Locus Robotics'.

Many RaaS systems come with dashboards and analytical tools that show you real-time statistics on "how well robots are working?', 'how productive they are?', and 'how much money they are saving?'. These insights help managers make smart decisions about how to run their businesses.

As RaaS models improve, they are expected to be used widely across various industries and areas, making robots available to businesses of all sizes and types in future.

5. CONCLUSION

(Wang et al., 2021) has put in the picture of the use of robots in logistics has gone beyond automating simple tasks to creating complex, interconnected systems that make every part of the supply chain more efficient than before. This chapter talks about how robots may be used for warehouse automation, transportation, inventory management, and last-mile delivery etc. Using them has led to measurable improvements in speed, accuracy, cost-effectiveness, and adaptability as reported by some of the researchers already.

Case studies from Amazon, DHL, JD.com, and other companies show how robots may have real influence on various performance measures. Despite the high costs of initial- investment, difficulties of integrating systems, need for trained workers, and the risks to cybersecurity, robots have valid and proven strategic benefits.

Also, emerging ideas like AI-driven predictive robotics, human-robot interaction, sustainable automation, and Robotics-as-a-Service point to a future where robots are smarter, easier to use, and provide for a sustainable environment.

The Strategic Importance of Robotics in Modern Logistics

In this era of things are rapidly evolving digital world, customers are imperative and thus the competitive world is hard to forecast; robots provide a strategic advantage that goes beyond just saving money. It lets businesses:

∇ Get Agile- Quickly adjust to changes in the market as per the wishes of the customers.
∇ Improve resilience- Use independent and backup systems to minimize disruptions if any.
∇ Increase Visibility- Get real-time information for better decision making.
∇ Improve sustainability- By aligning the businesses with environmental and legal goals.
∇ Reduce attrition- By making the workplace safer and less labor-intensive.
∇ Robotics has gone from being a minor hobby or future aspiration to an important skill that affects how logistics companies operate altogether.

Robots will be used more and more in logistics in the future; clubbing AI, the Internet of Things (IoT), cloud computing, and 5G will prove how intelligent and independent robots can be. Instead of being rare, human-robot cooperation will become the norm in the days to come. Robots will go from being tools to being part of logistical innovation thus becoming irreplaceable. Success takes planning ahead, making investments in people, and being ready to adopt the change. Governments, universities, and businesses all need to work together to make sure that robots can grow in a responsible and inclusive way by providing the right infrastructure, skills, and rules at once.

In short, the future of logistics will be defined by automation, the capacity to foresee, working together, being environmentally friendly, and putting the customer first. Robotics is a vital part of this alteration, moving the industry toward a smarter, faster, and a sustainable future.

REFERENCES

Bag, S., Telukdarie, A., & Gupta, S. (2021). Industry 4.0 maturity and supply chain performance: An empirical investigation. *International Journal of Production Economics, 231*, 107867. DOI: 10.1016/j.ijpe.2020.107867

Banur, O. M., Patle, B. K., & Pawar, S. (2024). Integration of robotics and automation in supply chain: A comprehensive review. *Robotic Systems and Applications, 4*(1), 1–19. DOI: 10.21595/rsa.2023.23349

Belter, D., & Skrzypczyński, P. (2019). Energy-efficient motion planning for mobile robots using sensor-driven adaptive control. *Robotics and Autonomous Systems, 111*, 199–216. DOI: 10.1016/j.robot.2018.10.015

Berkers, H. A., van der Heijden, B., & Bosch, C. (2023). The role of robotization in work design: A comparative case study of order pickers in eight logistic warehouses. *International Journal of Human Resource Management*. Advance online publication. DOI: 10.1080/09585192.2022.2043925

Bondar, M. (2023, July 13). Is your supply chain trustworthy? *Deloitte Insights*. https://www.deloitte.com/us/en/insights/topics/supply-chain/issues-in-global-supply-chain.html

Boone, T. (2023, January 23). Why you must digitize your products for today's complex global supply chain. *Supply Chain Dive*. https://www.supplychaindive.com/spons/why-you-must-digitize-your-products-for-todays-complex-global-supply-chain/640439/

Boysen, N., Briskorn, D., & Schwerdfeger, S. (2019). Drone delivery from a logistics perspective. *Computers & Operations Research, 104*, 105–117. DOI: 10.1016/j.cor.2018.12.002

Caputo, F., Greco, A., D'Aniello, G., & Notarnicola, F. (2021). Autonomous mobile robots in warehouse logistics: A literature review and future research agenda. *Computers & Industrial Engineering, 162*, 107783. DOI: 10.1016/j.cie.2021.107783

Carvalho, T. P., Soares, F. A., Vita, R., Francisco, R. P., Basto, J. P., & Alcalá, S. G. (2019). A systematic literature review of machine learning methods applied to predictive maintenance. *Computers & Industrial Engineering, 137*, 106024. DOI: 10.1016/j.cie.2019.106024

Cherubini, A., Passama, R., Crosnier, A., Lasnier, A., & Fraisse, P. (2016). Collaborative manufacturing with physical human–robot interaction. *Robotics and Computer-integrated Manufacturing, 40*, 1–13. DOI: 10.1016/j.rcim.2015.12.007

Deng, X., Li, H., Cheng, Y., & Zhang, Y. (2023). Deep learning-based predictive maintenance for industrial robotics: A comprehensive review. *Robotics and Computer-integrated Manufacturing*, *80*, 102517. DOI: 10.1016/j.rcim.2022.102517

FreightAmigo. (2023). *10 key supply-chain trends shaping global trade in 2023.* https://www.freightamigo.com/blog/10-key-supply-chain-trends-shaping-global-trade-in-2023

Guerra, M., Murciano-Brea, J., & Martín, D. (2023). Autonomous delivery robots in urban logistics: Current development and future trends. *Sensors (Basel)*, *23*(3), 1120. DOI: 10.3390/s23031120 PMID: 36772161

Gunasekaran, A., Subramanian, N., & Ngai, E. W. T. (2018). The future of operations management: An outlook and analysis. *International Journal of Production Economics*, *195*, 102–118. DOI: 10.1016/j.ijpe.2017.12.002

Krüger, J., Lien, T. K., & Verl, A. (2019). Cooperation of human and machines in assembly lines. *CIRP Annals*, *68*(2), 1–24. DOI: 10.1016/j.cirp.2019.04.097

Li, J., Luo, H., Xie, N., & Zhang, Y. (2022). Automation and robotics in smart warehousing: A systematic review. *Robotics and Computer-integrated Manufacturing*, *78*, 102386. DOI: 10.1016/j.rcim.2022.102386

Majeed, A., & Rupasinghe, T. (2023). Drones in last-mile delivery logistics: Applications, challenges, and future trends. *Drones (Basel)*, *7*(4), 260. DOI: 10.3390/drones7040260

Park, J., Kim, J., & Lee, D. (2021). Design and development of last-mile delivery robots: A review of autonomous ground vehicles. *IEEE Access : Practical Innovations, Open Solutions*, *9*, 13472–13490. DOI: 10.1109/ACCESS.2021.3051204

Sivakumar, A. I., & Li, W. (2020). Last-mile logistics: Key challenges and potential solutions. *Transportation Research Part E, Logistics and Transportation Review*, *141*, 102014. DOI: 10.1016/j.tre.2020.102014

Soteriades, E. S., & Psarommatis, F. (2021). Robotics in logistics: A detailed review of intelligent warehouse systems. *Computers & Industrial Engineering*, *159*, 107550. DOI: 10.1016/j.cie.2021.107550

Supply Chain Digital. (2024, August 7). *4 challenges in logistics industry in 2024.* https://www.dhl.com/discover/en-vn/logistics-advice/logistics-insights/challenges-in-logistics-and-supply-chain-management-best-practices

Villani, V., Pini, F., Leali, F., & Secchi, C. (2018). Survey on human–robot collaboration in industrial settings: Safety, interaction, and future directions. *Mechatronics*, *55*, 248–266. DOI: 10.1016/j.mechatronics.2018.02.009

Wamba, S. F., & Queiroz, M. M. (2020). Industry 4.0 and supply chain digitalization: A systematic review. *Production Planning and Control*, *33*(2–3), 91–104. DOI: 10.1080/09537287.2020.1810761

Wang, X. V., Kemény, Z., Váncza, J., & Wang, L. (2021). Human–robot collaboration in manufacturing: A survey of communication, safety, and control. *Robotics and Computer-integrated Manufacturing*, *68*, 102012. DOI: 10.1016/j.rcim.2020.102012

Wurman, P. R., Barrett, S., & Kawamoto, S. (2022). Coordinating autonomous robots in large-scale logistics environments. *AI Magazine*, *43*(1), 17–30. DOI: 10.1609/aimag.v43i1.20020

Yang, G., Lv, K., & Fan, S. (2023). Robots for sustainable development: Supporting SDGs through intelligent automation. *Nature Communications*, *14*, 2543. DOI: 10.1038/s41467-023-38215-y

Zhang, C., & Chen, Y. (2020). Sensor-based fault diagnosis of industrial robots: A review. *IEEE Transactions on Industrial Informatics*, *16*(7), 4446–4457. DOI: 10.1109/TII.2019.2958743

Zonta, T., da Costa, C. A., da Rosa Righi, R., de Lima, M. J., da Trindade, E. S., & Li, G. P. (2020). Predictive maintenance in the Industry4.0: A systematic review. *Computers & Industrial Engineering*, *150*, 106889. DOI: 10.1016/j.cie.2020.106889

Compilation of References

Abdollahi, H., Mahoor, M. H., Zandie, R., Siewierski, J., & Qualls, S. H. (2022). Artificial emotional intelligence in socially assistive robots for older adults: A pilot study. *IEEE Transactions on Affective Computing*, *13*(3), 1123–1137. DOI: 10.1109/TAFFC.2022.3143803 PMID: 37840968

Abdulhafedh, A. (2017). How to Detect and Remove Temporal Autocorrelation in Vehicular Crash Data. *Journal of Transportation Technologies*, *07*(02), 133–147. DOI: 10.4236/jtts.2017.72010

Abeliansky, A. L., Beulmann, M., & Prettner, K. (2024). Are they coming for us? Industrial robots and the mental health of workers. *Research Policy*, *53*(3), 104956. Advance online publication. DOI: 10.1016/j.respol.2024.104956

Acrelec United Kingdom. (2022, June 1). *Evolution of self-service technology*. LinkedIn. https://www.linkedin.com/pulse/evolution-self-service-technology-acrelec-united-kingdom/

Adhiatma, A., Fachrunnisa, O., & Tjahjono, H. K. (2021). A Value Creation Process for Sustainability of Knowledge Based-Society. In Barolli, L., Poniszewska-Maranda, A., & Enokido, T. (Eds.), *Complex, Intelligent and Software Intensive Systems* (pp. 307–314). Springer International Publishing., DOI: 10.1007/978-3-030-50454-0_29

Adil, M., Alghaswyneh, O. F. M., & Albkour, A. M. (2013, March 15). *SERVQUAL and SERVPERF: A Review of Measures in Services Marketing Research*. ResearchGate. https://www.researchgate.net/publication/304461403_SERVQUAL_and_SERVPERF_A_Review_of_Measures_in_Services_Marketing_Research

Agrawal, K., & Bhat, A. R. (2025). Advances in 3D printing with eco-friendly materials: A sustainable approach to manufacturing. *RSC Sustainability*, *3*(6), 2582–2604. DOI: 10.1039/D4SU00718B

Ahn, J. A., & Seo, S. (2018). Consumer responses to interactive restaurant self-service technology (IRSST): The role of gadget-loving propensity. *International Journal of Hospitality Management, 74*, 109–121. DOI: 10.1016/j.ijhm.2018.02.020

AITI - ICT Development Index 2023. (2023, December 29). Retrieved June 12, 2024, from https://aiti.gov.bn/news/2023/ict-development-index-2023/

AITI. (2019). Brunei Darussalam Information Communications Technology (ICT) Household Report 2019. In aiti.gov.bn

Alavi, M., & Leidner, D. E. (2001). Review: Knowledge Management and Knowledge Management Systems: Conceptual Foundations and Research Issues. *Management Information Systems Quarterly, 25*(1), 107–136. DOI: 10.2307/3250961

Alenjareghi, M. J., Keivanpour, S., Chinniah, Y. A., Jocelyn, S., & Oulmane, A. (2024). Safe human-robot collaboration: A systematic review of risk assessment methods with AI integration and standardization considerations. *International Journal of Advanced Manufacturing Technology, 133*(9-10), 4077–4110. Advance online publication. DOI: 10.1007/s00170-024-13948-3

Alin, A. (2010). Multicollinearity. *Wiley Interdisciplinary Reviews: Computational Statistics, 2*(3), 370–374. DOI: 10.1002/wics.84

Allenby, B. (2006). The ontologies of industrial ecology? *Progress in Industrial Ecology, 3*(1–2), 28–40. DOI: 10.1504/PIE.2006.010039

Allgor, R., Cezik, T., & Chen, D. (2023). Algorithm for robotic picking in Amazon fulfillment centers enables humans and robots to work together effectively. *INFORMS Journal on Applied Analytics, 53*(4), 266–282. DOI: 10.1287/inte.2022.1143

Almalki, M., Gray, K., & Martin-Sanchez, F. (2016). The use of self-quantification systems for personal health information: Big data management activities and prospects. *Health Information Science and Systems, 4*(1), 1–9. PMID: 26019809

Almokdad, E., & Lee, C. H. (2024). *Service Robots in the Workplace: Fostering Sustainable Collaboration by Alleviating Perceived Burdensomeness*. Sustainability., DOI: 10.3390/su16219518

Almunawar, M. N., & Anshari, M. (2022). Customer acceptance of online delivery platform during the COVID-19 pandemic: the case of Brunei Darussalam. *Journal of Science and Technology Policy Management/Journal of Science & Technology Policy Management, 15*(2), 288–310. DOI: 10.1108/JSTPM-04-2022-0073

Almunawar, M. N., & Almunawar, H. J. (2018). From Information Systems Outsourcing to Cloud Computing. In *Encyclopedia of Information Science and Technology* (4th ed., Vol. 1–10, pp. 1101–1115). IGI Global.

Almunawar, M. N., & Anshari, M. (2018). Moving to the Cloud. *Journal of Business and Economic Analysis, 01*(01), 65–80. DOI: 10.1142/S2737566821800049

Alolah, T., Stewart, R. A., Panuwatwanich, K., & Mohamed, S. (2014). Determining the causal relationships among balanced scorecard perspectives on school safety performance: Case of Saudi Arabia. *Accident; Analysis and Prevention, 68*, 57–74. DOI: 10.1016/j.aap.2014.02.002 PMID: 24589246

Alqahtani, M. F., & Afy-Shararah, M. (2025). Integrating Circular Economy Practices into Renewable Energy in the Manufacturing Sector: A Systematic Review of the Literature. *Sustainability (Basel), 17*(16), 7301. DOI: 10.3390/su17167301

Alshadoodee, H. A. A., Mansoor, M. S. G., Kuba, H. K., & Gheni, H. M. (2022). The role of artificial intelligence in enhancing administrative decision support systems by depend on knowledge management. *Bulletin of Electrical Engineering and Informatics, 11*(6), 3577–3589. DOI: 10.11591/eei.v11i6.4243

Alves, J., Lima, T. M., & Gaspar, P. D. (2023). Is Industry 5.0 a Human-Centred Approach? A Systematic Review. *Processes (Basel, Switzerland), 11*(1), 1. Advance online publication. DOI: 10.3390/pr11010193

Amodei, D., Olah, C., Steinhardt, J., Christiano, P., Schulman, J., & Mané, D. (2016). Concrete problems in AI safety. arXiv preprint arXiv:1606.06565.

Amu, A. (2021, September 16). Progresif debuts self-service kiosk at store front. *The Bruneian*. Retrieved August 28, 2023, from https://thebruneian.news/2021/09/16/progresif-debuts-self-service-kiosk-at-store-front/.

Anitha, M., Babel, A., Kumar, A., Rauniyar, A., & Zahid, K. (2019). Cloud-based secured QR code for self-service access control system at resort and hotels. Lecture Notes in Networks and Systems, 387–395. DOI: 10.1007/978-981-13-7150-9_41

Anshari, M., Almunawar, M. N., Lim, S. A., & Al-Mudimigh, A. (2019). Customer relationship management and big data enabled: Personalization & customization of services. *Applied Computing and Informatics, 15*(2), 2. Advance online publication. DOI: 10.1016/j.aci.2018.05.004

Anshari, M., Almunawar, M. N., & Low, P. K. C. (2012). CRM 2.0 Within E-Health Systems: Towards Achieving Health Literacy & Customer Satisfaction. *International Journal on Informatics for Development, 1*(2), 90476. DOI: 10.14421/ijid.2012.01201

Argyris, C., & Schön, D. A. (1978). *Organizational Learning: A Theory of Action Perspective*. Addison-Wesley.

Armbrust, M., Stoica, I., Zaharia, M., Fox, A., Griffith, R., Joseph, A. D., Katz, R., Konwinski, A., Lee, G., Patterson, D., & Rabkin, A. (2010). A view of cloud computing. *Communications of the ACM*, *53*(4), 4. Advance online publication. DOI: 10.1145/1721654.1721672

Ashraf, Z., Sohail, A., & Yousaf, M. (2019). Performance Analysis of Network Applications on IPv6 Cloud Connected Virtual Machine. *International Journal of Computer Network and Information Security*, *11*(12), 1–9. DOI: 10.5815/ijcnis.2019.12.01

Autor, D., Mindell, D., & Reynolds, E. (2020). *The Work of the Future: Building Better Jobs in an Age of Intelligent Machines*. MIT Task Force.

Ayoub, J., Avetisyan, L., Makki, M., & Zhou, F. (2021). An investigation of drivers' dynamic situational trust in conditionally automated driving. *arXiv preprint arXiv:2112.04095*. https://doi.org//arXiv.2112.04095DOI: 10.48550

Azahari, I. (2022, February 23). Riding the digitisation wave. *Borneo Bulletin*. Retrieved August 28, 2023, from https://borneobulletin.com.bn/riding-the-digitisation-wave/

Bag, S., Telukdarie, A., & Gupta, S. (2021). Industry 4.0 maturity and supply chain performance: An empirical investigation. *International Journal of Production Economics*, *231*, 107867. DOI: 10.1016/j.ijpe.2020.107867

Baltrusch, S. J., Krause, F., de Vries, A. W., van Dijk, W., & de Looze, M. P. (2022). What about the human in human robot collaboration?: A literature review on HRC's effects on aspects of job quality. In *Ergonomics* (Vol. 65, Issue 5). DOI: 10.1080/00140139.2021.1984585

Banasiewicz, A. (2021). Organizational Learning in the Age of Data. In *EAI/Springer Innovations in Communication and Computing*. Springer International Publishing., DOI: 10.1007/978-3-030-74866-1

Bank, D. B. S. (2022). *Annual Sustainability Report*. https://www.dbs.com/sustainability

Banur, O. M., Patle, B. K., & Pawar, S. (2024). Integration of robotics and automation in supply chain: A comprehensive review. *Robotic Systems and Applications*, *4*(1), 1–19. DOI: 10.21595/rsa.2023.23349

Barata, J., & Kayser, I. (2024). How will the digital twin shape the future of industry 5.0? *Technovation, 134*, 103025. DOI: 10.1016/j.technovation.2024.103025

Barnes, S., & Vidgen, R. (2001). Assessing the quality of auction Web sites. *Proceedings of the 34th Annual Hawaii International Conference on System Sciences*. DOI: 10.1109/HICSS.2001.927087

Bass, B. M. (1985). *Leadership and Performance Beyond Expectations*. Free Press.

BBC. (2020, August 17). A-level results: What went wrong with the algorithm? *BBC News*. https://www.bbc.com/news/education-53805156

Belter, D., & Skrzypczyński, P. (2019). Energy-efficient motion planning for mobile robots using sensor-driven adaptive control. *Robotics and Autonomous Systems, 111*, 199–216. DOI: 10.1016/j.robot.2018.10.015

Benyus, J. M. (2002). *Biomimicry: Innovation Inspired by Nature*. Harper Perennial.

Berkers, H. A., van der Heijden, B., & Bosch, C. (2023). The role of robotization in work design: A comparative case study of order pickers in eight logistic warehouses. *International Journal of Human Resource Management*. Advance online publication. DOI: 10.1080/09585192.2022.2043925

Bernoussi, R. E., & Rockinger, M. (2022). Rebalancing with transaction costs: Theory, simulations, and actual data. *Financial Markets and Portfolio Management, 37*(2), 121–160. DOI: 10.1007/s11408-022-00419-6

Bessen, J. E. (2019). *AI and Jobs: The Role of Demand*. NBER Working Paper No. 24235. DOI: 10.3386/w24235

Bhardwaj, P. (2019). Types of sampling in research. *Journal of the Practice of Cardiovascular Sciences, 5*(3), 157. DOI: 10.4103/jpcs.jpcs_62_19

Bhargava, A., Bester, M., & Bolton, L. (2021). Employees' perceptions of the implementation of robotics, artificial intelligence, and automation (RAIA) on job satisfaction, job security, and employability. *Journal of Technology in Behavioral Science, 6*(2), 106–113. DOI: 10.1007/s41347-020-00153-8

BIBD Debit Mastercard Contactless | Bank Islam Brunei Darussalam. (n.d.). https://www.bibd.com.bn/personal/cards/debit-card/debit-mastercard/#:~:text=Brunei%20Citizens%2C%20Permanent%20Residents%20and,BIBD%20Savings%20or%20Current%20Account

Binns, R. (2018). Fairness in machine learning: Lessons from political philosophy. *Proceedings of the 2018 Conference on Fairness, Accountability and Transparency*, 149–159.

BIPO Service. (2022, November). Understanding Brunei's Employment Laws & Regulations. www.biposervice.com. Retrieved June 27, 2024, from https://www.biposervice.com/wp-content/uploads/2022/11/BIPO-Times-November-2022-en.pdf

Bittencourt, V., Saakes, D., & Thiede, S. (2025). Surrogate modelling for continuous ergonomic assessment and adaptive configuration of industrial human-centered workplaces. *Journal of Manufacturing Systems*, *79*, 383–397. DOI: 10.1016/j.jmsy.2025.02.001

Blut, M., Wang, C., & Schoefer, K. (2016). Factors influencing the acceptance of self-service technologies. *Journal of Service Research*, *19*(4), 396–416. DOI: 10.1177/1094670516662352

Bobko, P., Hirshfield, L., Eloy, L., Spencer, C., Doherty, E., Driscoll, J., & Obolsky, H. (2022). Human-agent teaming and trust calibration: A theoretical framework, configurable testbed, empirical illustration, and implications for the development of adaptive systems. *Theoretical Issues in Ergonomics Science*, *24*(3), 310–334. DOI: 10.1080/1463922X.2022.2086644

Bogicevic, V., Bujisic, M., Bilgihan, A., Yang, W., & Cobanoglu, C. (2017). The impact of Traveler-focused airport technology on traveler satisfaction. *Technological Forecasting and Social Change*, *123*, 351–361. DOI: 10.1016/j.techfore.2017.03.038

Bogue, R. (2016). *Deleuze's way: Essays in transverse ethics and aesthetics*. Routledge. DOI: 10.4324/9781315576312

Boisrond, K., Tardif, P. M., & Jaafar, F. (2024). Ensuring the Integrity, Confidentiality, and Availability of IoT Data in Industry 5.0: A Systematic Mapping Study. *IEEE Access : Practical Innovations, Open Solutions*, *12*, 107017–107045. DOI: 10.1109/ACCESS.2024.3434618

Bolla, R., Bruschi, R., Davoli, F., Lombardo, C., Mohammadpour, A., Trivisonno, R., & Poe, W. Y. (2023). Adaptive Reliability for the Automated Control of Human-Robot Collaboration in Beyond-5G Networks. *IEEE Transactions on Network and Service Management*, *20*(3), 2489–2503. Advance online publication. DOI: 10.1109/TNSM.2023.3298171

Bommasani, R. (2021). *On the Opportunities and Risks of Foundation Models*. Stanford University.

Bondar, M. (2023, July 13). Is your supply chain trustworthy? *Deloitte Insights*. https://www.deloitte.com/us/en/insights/topics/supply-chain/issues-in-global-supply-chain.html

Bonett, D. G., & Wright, T. A. (2014). Cronbach's alpha reliability: Interval estimation, hypothesis testing, and sample size planning. *Journal of Organizational Behavior*, *36*(1), 3–15. DOI: 10.1002/job.1960

Boone, T. (2023, January 23). Why you must digitize your products for today's complex global supply chain. *Supply Chain Dive.* https://www.supplychaindive.com/spons/why-you-must-digitize-your-products-for-todays-complex-global-supply-chain/640439/

Borneo Bulletin. (2024, January 6). Brunei telco tells customers to be more vigilant after big surge in scams in the country. *The Star*. Retrieved June 21, 2024, from https://www.thestar.com.my/aseanplus/aseanplus-news/2024/01/06/brunei-telco-tells-customers-to-be-more-vigilant-after-big-surge-in-scams-in-the-country

Boysen, N., Briskorn, D., & Schwerdfeger, S. (2019). Drone delivery from a logistics perspective. *Computers & Operations Research*, *104*, 105–117. DOI: 10.1016/j.cor.2018.12.002

Braddock, D., & Parish, S. L. (2001). An institutional history of disability. Handbook of Disability Studies, 11(1), 11-68. DOI: 10.4135/9781412976251.n2

Breazeal, C. (2003). Emotion and sociable humanoid robots. *International Journal of Human-Computer Studies*, *59*(1–2), 119–155. DOI: 10.1016/S1071-5819(03)00018-1

Broadbent, E., Stafford, R., & MacDonald, B. (2009). Acceptance of healthcare robots for the older population: Review and future directions. *International Journal of Social Robotics*, *1*(4), 319–330. DOI: 10.1007/s12369-009-0030-6

Brocal, F., González, C., Komljenovic, D., Katina, P. F., & Sebastián, M. A. (2019). Emerging Risk Management in Industry 4.0: An Approach to Improve Organizational and Human Performance in the Complex Systems. *Complexity*, *2019*(1), 2089763. DOI: 10.1155/2019/2089763

Brown, T., & Martin, R. (2015). Action design. *Harvard Business Review*, *93*(9), 56–64.

Brunei Darussalam in 2019: Issues Revisited. (2020). In *ISEAS Publishing eBooks* (pp. 99–116). DOI: 10.1355/9789814881319-007

Brynjolfsson, E., & McAfee, A. (2014). *The second machine age: Work, progress, and prosperity in a time of brilliant technologies*. WW Norton & company.

Brynjolfsson, E., Li, D., & Raymond, L. (2023). Can AI improve worker performance? Evidence from a call center experiment. *Harvard Business School Working Paper*. https://www.hbs.edu

Brynjolfsson, E., & McAfee, A. (2017). The business of artificial intelligence: What it can—and cannot—do for your organization. *Harvard Business Review*.

Brynjolfsson, E., & McAfee, A. (2022). Human-compatible automation: Ethical and economic implications for the future workforce. *Journal of Artificial Intelligence Research*, *74*(2), 155–178. DOI: 10.1613/jair.1.13520

Bryson, J. J. (2018). Patiency is not a virtue: AI and the design of ethical systems. *Ethics and Information Technology*, *20*(1), 15–26. DOI: 10.1007/s10676-018-9448-6

Bughin, J., Catlin, T., Hirt, M., & Willmott, P. (2018). Why digital strategies fail. *The McKinsey Quarterly*, *1*(1), 14–25.

Bulmer, S., Elms, J., & Moore, S. (2018). Exploring the adoption of self-service checkouts and the associated social obligations of shopping practices. *Journal of Retailing and Consumer Services*, *42*, 107–116. DOI: 10.1016/j.jretconser.2018.01.016

Buolamwini, J., & Gebru, T. (2018). Gender shades: Intersectional accuracy disparities in commercial gender classification. *Proceedings of the 1st Conference on Fairness, Accountability and Transparency*, 77–91.

Burrell, J. (2016). How the machine 'thinks': Understanding opacity in machine learning algorithms. *Big Data & Society*, *3*(1), 2053951715622512. DOI: 10.1177/2053951715622512

Calo, R. (2015). Robotics and the lessons of cyberlaw. *California Law Review*, *103*, 513–563.

Calzavara, M., Faccio, M., Granata, I., & Trevisani, A. (2024). Achieving productivity and operator well-being: A dynamic task allocation strategy for collaborative assembly systems in Industry 5.0. *International Journal of Advanced Manufacturing Technology*. Advance online publication. DOI: 10.1007/s00170-024-14302-3

Campbell, C. S., Maglio, P. P., & Davis, M. M. (2010). From self-service to super-service: A resource mapping framework for co-creating value by shifting the boundary between provider and customer. *Information Systems and e-Business Management*, *9*(2), 173–191. DOI: 10.1007/s10257-010-0142-4

Cao, J., & Chen, N. (2024). The Influence of Robots' Fairness on Humans' Reward-Punishment Behaviors and Trust in Human-Robot Cooperative Teams. *Human Factors*, *66*(4), 1103–1117. Advance online publication. DOI: 10.1177/00187208221133272 PMID: 36218282

Caputo, F., Greco, A., D'Aniello, G., & Notarnicola, F. (2021). Autonomous mobile robots in warehouse logistics: A literature review and future research agenda. *Computers & Industrial Engineering*, *162*, 107783. DOI: 10.1016/j.cie.2021.107783

Carissoli, C., Negri, L., Bassi, M., Storm, F. A., & Delle Fave, A. (2023). Mental Workload and Human-Robot Interaction in Collaborative Tasks: A Scoping Review. *International Journal of Human-Computer Interaction*. Advance online publication. DOI: 10.1080/10447318.2023.2254639

Carvalho, T. P., Soares, F. A., Vita, R., Francisco, R. P., Basto, J. P., & Alcalá, S. G. (2019). A systematic literature review of machine learning methods applied to predictive maintenance. *Computers & Industrial Engineering*, *137*, 106024. DOI: 10.1016/j.cie.2019.106024

Cascio, W. F., & Montealegre, R. (2016). How technology is changing work and organizations. *Annual Review of Organizational Psychology and Organizational Behavior*, *3*(1), 349–375. DOI: 10.1146/annurev-orgpsych-041015-062352

Casner, S. M., & Geven, R. W. (2013). The trouble with automation: Learnings from aviation. *Human Factors*, *55*(2), 345–353. DOI: 10.1177/0018720812461372

Castillo, C., Otero-Romero, T., & Alvarez-Palau, E. J. (2025). Navigating the transition to industry 5.0: Advancing sustainability, resilience, and human-centricity in Spanish supply chain management. *Discover Sustainability*, *6*(1), 331. DOI: 10.1007/s43621-025-01190-0

Cepolina, F., & Razzoli, R. (2024). Review of robotic surgery platforms and end effectors. *Journal of Robotic Surgery*, *18*(1), 74. DOI: 10.1007/s11701-023-01781-x PMID: 38349595

Cerchione, R., Centobelli, P., Borin, E., Usai, A., & Oropallo, E. (2024, November 19). Piera Centobelli, Borin, E., Usai, A., & Oropallo, E. (2024). The WISED knowledge-creating company: Rethinking SECI model in light of the digital transition. *Journal of Knowledge Management*, *28*(10), 2997–3022. Advance online publication. DOI: 10.1108/JKM-02-2024-0133

Chan, C. Y. T., & Petrikat, D. (2022, November 30). Self-Service Technology: Benefits and Challenges. *Journal of Computer Science and Technology Studies*, *4*(2), 118–127. DOI: 10.32996/jcsts.2022.4.2.14

Chanchaichujit, J., Balasubramanian, S., & Charmaine, N. S. M. (2020, January 1). A systematic literature review on the benefit-drivers of RFID implementation in supply chains and its impact on organizational competitive advantage. *Cogent Business & Management*, *7*(1), 1818408. DOI: 10.1080/23311975.2020.1818408

Chand, S., & Lu, Y. (2023). Dual task scheduling strategy for personalized multi-objective optimization of cycle time and fatigue in human-robot collaboration. *Manufacturing Letters*, *35*, 88–95. Advance online publication. DOI: 10.1016/j.mfglet.2023.08.064

Chang, H. J., Wu, C. H., Ho, J. F., & Chen, P. Y. (2008). On sample size in using central limit theorem for gamma distribution. *Information and Management Sciences*, *19*(1), 153–174.

Chao, C. M. (2019, July 16). Factors Determining the Behavioral Intention to Use Mobile Learning: An Application and Extension of the UTAUT Model. *Frontiers in Psychology*, *10*, 1652. Advance online publication. DOI: 10.3389/fpsyg.2019.01652 PMID: 31379679

Chatterjee, S., Das, S., Ganguly, K., & Mandal, D. (2024). Advancements in robotic surgery: Innovations, challenges and future prospects. *Journal of Robotic Surgery*, *18*(1), 28. DOI: 10.1007/s11701-023-01801-w PMID: 38231455

Cheah, J. H., Magno, F., & Cassia, F. (2023). Reviewing the SmartPLS 4 software: The latest features and enhancements. *Journal of Marketing Analytics*, *12*(1), 97–107. DOI: 10.1057/s41270-023-00266-y

Chen, W., Botchie, D., Braganza, A., & Han, H. (2022, January). A transaction cost perspective on blockchain governance in global value chains. *Strategic Change*, *31*(1), 75–87. DOI: 10.1002/jsc.2487

Cherubini, A., Passama, R., Crosnier, A., Lasnier, A., & Fraisse, P. (2016). Collaborative manufacturing with physical human–robot interaction. *Robotics and Computer-integrated Manufacturing*, *40*, 1–13. DOI: 10.1016/j.rcim.2015.12.007

Chih-Hung Wang, M. (2012, March 16). Determinants and consequences of consumer satisfaction with self-service technology in a retail setting. *Managing Service Quality*, *22*(2), 128–144. DOI: 10.1108/09604521211218945

Chiou, E. K., & Lee, J. D. (2021). Trusting automation: Designing for responsivity and resilience. *Human Factors*, *63*(1), 5–28. DOI: 10.1177/00187208211009995 PMID: 33906505

Chock, S. (2020). *Design Justice: Community-Led Practices to Build the Worlds We Need*. MIT Press. DOI: 10.7551/mitpress/12255.001.0001

Chowdhury, R. (2023). Impact of perceived convenience, service quality and security on consumers' behavioural intention towards online food delivery services: The role of attitude as mediator. *SN Business & Economics*, *3*(1), 29. Advance online publication. DOI: 10.1007/s43546-023-00422-7 PMID: 36643187

Christoforakos, L., Gallucci, A., Surmava-Große, T., Ullrich, D., & Diefenbach, S. (2021). Can robots earn our trust the same way humans do? A systematic exploration of competence, warmth, and anthropomorphism as determinants of trust development in HRI. *Frontiers in Robotics and AI*, *8*, 640444. Advance online publication. DOI: 10.3389/frobt.2021.640444 PMID: 33898531

Chuang, S. (2024). Indispensable skills for human employees in the age of robots and AI. *European Journal of Training and Development*, *48*(1/2), 179–195. Advance online publication. DOI: 10.1108/EJTD-06-2022-0062

Chun, R. (2019, January 31). The Dark Art of Stealing from Self-Checkouts. The Atlantic. https://www.theatlantic.com/magazine/archive/2018/03/stealing-from-self-checkout/550940/

Chung, K. C., & Tan, P. J. B. (2025). Artificial intelligence and internet of things to improve smart hospitality services. *Internet of Things (The Netherlands)*. DOI: 10.1016/j.iot.2025.101544

Ciano, M. P., Peron, M., Panza, L., & Pozzi, R. (2025). Industry 4.0 technologies in support of circular Economy: A 10R-based integration framework. *Computers & Industrial Engineering*, *201*, 110867. Advance online publication. DOI: 10.1016/j.cie.2025.110867

Clark, A., & Chalmers, D. J. (1998). The extended mind. *Analysis*, *58*(1), 7–19. DOI: 10.1093/analys/58.1.7

Coeckelbergh, M. (2010). Health care, capabilities, and AI assistive technologies. *Ethics and Information Technology*, *12*(3), 205–216.

Coeckelbergh, M. (2020). *AI Ethics*. MIT Press. DOI: 10.7551/mitpress/12549.001.0001

Cohen, J. (2013). Statistical Power Analysis for the Behavioral Sciences. In *Routledge eBooks*. DOI: 10.4324/9780203771587

Colgate, J. E., Wannasuphoprasit, W., & Peshkin, M. A. (1996, November). Cobots: Robots for collaboration with human operators. In *ASME international mechanical engineering congress and exposition* (Vol. 15281, pp. 433-439). American Society of Mechanical Engineers.

Colgate, J. E., Wannasuphoprasit, W., & Peshkin, M. A. (1996). Cobots: Robots for collaboration with human operators. *Proceedings of the ASME Dynamic Systems and Control Division*, *58*(1), 433–439. DOI: 10.1115/IMECE1996-0367

Collier, J. E., & Kimes, S. E. (2012). Only if it is convenient: Understanding How Convenience Influences Self-Service Technology Evaluation. *Journal of Service Research, 16*(1), 39–51. DOI: 10.1177/1094670512458454

Collier, J. E., & Sherrell, D. L. (2010). Examining the influence of control and convenience in a self-service setting. *Journal of the Academy of Marketing Science, 38*(4), 490–509. DOI: 10.1007/s11747-009-0179-4

Commisso, D. (2022, June 21). *Checking out in 2022, from self-checkout to Amazon's "Just walk out" tech*. CivicScience. https://civicscience.com/checking-out-in-2022-from-self-checkout-to-amazons-just-walk-out-tech/

Considine, E., & Cormican, K. (2016). Self-service Technology Adoption: An Analysis of Customer to Technology Interactions. *Procedia Computer Science, 100*, 103–109. DOI: 10.1016/j.procs.2016.09.129

Coronado, E., Ueshiba, T., & Ramirez-Alpizar, I. G. (2024). A Path to Industry 5.0 Digital Twins for Human–Robot Collaboration by Bridging NEP+ and ROS. *Robotics (Basel, Switzerland), 13*(2), 28. Advance online publication. DOI: 10.3390/robotics13020028

Council, A. I. (2021). *AI Roadmap 2021: Making the UK a global center for AI excellence and innovation*. https://www.gov.uk/government/publications/artificial-intelligence-roadmap

Creative Dock. (2024). *The future of work aka is your position on the list of jobs most at risk from AI?* https://www.creativedock.com//blog/the-future-of-work

Cronin, J. J., & Taylor, S. A. (1994, January 1). *Servperf versus Servqual: Reconciling Performance-Based and Perceptions-Minus-Expectations Measurement of Service Quality*. Journal of Marketing; American Marketing Association. DOI: 10.1177/002224299405800110

Curran, J. M., & Meuter, M. L. (2005, March 1). Self-service technology adoption: Comparing three technologies. *Journal of Services Marketing, 19*(2), 103–113. DOI: 10.1108/08876040510591411

Cuypers, I. R. P., Hennart, J. F., Silverman, B. S., & Ertug, G. (2021, January). Transaction Cost Theory: Past Progress, Current Challenges, and Suggestions for the Future. *The Academy of Management Annals, 15*(1), 111–150. DOI: 10.5465/annals.2019.0051

D'souza, C. P., & Tapas, P. (2024). Diversity 5.0 framework: Managing innovation in Industry 5.0 through diversity and inclusion. *European Journal of Innovation Management*. Advance online publication. DOI: 10.1108/EJIM-02-2024-0139

Dagkinis, I. K., Spyrou, S., Georgantis, G. K., Psomas, P. M., Platis, A. N., & Tsoulfas, G. (2025). Impact of innovative technologies on quality and safety in surgery. *Journal of Robotic Surgery*. Advance online publication. DOI: 10.1007/s11701-024-02192-2 PMID: 39707082

Danjuma, I., Bawuro, F. A., Vassumu, M. A., & Habibu, S. A. (2018). The Service Quality Scale Debate: A Tri-Instrument Perspective for Higher Education Institutions. *Expert Journal of Business and Management, 6*(2), 127–133. https://business.expertjournals.com/ark:/16759/EJBM_612danjuma127-133.pdf

Daronnat, S., Azzopardi, L., Halvey, M., & Dubiel, M. (2021). Inferring trust from users' behaviours; agents' predictability positively affects trust, task performance and cognitive load in human-agent real-time collaboration. *Frontiers in Robotics and AI, 8*, 642201. Advance online publication. DOI: 10.3389/frobt.2021.642201 PMID: 34307467

Daugherty, P. R., & Wilson, H. J. (2018). *Human + Machine: Reimagining Work in the Age of AI*. Harvard Business Review Press.

Davenport, T. H., & Ronanki, R. (2018). Artificial intelligence for the real world. *Harvard Business Review, 96*(1), 108–116.

Davis, F. D. (1989). Perceived usefulness, perceived ease of use, and user acceptance of Information Technology. *Management Information Systems Quarterly, 13*(3), 319–340. DOI: 10.2307/249008

De Graaf, M. M., & Allouch, S. B. (2013). Exploring influencing variables for the acceptance of social robots. *Robotics and Autonomous Systems, 61*(12), 1476–1486. DOI: 10.1016/j.robot.2013.07.007

Definition of Customer Loyalty - Gartner Marketing Glossary. (n.d.). Gartner. https://www.gartner.com/en/marketing/glossary/customer-loyalty

Dellermann, D., Ebel, P., Söllner, M., & Leimeister, J. M. (2019). Hybrid intelligence. *Business & Information Systems Engineering, 61*(5), 637–643. DOI: 10.1007/s12599-019-00595-2

Demaître, E. (2019). Trust is key to service robot design, says SoftBank Robotics. *The Robot Report*. Retrieved from https://www.therobotreport.com/trust-key-service-robot-design-softbank-pepper/

Demirci Orel, F., & Kara, A. (2014, March). Supermarket self-checkout service quality, customer satisfaction, and loyalty: Empirical evidence from an emerging market. *Journal of Retailing and Consumer Services, 21*(2), 118–129. DOI: 10.1016/j.jretconser.2013.07.002

Demir, K. A., Döven, G., & Sezen, B. (2019). Industry 5.0 and Human-Robot Co-working. *Procedia Computer Science*, *158*, 688–695. DOI: 10.1016/j.procs.2019.09.104

Demoulin, N. T. M., & Djelassi, S. (2016, May 9). An integrated model of self-service technology (SST) usage in a retail context. *International Journal of Retail & Distribution Management*, *44*(5), 540–559. DOI: 10.1108/IJRDM-08-2015-0122

Deng, X., Li, H., Cheng, Y., & Zhang, Y. (2023). Deep learning-based predictive maintenance for industrial robotics: A comprehensive review. *Robotics and Computer-integrated Manufacturing*, *80*, 102517. DOI: 10.1016/j.rcim.2022.102517

Dennehy, D., Griva, A., Pouloudi, N., Dwivedi, Y. K., Mäntymäki, M., & Pappas, I. O. (2023). Artificial intelligence (AI) and information systems: Perspectives to responsible AI. *Information Systems Frontiers*, *25*(1), 1–7. DOI: 10.1007/s10796-022-10365-3

Department of Economic Planning and Statistics - Population. (n.d.). https://deps.mofe.gov.bn/SitePages/Population.aspx

Descriptive Research. (2021). Encyclopedia of Autism Spectrum Disorders, 1330–1330. DOI: 10.1007/978-3-319-91280-6_300489

Dieste, M., Orzes, G., Culot, G., Sartor, M., & Nassimbeni, G. (2023). The "dark side" of Industry 4.0: How can technology be made more sustainable? *International Journal of Operations &amp. International Journal of Operations & Production Management*, *44*(5), 900–933. DOI: 10.1108/IJOPM-11-2022-0754

Dignum, V. (2019). *Responsible Artificial Intelligence: How to Develop and Use AI in a Responsible Way*. Springer. DOI: 10.1007/978-3-030-30371-6

Ding, D. X., Hu, P. J. H., & Sheng, O. R. L. (2011). e-SELFQUAL: A scale for measuring online self-service quality. *Journal of Business Research*, *64*(5), 508–515. DOI: 10.1016/j.jbusres.2010.04.007

Dittig, J. (2023, January 13). *The rise of self-service - and how it's Reinventing Customer Service*. Samsung Business Insights. https://insights.samsung.com/2023/01/13/the-rise-of-self-service-and-how-its-reinventing-customer-service/

Djelassi, S., Diallo, M. F., & Zielke, S. (2018). How self-service technology experience evaluation affects waiting time and customer satisfaction? A moderated mediation model. *Decision Support Systems*, *111*, 38–47. DOI: 10.1016/j.dss.2018.04.004

Dodla, T. R., & Jones, L. A. (2023). Identifying knowledge management strategies for knowledge management systems. *Access Journal - Access to Science, Business. Innovation in the Digital Economy*, *4*(2), 261–277. DOI: 10.46656/access.2023.4.2(8)

Dogru, A. K., & Keskin, B. B. (2020). AI in operations management: Applications, challenges and opportunities. *Journal of Data, Information and Management*, *2*(2), 67–74. DOI: 10.1007/s42488-020-00023-1

Dormann, C. F., Elith, J., Bacher, S., Buchmann, C., Carl, G., Carré, G., Marquéz, J. R. G., Gruber, B., Lafourcade, B., Leitão, P. J., Münkemüller, T., McClean, C., Osborne, P. E., Reineking, B., Schröder, B., Skidmore, A. K., Zurell, D., & Lautenbach, S. (2012). Collinearity: A review of methods to deal with it and a simulation study evaluating their performance. *Ecography*, *36*(1), 27–46. DOI: 10.1111/j.1600-0587.2012.07348.x

Doshi-Velez, F., & Kim, B. (2017). Towards a rigorous science of interpretable machine learning. arXiv preprint arXiv:1702.08608.

Dwivedi, Y. K. (2023). Responsible AI and robotics: A review of ethical frameworks, challenges, and governance. *International Journal of Information Management*, *69*, 102605. DOI: 10.1016/j.ijinfomgt.2022.102605

Dwivedi, Y. K., Hughes, L., Ismagilova, E., Aarts, G., Coombs, C., Crick, T., Duan, Y., Dwivedi, R., Edwards, J., Eirug, A., Galanos, V., Ilavarasan, P. V., Janssen, M., Jones, P., Kar, A. K., Kizgin, H., Kronemann, B., Lal, B., Lucini, B., & Williams, M. D. (2021). Artificial Intelligence (AI): Multidisciplinary Perspectives on Emerging challenges, opportunities, and Agenda for research, Practice and Policy. *International Journal of Information Management*, *57*(101994), 101994. DOI: 10.1016/j.ijinfomgt.2019.08.002

Edmondson, A. C. (2019). *The Fearless Organization: Creating Psychological Safety in the Workplace for Learning, Innovation, and Growth*. Wiley.

Eisend, M., & Kuss, A. (2019). Testing Causal Relationships. *Research Methodology in Marketing*, 173–193. DOI: 10.1007/978-3-030-10794-9_8

Elendu, C., Elendu, T. C., & Elendu, I. D. (2024). *5G-enabled smart hospitals: Innovations in patient care and facility management*. Medicine., DOI: 10.1097/MD.0000000000038239

El-Hashash, E. F., & Shiekh, R. H. A. (2022). A Comparison of the Pearson, Spearman Rank and Kendall Tau Correlation Coefficients Using Quantitative Variables. *Asian Journal of Probability and Statistics*, 36–48. DOI: 10.9734/ajpas/2022/v20i3425

Elnadi, M., & Abdallah, Y. O. (2024). Industry 4.0: Critical investigations and synthesis of key findings. *Management Review Quarterly*, *74*(2), 711–744. DOI: 10.1007/s11301-022-00314-4

Emad-ul-Haq, Q., Aboalsamh, H., Belghith, A., Hussain, M., Abdul, W., Dahshan, M. H., & Ghouzali, S. (2015). Challenges and solutions for Internet of Things Driven by IPv6. [TIIS]. *KSII Transactions on Internet and Information Systems*, *9*(12), 4739–4758. DOI: 10.3837/tiis.2015.12.001

Engeström, Y. (1987). *Learning by Expanding: An Activity-Theoretical Approach to Developmental Research*. Orienta-Konsultit.

Eriksson, K. M., Olsson, A. K., & Carlsson, L. (2024). Beyond lean production practices and Industry 4.0 technologies toward the human-centric Industry 5.0. *Technological Sustainability*, *3*(3), 286–308. DOI: 10.1108/TECHS-11-2023-0049

Erro-Garcés, A., & Aramendia-Muneta, M. E. (2023). The role of human resource management practices on the results of digitalisation. From Industry 4.0 to Industry 5.0. *Journal of Organizational Change Management*, *36*(4), 585–602. Advance online publication. DOI: 10.1108/JOCM-11-2021-0354

Eswar Balachandar, G., & Bhaskar Reddy, C. (2024). *Mathematical Model for Multi-Objective Optimization of Facility Allocation in Flexible Manufacturing Systems with Collaborative Robots*. International Research Journal of Multidisciplinary Scope., DOI: 10.47857/irjms.2024.v05i03.0892

Eswaran, M., Inkulu, A. K., Tamilarasan, K., Bahubalendruni, M. V. A. R., Jaideep, R., Faris, M. S., & Jacob, N. (2024). Optimal layout planning for human robot collaborative assembly systems and visualization through immersive technologies. *Expert Systems with Applications*, *241*, 122465. Advance online publication. DOI: 10.1016/j.eswa.2023.122465

Eubanks, V. (2018). *Automating inequality: How high-tech tools profile, police, and punish the poor*. Macmillan+ ORM.

European Commission. (2019). *Ethics guidelines for trustworthy AI*. Publications Office of the European Union. https://digital-strategy.ec.europa.eu/en/library/ethics-guidelines-trustworthy-ai

European Commission. (2020). *AI White Paper: A European approach to excellence and trust*. https://ec.europa.eu/info/publications/white-paper-artificial-intelligence-european-approach-excellence-and-trust_en

European Commission. (2020). *Ethics guidelines for trustworthy AI*. https://digital-strategy.ec.europa.eu/en/library/ethics-guidelines-trustworthy-ai

European Commission. (2021). *Coordinated Plan on Artificial Intelligence 2021 Review*. https://digital-strategy.ec.europa.eu/en/library/coordinated-plan-artificial-intelligence-2021-review

European Commission. (2021). *Digital Europe Programme: 2021–2027*. https://digital-strategy.ec.europa.eu/en/activities/digital-programme

European Commission. (2021). *Proposal for a Regulation Laying Down Harmonised Rules on Artificial Intelligence (AI Act)*. https://eur-lex.europa.eu/legal-content/EN/TXT/?uri=CELEX%3A52021PC0206

European Commission. (2021). *Proposal for a Regulation Laying Down Harmonised Rules on Artificial Intelligence (Artificial Intelligence Act)*. https://eur-lex.europa.eu/legal-content/EN/TXT/?uri=CELEX%3A52021PC0206

European Commission. (2021). *Proposal for a Regulation laying down harmonized rules on artificial intelligence*. Artificial Intelligence Act.

Fachrunnisa, O., Tjahjono, H. K., & Palupi, M. (2018). Cognitive collective engagement in virtual collaborative team. *2018 7th International Conference on Industrial Technology and Management (ICITM)*, 390–393. DOI: 10.1109/ICITM.2018.8333981

Fachrunnisa, O., Adhiatma, A., & Tjahjono, H. K. (2019). Collective Engagement and Spiritual Wellbeing in Knowledge Based Community: A Conceptual Model. In Barolli, L., Hussain, F. K., & Ikeda, M. (Eds.), *Complex, Intelligent, and Software Intensive Systems* (pp. 899–906). Springer International Publishing., DOI: 10.1007/978-3-030-22354-0_83

Feddoul, Y., Ragot, N., Duval, F., Havard, V., Baudry, D., & Assila, A. (2023). Exploring human-machine collaboration in industry: A systematic literature review of digital twin and robotics interfaced with extended reality technologies. *International Journal of Advanced Manufacturing Technology*, *129*(5-6), 1917–1932. Advance online publication. DOI: 10.1007/s00170-023-12291-3

Fernandes, T., & Pedroso, R. (2016). The effect of self-checkout quality on customer satisfaction and repatronage in a retail context. *Service Business*, *11*(1), 69–92. DOI: 10.1007/s11628-016-0302-9

Flavián, C., Ibáñez-Sánchez, S., & Orús, C. (2019). The impact of virtual, augmented and mixed reality technologies on the customer experience. *Journal of Business Research*, *100*, 547–560. DOI: 10.1016/j.jbusres.2018.10.050

Floridi, L. (2018). Soft ethics and the governance of the digital. *Philosophy & Technology*, *31*(1), 1–8. DOI: 10.1007/s13347-018-0303-9 PMID: 30322997

Floridi, L., & Cowls, J. (2019). A Unified Framework of Five Principles for AI in Society. *Harvard Data Science Review, 1*(1). Advance online publication. DOI: 10.1162/99608f92.8cd550d1

Floridi, L., Cowls, J., Beltrametti, M., Chatila, R., Chazerand, P., Dignum, V., Luetge, C., Madelin, R., Pagallo, U., Rossi, F., Schafer, B., Valcke, P., & Vayena, E. (2018). AI4People—An ethical framework for a good AI society: Opportunities, risks, principles, and recommendations. *Minds and Machines, 28*(4), 689–707. DOI: 10.1007/s11023-018-9482-5 PMID: 30930541

Folgado, F. J., Calderón, D., González, I., & Calderón, A. J. (2024). Review of Industry 4.0 from the Perspective of Automation and Supervision Systems: Definitions, Architectures and Recent Trends. *Electronics (Basel), 13*(4), 4. Advance online publication. DOI: 10.3390/electronics13040782

Fornell, C., Johnson, M. D., Anderson, E. W., Cha, J., & Bryant, B. E. (1996). The American Customer Satisfaction Index: Nature, Purpose, and Findings. *Journal of Marketing, 60*(4), 7–18. DOI: 10.1177/002224299606000403

Fraboni, F., Brendel, H., & Pietrantoni, L. (2023). Evaluating Organizational Guidelines for Enhancing Psychological Well-Being, Safety, and Performance in Technology Integration. *Sustainability (Basel), 15*(10), 8113. Advance online publication. DOI: 10.3390/su15108113

Franzese, M., & Iuliano, A. (2019). Correlation Analysis. In *Elsevier eBooks* (pp. 706–721). DOI: 10.1016/B978-0-12-809633-8.20358-0

Fraunhofer, I. P. K. (2022). Bolting Cobots – Human and Machine in the Same Workspace. *FUTUR* (Fraunhofer IPK publication). Retrieved from https://www.ipk.fraunhofer.de/en/media/futur/futur-2022-2/bolting-cobots.html

FreightAmigo. (2023). *10 key supply-chain trends shaping global trade in 2023.* https://www.freightamigo.com/blog/10-key-supply-chain-trends-shaping-global-trade-in-2023

French Ministry of Education. (2021). *AI for Everyone: Promoting Digital Literacy.* https://www.education.gouv.fr

Frey, C. B., & Osborne, M. A. (2017). The future of employment: How susceptible are jobs to computerization? *Technological Forecasting and Social Change, 114*, 254–280. DOI: 10.1016/j.techfore.2016.08.019

Garvin, D. A., Wagonfeld, A. B., & Kind, L. (2013). Google's Project Oxygen: Do Managers Matter? Harvard Business School Case No. 313-110.

Geekturf. (2019, July 28). *Self-Ordering Kiosks. Coming soon to McDonald's Brunei Lambak*. Facebook. https://m.facebook.com/geekturf/photos/a.434705936971963/726222391153648/

Gelbrich, K. (2009). Beyond just being dissatisfied: How angry and helpless customers react to failures when using self-service technologies. *Schmalenbach Business Review*, *61*(1), 40–59. DOI: 10.1007/BF03396779

Gershman, S. J., Horvitz, E. J., & Tenenbaum, J. B. (2015). Computational rationality: A converging paradigm for intelligence in brains, minds, and machines. *Science*, *349*(6245), 273–278. DOI: 10.1126/science.aac6076 PMID: 26185246

Gervasi, R., Capponi, M., Mastrogiacomo, L., & Franceschini, F. (2024). Eye-tracking support for analyzing human factors in human-robot collaboration during repetitive long-duration assembly processes. *Production Engineering*. Advance online publication. DOI: 10.1007/s11740-023-01230-6

Ghasemaghaei, M. (2020). The role of positive and negative valence factors on the impact of bigness of data on big data analytics usage. *International Journal of Information Management*, *50*, 395–404. DOI: 10.1016/j.ijinfomgt.2018.12.011

Ghazali, F. B., Ramlee, S. N. S., Alwi, N., & Hizan, H. (2020). Content validity and test–retest reliability with principal component analysis of the translated Malay four-item version of Paffenbarger physical activity questionnaire. *Journal of Health Research. Warasan Wichai Witthayasat Kanphaet*, *35*(6), 493–505. DOI: 10.1108/JHR-11-2019-0269

Ghobakhloo, M., Iranmanesh, M., Fathi, M., Rejeb, A., Foroughi, B., & Nikbin, D. (2024). Beyond Industry 4.0: A systematic review of Industry 5.0 technologies and implications for social, environmental and economic sustainability. *Asia-Pacific Journal of Business Administration*. Advance online publication. DOI: 10.1108/APJBA-08-2023-0384

Ghosh, M. (2020). Understanding the relationship of self-service technology quality with user adoption using SSTQUAL. *The TQM Journal*, *33*(2), 293–314. DOI: 10.1108/TQM-12-2019-0291

Glikson, E., & Woolley, A. W. (2020). Human Trust in Artificial Intelligence: Review of Empirical Research. *The Academy of Management Annals*, *14*(2), 627–660. DOI: 10.5465/annals.2018.0057

Glushko, R. J., & Nomorosa, K. J. (2012). Substituting Information for Interaction. *Journal of Service Research*, *16*(1), 21–38. DOI: 10.1177/1094670512463967

Goleman, D. (1995). *Emotional Intelligence: Why It Can Matter More Than IQ*. Bantam Books.

Grand View Research. (2020, July 7). Self-service Technology Market Size Report, 2020-2027. https://www.grandviewresearch.com/industry-analysis/self-service-technology-market#

Group, B. M. W. (2013, September 10). *Innovative human-robot cooperation in BMW Group production*. Press release. https://www.press.bmwgroup.com/global/article/attachment/T0209722EN/301973

Gubeladze, D., & Pavliashvili, S. (2020). LINEAR ECONOMY AND CIRCULAR ECONOMY - CURRENT STATE ASSESSMENT AND FUTURE VISION. *International Journal of Innovative Technologies in Economy, 5(32). Article*, *5*(32). Advance online publication. DOI: 10.31435/rsglobal_ijite/30122020/7286

Guerra, M., Murciano-Brea, J., & Martín, D. (2023). Autonomous delivery robots in urban logistics: Current development and future trends. *Sensors (Basel)*, *23*(3), 1120. DOI: 10.3390/s23031120 PMID: 36772161

Gummerus, J., Lipkin, M., Dube, A., & Heinonen, K. (2019, February 11). Technology in use – characterizing customer self-service devices (SSDS). *Journal of Services Marketing*, *33*(1), 44–56. DOI: 10.1108/JSM-10-2018-0292

Gunasekaran, A., Subramanian, N., & Ngai, E. W. T. (2018). The future of operations management: An outlook and analysis. *International Journal of Production Economics*, *195*, 102–118. DOI: 10.1016/j.ijpe.2017.12.002

Hajli, N. (2019). The impact of positive valence and negative valence on social commerce purchase intention. *Information Technology & People*, *33*(2), 774–791. DOI: 10.1108/ITP-02-2018-0099

Halton, C. (2023, June 17). *Diffusion of Innovations Theory: Definition and Examples*. Investopedia. https://www.investopedia.com/terms/d/diffusion-of-innovations-theory.asp#:~:text=What%20Are%20the%20Steps%20In,later%20editions%20of%20his%20book

Hamid, M. A., Salleh, S., & Laxman, K. (2020). A Study on the Factors Influencing Students' Acceptance of Learning Management Systems (LMS): A Brunei Case Study. *International Journal of Technology in Education and Science*, *4*(3), 203–217. DOI: 10.46328/ijtes.v4i3.101

Haming, M., Syaiful, A. Z., Putra, A. H. P. K., & Murdifin, I. (2019, February). The Application of SERVQUAL Distribution In Measuring Customer Satisfaction of Retails Company. *Journal of Distribution Science*, *17*(2), 25–31. DOI: 10.15722/jds.17.2.201902.25

Hancock, P. A., Billings, D. R., & Schaefer, K. E. (2011). Can You Trust Your Robot? *Ergonomics in Design*, *19*(3), 24–29. DOI: 10.1177/1064804611415045

Hancock, P. A., Billings, D. R., Schaefer, K. E., Chen, J. Y. C., de Visser, E. J., & Parasuraman, R. (2011). A Meta-Analysis of Factors Affecting Trust in Human-Robot Interaction. *Human Factors*, *53*(5), 517–527. DOI: 10.1177/0018720811417254 PMID: 22046724

Hancock, P. A., Kessler, T., Kaplan, A. D., Brill, J. C., & Szalma, J. L. (2020). Evolving trust in robots: Specification through sequential and comparative meta-analyses. *Human Factors*, *63*(7), 1196–1229. DOI: 10.1177/0018720820922080 PMID: 32519902

Han, J., Forbes, H., & Schaefer, D. (2019). An Exploration of the Relations between Functionality, Aesthetics and Creativity in Design. Proceedings of the . . . *International Conference on Engineering Design*, 1(1), 259–268. DOI: 10.1017/dsi.2019.29

Hao, K. (2020, December 4). We read the paper that led to Timnit Gebru's departure from Google. Here is what it says. *MIT Technology Review*. https://www.technologyreview.com/2020/12/04/1013294/

Haradhan, M. (2020). Quantitative Research: A Successful Investigation in Natural and Social Sciences. *Journal of Economic Development*, *9*(4), 52–79. https://mpra.ub.uni-muenchen.de/105149/

Haselberger, D. (2020). Symbiotic Intelligence: Merging AI and Human Capabilities for Better Decision-Making. *Journal of Artificial Intelligence Research*, *69*, 45–62.

Hattula, J. D., Herzog, W., & Dhar, R. (2022). The impact of touchscreen devices on consumers' choice confidence and purchase likelihood. *Marketing Letters*, *34*(1), 35–53. DOI: 10.1007/s11002-022-09623-w

Hazzi, O. A., & Maldaon, I. S. (2015). A Pilot Study: Vital Methodological Issues. *Business: Theory and Practice*, *16*(1), 53–62. DOI: 10.3846/btp.2015.437

Heifetz, R. A. (1994). *Leadership Without Easy Answers*. Harvard University Press. DOI: 10.4159/9780674038479

Heifetz, R., & Linsky, M. (2002). *Leadership on the Line: Staying Alive Through the Dangers of Leading*. Harvard Business Press.

Heifetz, R., & Linsky, M. (2002). *Leadership on the Line: Staying Alive through the Dangers of Leading*. Harvard Business Review Press.

Hennart, J. F., & Verbeke, A. (2022, September 27). Actionable and enduring implications of Oliver Williamson's transaction cost theory. *Journal of International Business Studies*, *53*(8), 1557–1575. DOI: 10.1057/s41267-022-00558-y

Henseler, J., Ringle, C. M., & Sinkovics, R. R. (2009). The use of partial least squares path modeling in international marketing. In *Advances in international marketing* (pp. 277–319). DOI: 10.1108/S1474-7979(2009)0000020014

Henseler, J., Hubona, G., & Ray, P. A. (2016). Using PLS path modeling in new technology research: Updated guidelines. *Industrial Management + Data Systems. Industrial Management & Data Systems*, *116*(1), 2–20. DOI: 10.1108/IMDS-09-2015-0382

He, P., Jiang, X., Wang, Q., & Zhang, B. (2025). Multi-objective Human-robot collaborative batch scheduling in distributed hybrid flowshop via automatic design of local search-reconstruction-feedback algorithm. *Computers & Industrial Engineering*, *203*, 110983. Advance online publication. DOI: 10.1016/j.cie.2025.110983

Herzberg, F. (1966). *Work and the Nature of Man*. World Publishing.

Hevelke, A., & Nida-Rümelin, J. (2023). Ethical decision-making in autonomous robotic systems: Human oversight and accountability. *Robotics and Autonomous Systems*, *162*, 104394. DOI: 10.1016/j.robot.2023.104394

Hilton, T., Hughes, T., Little, E., & Marandi, E. (2013). Adopting self-service technology to do more with less. *Journal of Services Marketing*, *27*(1), 3–12. DOI: 10.1108/08876041311296338

Hjorth, S., & Chrysostomou, D. (2022). Human–robot collaboration in industrial environments: A literature review on non-destructive disassembly. *Robotics and Computer-integrated Manufacturing*, *73*, 102208. Advance online publication. DOI: 10.1016/j.rcim.2021.102208

Hoff, K. A., & Bashir, M. (2015). Trust in Automation: Integrating Empirical Evidence on Factors That Influence Trust. *Human Factors*, *57*(3), 407–434. DOI: 10.1177/0018720814547570 PMID: 25875432

Hoffman, K., & Bateson, J. (2016, January 1). *Services Marketing: Concepts, Strategies, & Cases*. Cengage Learning.

Holland, J., Kingston, L., McCarthy, C., Armstrong, E., O'dwyer, P., Merz, F., & McConnell, M. (2021). Service robots in the healthcare sector. *Robotics (Basel, Switzerland), 10*(1), 47. Advance online publication. DOI: 10.3390/robotics10010047

Hollanek, T. (2023). AI transparency: A matter of reconciling design with critique. *AI & Society, 38*(5), 2071–2079. DOI: 10.1007/s00146-020-01110-y

Hsu, C.-H., Liu, J.-C., Cai, X.-Q., Zhang, T.-Y., & Lv, W.-Y. (2024). Enabling Sustainable Diffusion in Supply Chains Through Industry 5.0: An Impact Analysis of Key Enablers for SMEs in Emerging Economies. *Mathematics, 12*(24), 3938. Advance online publication. DOI: 10.3390/math12243938

Huang, Z., Ge, S., He, Y., Wang, D., & Zhang, S. (2024). Research on the Intelligent System Architecture and Control Strategy of Mining Robot Crowds. *Energies, 17*(8), 1834. Advance online publication. DOI: 10.3390/en17081834

Ietto, B., Ancillai, C., Sabatini, A., Carayannis, E. G., & Gregori, G. L. (2024). The Role of External Actors in SMEs' Human-Centered Industry 4.0 Adoption: An Empirical Perspective on Italian Competence Centers. *IEEE Transactions on Engineering Management, 71*, 1057–1072. Advance online publication. DOI: 10.1109/TEM.2022.3144881

IFR. (2022). World [International Federation of Robotics. https://ifr.org]. *Robotics Reports, 2022*, •••.

Independent Digital News and Media. (2011, March 15). *How Britain became a self-service nation*. The Independent. https://www.independent.co.uk/news/business/analysis-and-features/how-britain-became-a-selfservice-nation-2241830.html

Inkulu, A. K., & Bahubalendruni, M. V. A. R. (2024). *Human-robot collaborative task planning for assembly system productivity enhancement*. Robotic Intelligence and Automation., DOI: 10.1108/RIA-05-2023-0067

Inman, J. J., & Nikolova, H. (2017). Shopper-Facing Retail Technology: A Retailer Adoption Decision Framework Incorporating Shopper Attitudes and Privacy Concerns. *Journal of Retailing, 93*(1), 7–28. DOI: 10.1016/j.jretai.2016.12.006

Isaac, S. (2025, April 28). *XPENG Iron: Advanced Humanoid Robot Unveiled - Qviro Blog*. https://qviro.com/blog/xpeng-launches-iron-advanced-humanoid-robot/

Isaak, J., & Hanna, M. J. (2018). User data privacy: Facebook, Cambridge Analytica, and privacy protection. *Computer, 51*(8), 56–59. DOI: 10.1109/MC.2018.3191268

Islam, S. O. B., & Lughmani, W. A. (2023). A Connective Framework for Safe Human–Robot Collaboration in Cyber-Physical Production Systems. *Arabian Journal for Science and Engineering*, *48*(9), 11621–11644. Advance online publication. DOI: 10.1007/s13369-022-07490-1

Ivanov, D., Blackhurst, J., & Das, A. (2021). Supply chain resilience and its interplay with digital technologies: Making innovations work in emergency situations. *International Journal of Physical Distribution & Logistics Management*, *51*(2), 97–103. DOI: 10.1108/IJPDLM-03-2021-409

Jaeger, P. T. (2012). *Disability and the Internet: Confronting a digital divide*. Lynne Rienner Publishers.

Jain, S. K., & Gupta, G. (2004, April). Measuring Service Quality: Servqual vs. Servperf Scales. *Vikalpa*, *29*(2), 25–38. DOI: 10.1177/0256090920040203

Javaid, M., Haleem, A., Singh, R. P., Haq, M. I. U., Raina, A., & Suman, R. (2020). Industry 5.0: Potential Applications in COVID-19. *Journal of Industrial Integration and Management*, *05*(04), 507–530. DOI: 10.1142/S2424862220500220

Jia, Z., Xie, S., & Zhang, W. (2025). Flexible task assignment and assembly scheduling for human-robot collaboration cell considering uncertainty. *International Journal of Production Research*, *63*(16), 6134–6154. Advance online publication. DOI: 10.1080/00207543.2025.2469286

Jie, N. X., & Kamsin, I. F. (2021). *Self- checkout service with RFID technology in supermarket*. Atlantis Highlights in Computer Sciences., DOI: 10.2991/ahis.k.210913.062

Ji, H., Li, S., Chen, J., & Zhou, S. (2024). On-site human-robot collaboration for lunar exploration based on shared mixed reality. *Multimedia Tools and Applications*.

Jobin, A., Ienca, M., & Vayena, E. (2019). The global landscape of AI ethics guidelines. *Nature Machine Intelligence*, *1*(9), 389–399. DOI: 10.1038/s42256-019-0088-2

Joe, S., Kim, J. S., & Zemke, D. M. V. (2020, April 8). Effects of Social Influence and Perceived Enjoyment on Kiosk Acceptance: A Moderating Role of Gender. *International Journal of Hospitality & Tourism Administration*, *23*(2), 289–316. DOI: 10.1080/15256480.2020.1746217

Johanson, G. A., & Brooks, G. P. (2009). Initial Scale Development: Sample Size for Pilot Studies. *Educational and Psychological Measurement*, *70*(3), 394–400. DOI: 10.1177/0013164409355692

John, B., Alsamarra'l, Z., & Panteli, N. (2022). Enhancing Employee Experience in the Era of Hybrid Work: The Case of Microsoft Viva. *IEEE Software*, *40*(2), 1–9. DOI: 10.1109/MS.2022.3229956

Johnson, E. (2021). Cronbach's Alpha. *Encyclopedia of Autism Spectrum Disorders*, 1248–1249. DOI: 10.1007/978-3-319-91280-6_307

Kahneman, D., & Tversky, A. (1979). Prospect theory: An analysis of decision under risk. *Econometrica*, *47*(2), 263–291. DOI: 10.2307/1914185

Kaliyadan, F., & Kulkarni, V. (2019). Types of variables, descriptive statistics, and sample size. *Indian Dermatology Online Journal*, *10*(1), 82. DOI: 10.4103/idoj.IDOJ_468_18 PMID: 30775310

Kaminski, J. (2011). Diffusion of Innovation Theory. *Canadian Journal of Nursing Informatics*, *6*(2). https://cjni.net/journal/?p=1444

Kanda, W., Geissdoerfer, M., & Hjelm, O. (2021). From circular business models to circular business ecosystems. *Business Strategy and the Environment*, *30*(6), 2814–2829. DOI: 10.1002/bse.2895

Kane, A. (2025). Navigating the Transition: Challenges and Benefits of Shifting from IPv4 to IPv6 in a Rapidly Evolving Internet Landscape. *International Journal of Internet and Distributed Systems*, *7*(2), 2. Advance online publication. DOI: 10.4236/ijids.2025.72002

Kapur, A., & Graedel, T. E. (2004). Industrial Ecology. In Cleveland, C. J. (Ed.), *Encyclopedia of Energy* (pp. 373–382). Elsevier., DOI: 10.1016/B0-12-176480-X/00533-7

Kaushik, A. K., & Rahman, Z. (2017). An empirical investigation of Tourist's choice of Service Delivery Options. *International Journal of Contemporary Hospitality Management*, *29*(7), 1892–1913. DOI: 10.1108/IJCHM-08-2015-0438

Kavvathas, K., Kampourakis, E., Andronas, D., Fousekis, N., & Makris, S. (2024). A service-oriented orchestration and planning tool for plug and produce manufacturing: A deformable object handling supervision paradigm. *International Journal of Computer Integrated Manufacturing*, *37*(12), 1626–1649. Advance online publication. DOI: 10.1080/0951192X.2024.2331533

Kennedy, S., & Linnenluecke, M. K. (2022). Circular economy and resilience: A research agenda. *Business Strategy and the Environment*, *31*(6), 2754–2765. DOI: 10.1002/bse.3004

Kermenov, R., Di Biase, A., Pellicani, I., Longhi, S., & Bonci, A. (2025). Near Time-Optimal Trajectories with ISO Standard Constraints for Human–Robot Collaboration in Fabric Co-Transportation. *Robotics (Basel, Switzerland), 14*(2), 10. Advance online publication. DOI: 10.3390/robotics14020010

Keshvarparast, A., Battini, D., Battaia, O., & Pirayesh, A. (2024). Collaborative robots in manufacturing and assembly systems: Literature review and future research agenda. *Journal of Intelligent Manufacturing, 35*(5), 2065–2118. DOI: 10.1007/s10845-023-02137-w

Keshvarparast, A., Berti, N., Chand, S., Guidolin, M., Lu, Y., Battaia, O., Xu, X., & Battini, D. (2024). Ergonomic design of Human-Robot collaborative workstation in the Era of Industry 5.0. *Computers & Industrial Engineering, 198*, 110729. Advance online publication. DOI: 10.1016/j.cie.2024.110729

Khalufi, M. N. A., & Mohammad Shah, K. A. (2022). Factors Influencing Consumer's Intention to Use Self Service Technology in Retail. *Global Business and Management Research: An International Journal, 14*(3s), 1044–1052. http://www.gbmrjournal.com/pdf/v14n3s/V14N3s-71.pdf

Khoa, D. T., Gip, H. Q., Guchait, P., & Wang, C. Y. (2023). Competition or collaboration for human–robot relationship: A critical reflection on future cobotics in hospitality. *International Journal of Contemporary Hospitality Management, 35*(6), 2202–2215. Advance online publication. DOI: 10.1108/IJCHM-04-2022-0434

Kiffle, K. (2019, November 29). Brunei Information & Communications Technology Sector Initiative: The next step on the path to a Smart Nation - Brunei NetCo. Brunei NetCo. Retrieved January 14, 2024, from https://www.bruneinetco.com/brunei-information-communications-technology-sector-initiative-the-next-step-on-the-path-to-a-smart-nation/

Kim, H. S., Oh, B.-Y., Cheong, C., Park, M. H., Chung, S. S., Lee, R.-A., Kim, K. H., & Noh, G. T. (2023). Single-incision robotic colorectal surgery with the da Vinci SP® surgical system: Initial results of 50 cases. *Techniques in Coloproctology, 27*(7), 589–599. DOI: 10.1007/s10151-023-02791-5 PMID: 36971849

Kim, H.-J., & Lee, J.-M. (2020). Consumers' resistance and continued use intention of self-service kiosk. *Family and Environment Research, 58*(3), 401–416. DOI: 10.6115/fer.2020.029

Kim, I., & Min, J. (2023). Working hours and the regulations in Korea. *Annals of Occupational and Environmental Medicine, 35*(1), e18. Advance online publication. DOI: 10.35371/aoem.2023.35.e18 PMID: 37928375

Kim, S. (2022). Working With Robots: Human Resource Development Considerations in Human–Robot Interaction. *Human Resource Development Review*, *21*(1), 153448432110688. DOI: 10.1177/15344843211068810

Kim, T. K., & Park, J. H. (2019). More about the basic assumptions of t-test: Normality and sample size. *Korean Journal of Anesthesiology*, *72*(4), 331–335. DOI: 10.4097/kja.d.18.00292 PMID: 30929413

Kim, Y. H., Ye, W., Kumar, R., Bail, F., Dvorak, J., Tan, Y., May, M. C., Chang, Q., Athinarayanan, R., Lanza, G., Sutherland, J. W., Li, X., & Nath, C. (2024). Unlocking the Potential of Remanufacturing Through Machine Learning and Data-Driven Models—A Survey. *Algorithms*, *17*(12), 562. Advance online publication. DOI: 10.3390/a17120562

Kim, Y., Kim, H., Murphy, R., Lee, S., & Ahn, C. R. (2022). Delegation or Collaboration: Understanding Different Construction Stakeholders' Perceptions of Robotization. *Journal of Management Engineering*, *38*(1), 04021084. Advance online publication. DOI: 10.1061/(ASCE)ME.1943-5479.0000994

Kiosk Group. (n.d.). *Self Service kiosks 101*. https://www.kioskgroup.com/pages/self-service-kiosks-101

Koç, M. (2023). ARTIFICIAL INTELLIGENCE IN GERIATRICS. *Turk Geriatri Dergisi*.

Kokkinou, A., & Cranage, D. A. (2013, June). Using self-service technology to reduce customer waiting times. *International Journal of Hospitality Management*, *33*, 435–445. DOI: 10.1016/j.ijhm.2012.11.003

Kolasani, S. (2023). Optimizing Natural Language Processing, Large Language Models (LLMs) for Efficient Customer Service, and hyper-personalization to enable sustainable growth and revenue. *Transactions on Latest Trends in Artificial Intelligence, 4*(4). Retrieved from https://ijsdcs.com/index.php/TLAI/article/view/476/196

Korivand, S., Galvani, G., Ajoudani, A., Gong, J., & Jalili, N. (2024). Optimizing Human–Robot Teaming Performance through Q-Learning-Based Task Load Adjustment and Physiological Data Analysis. *Sensors (Basel)*, *24*(9), 2817. Advance online publication. DOI: 10.3390/s24092817 PMID: 38732923

Krishen, A. S. (2008). Perceived Versus Actual Complexity for Websites: Their Relationship to Consumer Satisfaction. *Journal of Consumer Satisfaction, Dissatisfaction & Complaining Behavior*, •••, 21.

Kristiawan, M., Suhono, S., Yussof, M. H. B., & Muslimah, M. (2021). The International School's Culture in Indonesia and Brunei Darussalam. *Jurnal Iqra, 6*(1), 180–191. DOI: 10.25217/ji.v6i1.1263

Krithikadatta, J. (2014). Normal Distribution. *Journal of Conservative Dentistry. Journal of Conservative Dentistry : JCD, 17*(1), 96. DOI: 10.4103/0972-0707.124171 PMID: 24554873

Krüger, J., Lien, T. K., & Verl, A. (2019). Cooperation of human and machines in assembly lines. *CIRP Annals, 68*(2), 1–24. DOI: 10.1016/j.cirp.2019.04.097

Krüger, R., Klucken, J., Weiss, D., Tönges, L., Kolber, P., Unterecker, S., Lorrain, M., Baas, H., Müller, T., & Riederer, P. (2017). Classification of advanced stages of Parkinson's disease: Translation into stratified treatments. *Journal of Neural Transmission (Vienna, Austria), 124*(8), 1015–1027. DOI: 10.1007/s00702-017-1707-x PMID: 28342083

Ku, E. C., & Chen, C. D. (2013). Fitting facilities to self-service technology usage: Evidence from kiosks in Taiwan airport. *Journal of Air Transport Management, 32*, 87–94. DOI: 10.1016/j.jairtraman.2013.07.001

Kwak, J. K. (2020). Empirical Analysis on Comparison between Self-checkout and Regular Staffed-checkout lanes in a Poland Retail Store, 8(1), pp. 56-61

Langer, M., Baum, K., & Schlicker, N. (2024). Effective Human Oversight of AI-Based Systems: A Signal Detection Perspective on the Detection of Inaccurate and Unfair Outputs. *Minds and Machines, 35*(1), 1. DOI: 10.1007/s11023-024-09701-0

Lario, J., Mateos, J., Psarommatis, F., & Ortiz, Á. (2025). Towards Zero Defect and Zero Waste Manufacturing by Implementing Non-Destructive Inspection Technologies. *Journal of Manufacturing and Materials Processing*. DOI: 10.3390/jmmp9020029

Larivière, B., Bowen, D., Andreassen, T. W., Kunz, W., Sirianni, N. J., Voss, C., Wünderlich, N. V., & De Keyser, A. (2017). "Service encounter 2.0": An investigation into the roles of technology, employees and customers. *Journal of Business Research, 79*, 238–246. DOI: 10.1016/j.jbusres.2017.03.008

Latour, B. (2005). *Reassembling the Social: An Introduction to Actor-Network-Theory*. Oxford University Press. DOI: 10.1093/oso/9780199256044.001.0001

Lawton, G. (2022, August 30). *9 disadvantages of self-service options: TechTarget*. Customer Experience. https://www.techtarget.com/searchcustomerexperience/tip/6-reasons-why-self-service-options-fail#:~:text=However%2C%20self%2Dservice%20initiatives%20can,before%20investing%20in%20this%20strategy

Lazar, J. (2017). *Ensuring digital accessibility through process and policy*. Morgan Kaufmann.

Lee, J. D., & See, K. A. (2004). Trust in Automation: Designing for Appropriate Reliance. *Human Factors*. https://journals.sagepub.com/doi/abs/10.1518/hfes.46.1.50_30392

Lee, J. D., & See, K. A. (2004). Trust in automation: Designing for appropriate reliance. *Human Factors*, *46*(1), 50–80. DOI: 10.1518/hfes.46.1.50.30392 PMID: 15151155

Lee, S. M., & Lee, D. (2019). "UNTACT": A new customer service strategy in the Digital age. *Service Business*, *14*(1), 1–22. DOI: 10.1007/s11628-019-00408-2

Leng, J., Sha, W., Wang, B., Zheng, P., Zhuang, C., Liu, Q., Wuest, T., Mourtzis, D., & Wang, L. (2022). Industry 5.0: Prospect and retrospect. *Journal of Manufacturing Systems*, *65*, 279–295. DOI: 10.1016/j.jmsy.2022.09.017

Leung, X. Y., Torres, B., & Fan, A. (2021). Do kiosks outperform cashiers? an S-O-R framework of restaurant ordering experiences. *Journal of Hospitality and Tourism Technology*, *12*(3), 580–592. DOI: 10.1108/JHTT-03-2020-0065

Lian, J. W. (2018, August 17). Why is self-service technology (SST) unpopular? Extending the IS success model. *Library Hi Tech*, *39*(4), 1154–1173. DOI: 10.1108/LHT-01-2018-0015

Liao, D., & Valliant, R. (2012). Variance inflation factors in the analysis of complex survey data. *Survey Methodology*, *38*, 53–62.

Liau, Y. Y., & Ryu, K. (2024). Framework of knowledge management for human–robot collaborative mold assembly using heterogeneous cobots. *Journal of Intelligent Manufacturing*, *35*(8), 3713–3729. DOI: 10.1007/s10845-024-02439-7

Li, J., Gong, X., Li, P., Xiao, L., Chang, X., Ouyang, X., & Tang, J. (2022). Application of Da Vinci robotic surgery system in cervical cancer: A single institution experience of 557 cases. *Asian Journal of Surgery*, *45*(2), 707–711. DOI: 10.1016/j.asjsur.2021.07.052 PMID: 34373166

Li, J., Luo, H., Xie, N., & Zhang, Y. (2022). Automation and robotics in smart warehousing: A systematic review. *Robotics and Computer-integrated Manufacturing*, *78*, 102386. DOI: 10.1016/j.rcim.2022.102386

Li, J.-M., Liu, K.-X., Xie, J.-F., & Wu, H.-Y. (2024). How Does Human-Robot Collaboration Affect Hotel Employees' Proactive Behavior? *International Journal of Human-Computer Interaction*. Advance online publication. DOI: 10.1080/10447318.2024.2431769

Li, J.-M., Zhang, R.-X., Wu, T.-J., & Mao, M. (2024). How does work autonomy in human-robot collaboration affect hotel employees' work and health outcomes? Role of job insecurity and person-job fit. *International Journal of Hospitality Management, 117*, 103654. DOI: 10.1016/j.ijhm.2023.103654

Lin, P. (2018). The ethics of autonomous cars. *The Atlantic*. https://www.theatlantic.com/technology/archive/2013/10/the-ethics-of-autonomous-cars/280360/

Lin, C. H., Shih, H. Y., & Sher, P. J. (2007). Integrating technology readiness into technology acceptance: The TRAM model. *Psychology and Marketing, 24*(7), 641–657. DOI: 10.1002/mar.20177

Lin, C.-Y., Liu, Y.-C., Chen, M.-C., Chiang, F.-F., & Lee, T.-Y. (2022). Learning curve and surgical outcome of robotic assisted colorectal surgery with ERAS program. *Scientific Reports, 12*(1), 20566. Advance online publication. DOI: 10.1038/s41598-022-24665-w PMID: 36446802

Lin, F. R., Ferrucci, L., Metter, E. J., An, Y., Zonderman, A. B., & Resnick, S. M. (2011). Hearing loss and cognition in the Baltimore Longitudinal Study of Aging. *Neuropsychology, 25*(6), 763–770. DOI: 10.1037/a0024238 PMID: 21728425

Lin, J.-S. C., & Hsieh, P.-L. (2011). Assessing the self-service technology encounters: Development and validation of SSTQUAL scale. *Journal of Retailing, 87*(2), 194–206. DOI: 10.1016/j.jretai.2011.02.006

Lin, S., & Lin, J. (2023). How organizations leverage digital technology to develop customization and enhance customer relationship performance: An empirical investigation. *Technological Forecasting & Social Change. Technological Forecasting and Social Change, 188*, 122254. DOI: 10.1016/j.techfore.2022.122254

Li, S., Wang, R., Zheng, P., & Wang, L. (2021). Towards proactive human–robot collaboration: A foreseeable cognitive manufacturing paradigm. *Journal of Manufacturing Systems, 60*, 547–552. Advance online publication. DOI: 10.1016/j.jmsy.2021.07.017

Liu, R., Natarajan, M., & Gombolay, M. C. (2022). Coordinating Human-Robot Teams with Dynamic and Stochastic Task Proficiencies. *ACM Transactions on Human-Robot Interaction, 11*(1), 1–42. Advance online publication. DOI: 10.1145/3477391

Loeb, S., Dynarski, S., McFarland, D., Morris, P., Reardon, S., & Reber, S. (2017). *Descriptive Analysis in Education: A Guide for Researchers. NCEE 2017-4023.* https://eric.ed.gov/?id=ED573325

Lu, C., Gao, R., Yin, L., & Zhang, B. (2024). Human-Robot Collaborative Scheduling in Energy-Efficient Welding Shop. *IEEE Transactions on Industrial Informatics, 20*(1), 963–971. Advance online publication. DOI: 10.1109/TII.2023.3271749

Lucia-Palacios, L., Pérez-López, R., & Polo-Redondo, Y. (2020). Does stress matter in mall experience and customer satisfaction? *Journal of Services Marketing/the Journal of Services Marketing, 34*(2), 177–191. DOI: 10.1108/JSM-03-2019-0134

Luhmann, N. (1979). *Trust and power.* Wiley.

Lu, R., Zhao, X., Li, J., Niu, P., Yang, B., Wu, H., Wang, W., Song, H., Huang, B., Zhu, N., Bi, Y., Ma, X., Zhan, F., Wang, L., Hu, T., Zhou, H., Hu, Z., Zhou, W., Zhao, L., & Tan, W. (2020). Genomic characterisation and epidemiology of 2019 novel coronavirus: Implications for virus origins and receptor binding. *Lancet, 395*(10224), 565–574. DOI: 10.1016/S0140-6736(20)30251-8 PMID: 32007145

Maddikunta, P. K. R., Pham, Q.-V., B, P., Deepa, N., Dev, K., Gadekallu, T. R., Ruby, R., & Liyanage, M. (2022). Industry 5.0: A survey on enabling technologies and potential applications. *Journal of Industrial Information Integration, 26*, 100257. DOI: 10.1016/j.jii.2021.100257

Mahmood. (2025). AI-Driven Knowledge Ecosystems: Transforming Continuous Innovation in Cloud Data Engineering. *Journal of Computer Science and Technology Studies, 7*(8), 401–407. https://doi.org/DOI: 10.32996/jcsts.2025.7.8.44

Majeed, A., & Rupasinghe, T. (2023). Drones in last-mile delivery logistics: Applications, challenges, and future trends. *Drones (Basel), 7*(4), 260. DOI: 10.3390/drones7040260

Mao, Z., Zhang, J., Sun, Y., Fang, K., & Huang, D. (2025). Balancing parallel assembly lines with human-robot collaboration: Problem definition, mathematical model and tabu search approach. *International Journal of Production Research, 63*(1), 51–85. Advance online publication. DOI: 10.1080/00207543.2024.2356627

Marcoulides, K. M., & Raykov, T. (2018). Evaluation of Variance Inflation Factors in Regression Models Using Latent Variable Modeling Methods. *Educational and Psychological Measurement, 79*(5), 874–882. DOI: 10.1177/0013164418817803 PMID: 31488917

Marston, S., Li, Z., Bandyopadhyay, S., Zhang, J., & Ghalsasi, A. (2011). Cloud computing—The business perspective. *Decision Support Systems*, *51*(1), 1. Advance online publication. DOI: 10.1016/j.dss.2010.12.006

Martínez-Navalón, J. G., Gelashvili, V., & Gómez-Ortega, A. (2021). Evaluation of User Satisfaction and Trust of Review Platforms: Analysis of the Impact of Privacy and E-WOM in the Case of TripAdvisor. *Frontiers in Psychology*, *12*, 750527. Advance online publication. DOI: 10.3389/fpsyg.2021.750527 PMID: 34603166

Martin, K. D., & Palmatier, R. W. (2020, December). Data Privacy in Retail: Navigating Tensions and Directing Future Research. *Journal of Retailing*, *96*(4), 449–457. DOI: 10.1016/j.jretai.2020.10.002

Mastercard. (2023). *Mastercard leverages its AI capabilities to fight real-time payment scams*. https://newsroom.mastercard.com/news/press/2023/july/mastercard-leverages-its-ai-capabilities-to-fight-real-time-payment-scams/

Mastercard. (2024). *Mastercard supercharges consumer protection with gen AI*. https://newsroom.mastercard.com/news/press/2024/february/mastercard-supercharges-consumer-protection-with-gen-ai/

Mateus, J. C., Claeys, D., Limère, V., Cottyn, J., & Aghezzaf, E.-H. (2019). A structured methodology for the design of a human-robot collaborative assembly workplace. *International Journal of Advanced Manufacturing Technology*, *102*(5-8), 2663–2681. Advance online publication. DOI: 10.1007/s00170-019-03356-3

Mayer, R. C., Davis, J. H., & Schoorman, F. D. (1995). An integrative model of organizational trust. *Academy of Management Review*, *20*(3), 709–734. DOI: 10.2307/258792

Mbunge, E., Chitungo, I., & Dzinamarira, T. (2021a). *Unbundling the significance of cognitive robots and drones deployed to tackle COVID-19 pandemic: A rapid review to unpack emerging opportunities to improve healthcare in sub-Saharan Africa*. Cognitive Robotics., DOI: 10.1016/j.cogr.2021.11.001

McBride, N., Stahl, B. C., & Flick, C. (2021). Teaching responsible AI: Integrating ethics into technology education. *Communications of the ACM*, *64*(4), 33–35.

McDonough, W., & Braungart, M. (2002). *Cradle to Cradle: Remaking the Way We Make Things* (1st ed.). North Point Press.

McGrath, C., & Astell, A. (2016). The benefits and barriers to technology acquisition: Understanding the decision-making processes of older adults with age-related vision loss (ARVL). *British Journal of Occupational Therapy*, *80*(2), 123–131. DOI: 10.1177/0308022616667959

McKinsey & Company. (2020). *The State of AI in 2020*. https://www.mckinsey.com/business-functions/mckinsey-analytics/our-insights

McKinsey & Company. (2022). *The State of AI in 2022*. https://www.mckinsey.com

McKinsey Global Institute. (2021). *The state of AI in 2021*. https://www.mckinsey.com

Mehrabi, N., Morstatter, F., Saxena, N., Lerman, K., & Galstyan, A. (2021). A survey on bias and fairness in machine learning. *ACM Computing Surveys, 54*(6). *ACM Computing Surveys*, *115*(6), 1–35. DOI: 10.1145/3457607

MEITY. (2022). *FutureSkills PRIME: Empowering India's Workforce for the Digital Economy*. https://futureskillsprime.in

Meuter, M. L., Ostrom, A. L., Roundtree, R. I., & Bitner, M. J. (2000). Self-Service Technologies: Understanding Customer Satisfaction with Technology-Based Service Encounters. *Journal of Marketing*, *64*(3), 50–64. https://www.jstor.org/stable/3203487. DOI: 10.1509/jmkg.64.3.50.18024

Mia, M. R., & Shuford, J. (2024). Exploring the Synergy of Artificial Intelligence and Robotics in Industry 4.0 Applications. *Journal of Artificial Intelligence General Science (JAIGS) ISSN:3006-4023, 1*(1). DOI: 10.60087/jaigs.v1i1.31

Michalos, G., Kousi, N., Karagiannis, P., Gkournelos, C., Dimoulas, K., Koukas, S., Mparis, K., Papavasileiou, A., & Makris, S. (2018). Seamless human robot collaborative assembly – An automotive case study. *Mechatronics*, *55*, 194–211. DOI: 10.1016/j.mechatronics.2018.08.006

Microsoft. (2019). *Microsoft AI Principles*. https://www.microsoft.com/en-us/ai/responsible-ai

Min, B. H., & Borch, C. (2022). Systemic failures and organizational risk management in algorithmic trading: Normal accidents and high reliability in financial markets. *Social Studies of Science*, *52*(2), 277–302. DOI: 10.1177/03063127211048515 PMID: 34612758

Mishra, P., Singh, U., Pandey, C., Mishra, P., & Pandey, G. (2019). Application of student's t-test, analysis of variance, and covariance. *Annals of Cardiac Anaesthesia*. *Annals of Cardiac Anaesthesia*, *22*(4), 407–411. DOI: 10.4103/aca.ACA_94_19 PMID: 31621677

MIT Sloan Management Review & BCG. (2020). *The Cultural Benefits of Artificial Intelligence in the Enterprise*. https://sloanreview.mit.edu

MIT Sloan Management Review & BCG. (2020). *The Cultural Benefits of Artificial Intelligence in the Enterprise*. https://sloanreview.mit.edu/projects/the-cultural-benefits-of-artificial-intelligence-in-the-enterprises

Mitchell, M. (2019). Model cards for model reporting. *Proceedings of the Conference on Fairness, Accountability, and Transparency*, 220–229. DOI: 10.1145/3287560.3287596

Mohamad, L. (2022, November 29). Cashless initiative empowers smart nation. *Borneo Bulletin*. Retrieved August 28, 2023.

Mukherjee, S., Baral, M. M., Pal, S. K., Chittipaka, V., Roy, R., & Alam, K. (2022). Humanoid robot in healthcare: A Systematic Review and Future Research Directions. *2022 International Conference on Machine Learning, Big Data, Cloud and Parallel Computing, COM-IT-CON 2022*. DOI: 10.1109/COM-IT-CON54601.2022.9850577

Müller, V. C., & Bostrom, N. (2022). Future progress in artificial intelligence: Extent and limits of automation. *AI & Society*, *37*(4), 1239–1254. DOI: 10.1007/s00146-021-01191-0

Murakami, S., Shimizu, K., Tokoro, C., & Nakamura, T. (2022). Role of Resource Circularity in Carbon Neutrality. *Sustainability (Basel)*, *14*(24), 16408. DOI: 10.3390/su142416408

Na'imah, T., Tjahjono, H. K., & Madjid, A. (2023). Workplace Well-Being: The Roles of Perceived Organizational Support, Organizational Justice and Workplace Spirituality. *Quality - Access to Success*, *24*(193). Advance online publication. DOI: 10.47750/QAS/24.193.29

Nahavandi, S. (2019). Industry 5.0—A Human-Centric Solution. *Sustainability (Basel)*, *11*(16), 16. Advance online publication. DOI: 10.3390/su11164371

Nasution, M. I., Fahmi, M., Jufrizen, N., Muslih, N., & Prayogi, M. A. (2020). The Quality of Small and Medium Enterprises Performance Using the Structural Equation Model-Part Least Square (SEM-PLS). *Journal of Physics: Conference Series*, *1477*(5), 052052. DOI: 10.1088/1742-6596/1477/5/052052

National Transportation Safety Board. (2020). *Collision between a sport utility vehicle operating with partial driving automation and a crash attenuator—Mountain View, California*. Washington, DC: U.S. Government Printing Office. https://www.ntsb.gov/investigations/Pages/HWY18FH011.aspx

NIST. (2023). *AI Risk Management Framework (AI RMF 1.0)*. U.S. Department of Commerce. https://www.nist.gov/itl/ai-risk-management-framework

NIST. (2023). *AI Risk Management Framework 1.0*. U.S. Department of Commerce. https://www.nist.gov/itl/ai-risk-management-framework

Noor, A. (2022, June 9). Digital payment, data storage, information key to vibrant digital ecosystem. *Borneo Bulletin*. Retrieved August 28, 2023, from https://borneobulletin.com.bn/digital-payment-data-storage-information-key-to-vibrant-digital-ecosystem/

Norman, D. A. (2013). *The Design of Everyday Things* (Revised ed.). Basic Books.

O'Reilly, C. A.III, & Tushman, M. L. (2013). Organizational Ambidexterity: Past, Present, and Future. *The Academy of Management Perspectives*, *27*(4), 324–338. DOI: 10.5465/amp.2013.0025

Obermeier, G., Klingersberger, J., & Auinger, A. (2022, January 4). *Factors influencing usage intentions towards a self-service kiosk with biometric authentication*. Handle Proxy. http://hdl.handle.net/10125/79915

OECD. (2020). *The Future of Education and Skills 2030*. https://www.oecd.org/education/2030/

Ohira, T. (2018). On statistical independence and no-correlation for a pair of random variables taking two values: Classical and quantum. *Progress of Theoretical and Experimental Physics, 2018*(8). DOI: 10.1093/ptep/pty086

Orlikowski, W. J. (2007). Sociomaterial practices: Exploring technology at work. *Organization Studies*, *28*(9), 1435–1448. DOI: 10.1177/0170840607081138

Othman, A. K., Hamzah, M. I., & Abu Hassan, L. F. (2020). Modeling the contingent role of technological optimism on customer satisfaction with self-service technologies. *Journal of Enterprise Information Management*, *33*(3), 559–578. DOI: 10.1108/JEIM-09-2019-0295

Otto, A. S., Szymanski, D. M., & Varadarajan, R. (2019, May 31). Customer satisfaction and firm performance: Insights from over a quarter century of empirical research. *Journal of the Academy of Marketing Science*, *48*(3), 543–564. DOI: 10.1007/s11747-019-00657-7

Pai, R. Y., Shetty, A., Shetty, A. D., Bhandary, R., Shetty, J., Nayak, S., Dinesh, T. K., & D'souza, K. J. (2022). Integrating artificial intelligence for knowledge management systems – synergy among people and technology: A systematic review of the evidence. *Ekonomska Istrazivanja*, *35*(1), 1–23. DOI: 10.1080/1331677X.2022.2058976

Pandita, A., & Kiran, R. (2023). The Technology Interface and Student Engagement Are Significant Stimuli in Sustainable Student Satisfaction. *Sustainability (Basel), 15*(10), 7923. DOI: 10.3390/su15107923

Parasuraman, A. P., Zeithaml, V. A., & Berry, L. L. (1988, January 1). *SERVQUAL A Multiple-item Scale for Measuring Consumer Perceptions of Service Quality*. ResearchGate. https://www.researchgate.net/publication/200827786_SERVQUAL_A_Multiple-item_Scale_for_Measuring_Consumer_Perceptions_of_Service_Quality

Parasuraman, A. (2000). Technology Readiness Index (Tri). *Journal of Service Research, 2*(4), 307–320. DOI: 10.1177/109467050024001

Parasuraman, A., Ball, J., Aksoy, L., Keiningham, T. L., & Zaki, M. (2020). More than a feeling? Toward a theory of customer delight. *Journal of Service Management, 32*(1), 1–26. DOI: 10.1108/JOSM-03-2019-0094

Parasuraman, A., Zeithaml, V. A., & Berry, L. L. (1985). A Conceptual Model of Service Quality and Its Implications for Future Research. *Journal of Marketing, 49*(4), 41–50. DOI: 10.1177/002224298504900403

Parasuraman, A., Zeithaml, V. A., & Malhotra, A. (2005). E-S-QUAL. *Journal of Service Research, 7*(3), 213–233. DOI: 10.1177/1094670504271156

Parise, S., Guinan, P. J., & Kafka, R. (2016). Solving the crisis of immediacy: How Digital Technology can transform the customer experience. *Business Horizons, 59*(4), 411–420. DOI: 10.1016/j.bushor.2016.03.004

Park, J., Kim, J., & Lee, D. (2021). Design and development of last-mile delivery robots: A review of autonomous ground vehicles. *IEEE Access : Practical Innovations, Open Solutions, 9*, 13472–13490. DOI: 10.1109/ACCESS.2021.3051204

Park, J.-S., Ha, S., & Jeong, S. W. (2020). Consumer acceptance of self-service technologies in Fashion Retail Stores. *Journal of Fashion Marketing and Management, 25*(2), 371–388. DOI: 10.1108/JFMM-09-2019-0221

Park, S., Lehto, X., & Lehto, M. (2021). Self-service technology kiosk design for restaurants: An QFD application. *International Journal of Hospitality Management, 92*, 102757. DOI: 10.1016/j.ijhm.2020.102757

Pasquale, F. (2015). *The Black Box Society: The Secret Algorithms That Control Money and Information*. Harvard University Press. DOI: 10.4159/harvard.9780674736061

Pauli, G. (2010). *Blue Economy-10 Years, 100 Innovations, 100 Million Jobs*. Paradigm Pubns.

Pearce, D. W., & Turner, R. K. (1989). *Economics of Natural Resources and the Environment*. JHU Press. DOI: 10.56021/9780801839863

Pech, M., & Vrchota, J. (2022). The Product Customization Process in Relation to Industry 4.0 and Digitalization. *Processes (Basel, Switzerland)*, *10*(3), 539. DOI: 10.3390/pr10030539

Pedersen, H. K., Gudmundsdottir, V., Nielsen, H. B., Hyotylainen, T., Nielsen, T., Jensen, B. A., Forslund, K., Hildebrand, F., Prifti, E., Falony, G., Le Chatelier, E., Levenez, F., Doré, J., Mattila, I., Plichta, D. R., Pöhö, P., Hellgren, L. I., Arumugam, M., Sunagawa, S., & Pedersen, O. (2016). Human gut microbes impact host serum metabolome and insulin sensitivity. *Nature*, *535*(7612), 376–381. DOI: 10.1038/nature18646 PMID: 27409811

Pelikan, H. R. M., Cheatle, A., Jung, M. F., & Jackson, S. J. (2018). Operating at a distance – How a teleoperated surgical robot reconfigures teamwork in the operating room. *Proceedings of the ACM on Human-Computer Interaction, 2*(CSCW), Article 54. https://doi.org/DOI: 10.1145/3274407

Peng, L. P., & Wu, Y. T. (2021). Assessing Multiple Stakeholders' Collaborative Network in Promoting Eco-agriculture. *Taiwan Shuili.*

Pereira, A. G., & Liu, S. (2024). Human-robot collaboration in digital workplaces: Ethical risks and organizational safeguards. *Journal of Technology Ethics*, *5*(1), 22–39.

Petersen, L., Robert, L. P., Yang, X. J., & Tilbury, D. (2019). Situational awareness, driver's trust in automated driving systems, and secondary task performance. *SAE International Journal of Connected and Automated Vehicles*, *2*(2), 65–76. DOI: 10.4271/12-02-02-0009

Pinho, I., Rego, A., & Pina e Cunha, M. (2012). Improving knowledge management processes: A hybrid positive approach. *Journal of Knowledge Management*, *16*(2), 215–242. DOI: 10.1108/13673271211218834

Poister, T. H., & Thomas, J. C. (2011, April 12). The Effect of Expectations and Expectancy Confirmation/Disconfirmation on Motorists' Satisfaction with State Highways. *Journal of Public Administration: Research and Theory*, *21*(4), 601–617. DOI: 10.1093/jopart/mur004

Poushneh, A., & Vasquez-Parraga, A. Z. (2018). The role of customer readiness and participation in non-technology-based Service Delivery. *Journal of Consumer Marketing*, *35*(6), 588–600. DOI: 10.1108/JCM-11-2016-2006

Powering the new world of commerce. Ingenico. (n.d.). https://ingenico.com/en

Probst, P. (2023). A Review of the Role of Robotics in Surgery: To DaVinci and Beyond! *Missouri Medicine*, *120*(5), 389–396. PMID: 37841561

Purba, R., Oesman, Y. M., & Komaladewi, R. (2022). The Effect of Self-Service Technology on Customers' Trust and Loyalty (Case Study at Terminal 3 Self-Service Technology Soekarno-Hatta International Airport). *Budapest International Research and Critics Institute-Journal (BIRCI-Journal),* *5*(1), 6527–6542. https://doi.org/ https://doi.org/10.33258/birci.v5i1.4380

Radomir, L., & Nistor, C. V. (2014). Comparing the Original and the Revised SSTQUAL Scale among High-educated Consumers in Romania. *Procedia Economics and Finance*, *15*, 926–934. DOI: 10.1016/S2212-5671(14)00564-4

Rahwan, I. (2018). Society-in-the-loop: Programming the algorithmic social contract. *Ethics and Information Technology*, *20*(1), 5–14. DOI: 10.1007/s10676-017-9430-8

Rahwan, I., Cebrian, M., Obradovich, N., Bongard, J., Bonnefon, J. F., Breazeal, C., Crandall, J. W., Christakis, N. A., Couzin, I. D., Jackson, M. O., Jennings, N. R., Kamar, E., Kloumann, I. M., Larochelle, H., Lazer, D., McElreath, R., Mislove, A., Parkes, D. C., Pentland, A. S., & Wellman, M. (2019). Machine behaviour. *Nature*, *568*(7753), 477–486. DOI: 10.1038/s41586-019-1138-y PMID: 31019318

Rai, A. (2019). Explainable AI: From black box to glass box. *Journal of the Academy of Marketing Science*, *48*(1), 137–141. DOI: 10.1007/s11747-019-00710-5

Raji, I. D., Scheuerman, M. K., & Amironesei, R. (2021). You can't sit with us: Exclusionary pedagogy in AI ethics education. In *Proceedings of the 2021 ACM Conference on Fairness, Accountability, and Transparency* (FAccT '21), March 3–10, 2021, Virtual Event, Canada. ACM, 515–525. https://doi.org/DOI: 10.1145/3442188.3445914

Raji, I. D. (2020). Closing the AI accountability gap: Defining an end-to-end framework for internal algorithmic auditing. *Proceedings of the 2020 Conference on Fairness, Accountability, and Transparency*, 33–44. DOI: 10.1145/3351095.3372873

Rakholia, R., Suarez-Cetrulo, A. L., Singh, M., & Simon, C. R. (2024). *Advancing Manufacturing Through Artificial Intelligence: Current Landscape, Perspectives, Best Practices, Challenges, and Future Direction*. IEEE., DOI: 10.1109/ACCESS.2024.3458830

Ram, C., Pitt, J. B., Wallace, M. W., Goldstein, S. D., & Zamora, I. J. (2024). Framework for pediatric robotic surgery program development. *Seminars in Pediatric Surgery*, *33*(1), 151389. Advance online publication. DOI: 10.1016/j.sempedsurg.2024.151389 PMID: 38245993

Rana, M., Akter, J., & Hanif, M. A. (2025). Next-gen vision: A systematic review on robotics transforming ophthalmic surgery. *Journal of Robotic Surgery*, *19*(1), 452. DOI: 10.1007/s11701-025-02623-8 PMID: 40762731

Rastgoftar, H. (2024). Safe Human-UAS Collaboration from High-Level Planning to Low-Level Tracking. *IEEE Transactions on Automation Science and Engineering*. Advance online publication. DOI: 10.1109/TASE.2024.3461726

Rayhan, S., & Rayhan, A. (2023). The psychological impact of AI: Adapting to a world of smart machines. *ResearchGate*. https://www.researchgate.net/publication/372724209_The_Psychological_Impact_of_AI_Adapting_to_a_World_of_Smart_Machines

RBPF issues warning against phishing scams. (2024, January 30). RBPF Issues Warning Against Phishing Scams. https://borneobulletin.com.bn/rbpf-issues-warning-against-phishing-scams/

Rejeb, A., Rejeb, K., Keogh, J. G., & Süle, E. (2025). When Industry 5.0 Meets the Circular Economy: A Systematic Literature Review. *Circular Economy and Sustainability*, *5*(4), 2621–2652. Advance online publication. DOI: 10.1007/s43615-025-00570-y

Rinaldi, M., Di Pasquale, V., Farina, P., Iannone, R., Macchiaroli, R., & Grosse, E. H. (2025). Human–robot interaction in industry: A tertiary study. *Procedia Computer Science*, *253*, 1691–1701. DOI: 10.1016/j.procs.2025.01.231

Rindfleisch, A. (2019, August 5). Transaction cost theory: past, present and future. DOI: 10.1007/s13162-019-00151-x

Robert, L. P., Dennis, A. R., & Ahuja, M. (2018). Social robots that tease: Understanding and leveraging humor in human–robot interaction. *Information Systems Research*, *29*(3), 525–545. DOI: 10.1287/isre.2018.0773

Robertson, J. (2018). *Robo Sapiens Japanicus: Robots, Gender, Family, and the Japanese Nation*. University of California Press.

Robins, B., & Dautenhahn, K. (2014). Tactile interactions with a humanoid robot: Novel play scenario implementations with children with autism. *International Journal of Social Robotics*, *6*(3), 397–415. DOI: 10.1007/s12369-014-0228-0

Robson, K., & Pitt, L. (2018). Tesco in South Korea: Strategic Localization. DOI: 10.4135/9781526439666

Rodrigues, L. L., Barkur, G., Varambally, K., & Golrooy Motlagh, F. (2011, October 4). Comparison of SERVQUAL and SERVPERF metrics: An empirical study. *The TQM Journal*, *23*(6), 629–643. DOI: 10.1108/17542731111175248

Roopa, S., & Rani, M. (2012). Questionnaire Designing for a Survey. *The Journal of Indian Orthodontic Society*, *46*(6), 273–277. DOI: 10.1177/0974909820120509S

Rose, D. C. (2021). Policy windows for the future of agriculture: Applying the multiple streams framework to the adoption of digital agriculture. *Journal of Rural Studies*, *82*, 226–236.

Sachdev, R. (2025, July 2). How AI & Robotics Are Transforming Laser Eye Surgery. *Planet Lasik - Best Eye Care Center | Lasik & Refractive Surgeries*. https://planetlasik.com/blog/ai-robotics-in-laser-eye-surgery

Sakrabani, P., Teoh, A. P., & Amran, A. (2019). Strategic impact of retail 4.0 on retailers' performance in Malaysia. *Strategic Direction*, *35*(11), 1–3. DOI: 10.1108/SD-05-2019-0099

Samek, W., & Müller, K. R. (2021). Towards explainable artificial intelligence. In *Explainable AI: Interpreting, Explaining and Visualizing Deep Learning* (pp. 5–22). Springer. https://doi.org/DOI: 10.1007/978-3-030-28954-6_1

Santos, B., Costa, R. L. C., & Santos, L. (2024). Cybersecurity in Industry 5.0: Open Challenges and Future Directions. *2024 21st Annual International Conference on Privacy, Security and Trust (PST)*, 1–6. DOI: 10.1109/PST62714.2024.10788065

Sariatli, F. (2017). Linear Economy Versus Circular Economy: A Comparative and Analyzer Study for Optimization of Economy for Sustainability. *Visegrad Journal on Bioeconomy and Sustainable Development*, *6*(1), 31–34. DOI: 10.1515/vjbsd-2017-0005

Sarikonda, A., Tecce, E., Leibold, A., Mansoor, A. D., Thalheimer, S., Heller, J., Prasad, S., Sharan, A., Jallo, J., Harrop, J., Vaccaro, A. R., & Sivaganesan, A. (2024). What is the Marginal Cost of Using Robot Assistance or Navigation for Transforaminal Lumbar Interbody Fusion? A Time-Driven Activity-Based Cost Analysis. *Neurosurgery*, *95*(3), 556–565. Advance online publication. DOI: 10.1227/neu.0000000000002899 PMID: 38465927

Schaefer, K. E., Billings, D. R., & Hancock, P. A. (2016). A meta-analysis of factors influencing the development of trust in automation: Implications for understanding autonomy in future systems. *Human Factors*, *58*(3), 377–400. DOI: 10.1177/0018720816634228 PMID: 27005902

Scherer, A., Wünderlich, N. V., & von Wangenheim, F. (2015). The value of self-service: Long-term effects of technology-based self-service usage on customer retention. *Management Information Systems Quarterly*, *39*(1), 177–200. DOI: 10.25300/MISQ/2015/39.1.08

Schmiedchen, R., & Beutnagel, W. (2024, September 24). *How cobots can support automotive production*. https://www.automotivemanufacturingsolutions.com/digitalisation/how-cobots-can-support-automotive-production/533015

Schneider, R. J., Lyons, J. B., & Khazon, S. (2015). Emotional intelligence and personality: Predicting transformational, transactional, and laissez-faire leadership styles. *Leadership and Organization Development Journal*, *36*(3), 396–414.

Schwab, K., Samans, R., Zahidi, S., Leopold, T. A., Ratcheva, V., Hausmann, R., & Tyson, L. D. (2017, November). The global gender gap report 2017. World Economic Forum.

Schweizer, A., Knoll, P., Urbach, N., von der Gracht, H. A., & Hardjono, T. (2020, November). To What Extent Will Blockchain Drive the Machine Economy? Perspectives From a Prospective Study. *IEEE Transactions on Engineering Management*, *67*(4), 1169–1183. DOI: 10.1109/TEM.2020.2979286

Selvi, A. F. (2019). Qualitative content analysis. *The Routledge Handbook of Research Methods in Applied Linguistics*, 440–452. https://doi.org/DOI: 10.4324/9780367824471-37

Seo, K. H., & Lee, J. H. (2021). The emergence of service robots at restaurants: Integrating trust, perceived risk, and satisfaction. *Sustainability (Basel)*, *13*(8), 4431. DOI: 10.3390/su13084431

Shabur, M., Shahriar, A., & Ara, M. A. (2025). From automation to collaboration: Exploring the impact of industry 5.0 on sustainable manufacturing. *Discover Sustainability*, *6*(1), 341. DOI: 10.1007/s43621-025-01201-0

Shahid Iqbal, M., Ul Hassan, M., & Habibah, U. (2018, January 1). Impact of self-service technology (SST) service quality on customer loyalty and behavioral intention: The mediating role of customer satisfaction. *Cogent Business & Management*, *5*(1), 1. DOI: 10.1080/23311975.2018.1423770

Shahid, S., Kaur, K., Mohyuddin, S. M., Prikshat, V., & Patel, P. (2025). Revolutionizing HRM: A review of human-robot collaboration in HRM functions and the imperative of change readiness. *Business Process Management Journal*, *31*(7), 2892–2928. Advance online publication. DOI: 10.1108/BPMJ-12-2023-0951

Shahril, Z., Zulkafly, H. A., Ismail, N. S., & Sharif, N. U. N. M. (2021, January 28). Customer Satisfaction Towards Self-Service Kiosks for Quick Service Restaurants (QSRs) in Klang Valley. *International Journal of Academic Research in Business & Social Sciences, 11*(13). Advance online publication. DOI: 10.6007/IJARBSS/v11-i13/8502

Shan, S., Wenger, E., Zhang, J., Li, H., Zheng, H., & Y. Zhao, B. (2020). Fawkes: Protecting Privacy against Unauthorized Deep Learning Models. USENIX, 1589–1604.

Sharbawi, S., & Mabud, S. A. (2021). Malay, Muslim and Monarchy: An Introduction to Brunei Darussalam and Its National Identity. In Springer eBooks (pp. 45–66). DOI: 10.1007/978-3-030-77119-5_3

Sharma, S., Sharma, C., Batra, I., Malik, A., Kaswan, M. S., Du, D., & Kumar, V. (2025). Navigating the Frontiers of Industry 5.0: Predictive Analysis Using Natural Language Processing. *Operations Research Forum, 6*(3), 93. DOI: 10.1007/s43069-025-00501-5

Shim, H. S., Han, S. L., & Ha, J. (2020). The Effects of Consumer Readiness on the Adoption of Self-Service Technology: Moderating Effects of Consumer Traits and Situational Factors. *Sustainability (Basel), 13*(1), 95. DOI: 10.3390/su13010095

Shirakura, N., Maruyama, T., Makihara, K., Ueshiba, T., Itadera, S., Endo, Y., Tada, M., & Domae, Y. (2025). Synergistic effects of robot performance on human–robot mutual assistance systems in manufacturing. *Advanced Robotics, 39*(4), 208–221. Advance online publication. DOI: 10.1080/01691864.2024.2445600

Shi, W., Cao, J., Zhang, Q., Li, Y., & Xu, L. (2016). Edge computing: Vision and challenges. *IEEE Internet of Things Journal, 3*(5), 637–646. DOI: 10.1109/JIOT.2016.2579198

Siau, K., & Wang, W. (2018). Building trust in artificial intelligence, machine learning, and robotics. *Cutter Business Technology Journal, 31*(2), 47–53.

Siemens. (2025). *AI-based predictive maintenance*. Retrieved from https://www.siemens.com/global/en/products/automation/topic-areas/industrial-ai/usecases/ai-based-predictive-maintenance.html

Sima, V., Gheorghe, I. G., Subić, J., & Nancu, D. (2020). Influences of the Industry 4.0 Revolution on the Human Capital Development and Consumer Behavior: A Systematic Review. *Sustainability (Basel), 12*(10), 10. Advance online publication. DOI: 10.3390/su12104035

Simon, H. A. (1957). *Models of Man: Social and Rational*. Wiley.

Singapore SkillsFuture. (2021). *Annual Report on Skills Transformation and Workforce Development*. https://www.skillsfuture.gov.sg

Singh, A. (2018). *Evaluating Passengers' Perceived Service Quality Towards Self-Service Luggage Check-In Technologies at Airports Using SSTQUAL Scale.* (Publication No. 10810972)[Master's thesis, Arizona State University]. ProQuest Dissertations and Theses Global.

Singh, K. D., & Singh, P. D. (2023). *Machine Learning in Robotics with Fog/ Cloud Computing and IoT*. EAI Endorsed Transactions on AI and Robotics., DOI: 10.4108/airo.3621

Siva Durga Prasad Nayak, M., & Narayan, K. A. (2019). Strengths and weaknesses of online surveys. *IOSR Journal of Humanities and Social Sciences*, *24*(5), 31–38.

Sivakumar, A. I., & Li, W. (2020). Last-mile logistics: Key challenges and potential solutions. *Transportation Research Part E, Logistics and Transportation Review*, *141*, 102014. DOI: 10.1016/j.tre.2020.102014

Skiera, B., Reiner, J., & Albers, S. (2021). Regression Analysis. In *Springer eBooks* (pp. 299–327). DOI: 10.1007/978-3-319-57413-4_17

Smart Nation and Digital Government Office. (2019). *National AI Strategy of Singapore*. https://www.smartnation.gov.sg

Sohail, M. S., & Hasan, M. (2021, March 4). Students' perceptions of service quality in Saudi universities: The SERVPERF model. *Learning and Teaching in Higher Education: Gulf Perspectives*, *17*(1), 54–66. DOI: 10.1108/LTHE-08-2020-0016

Somashekhar, S. P., Sepúlveda, M. J., Puglielli, S., Norden, A. D., Rauthan, A., Arun, K. N., & Raghunath, S. (2018). Watson for oncology and breast cancer treatment recommendations: Agreement with an expert multidisciplinary tumor board. *Annals of Oncology : Official Journal of the European Society for Medical Oncology*, *29*(2), 418–423. DOI: 10.1093/annonc/mdx781 PMID: 29324970

Soteriades, E. S., & Psarommatis, F. (2021). Robotics in logistics: A detailed review of intelligent warehouse systems. *Computers & Industrial Engineering*, *159*, 107550. DOI: 10.1016/j.cie.2021.107550

Sreejesh, S., Mohapatra, S., & Anusree, M. R. (2013). Exploratory Factor and Principal Component Analysis. In *Springer eBooks* (pp. 207–228). DOI: 10.1007/978-3-319-00539-3_9

Stasevych, M., & Zvarych, V. (2023). Innovative Robotic Technologies and Artificial Intelligence in Pharmacy and Medicine: Paving the Way for the Future of Health Care—A Review. In *Big Data and Cognitive Computing* (Vol. 7, Issue 3). DOI: 10.3390/bdcc7030147

Sternberg, R. J. (1985). *Beyond IQ: A Triarchic Theory of Human Intelligence*. Cambridge University Press.

Story, M. F., Mueller, J. L., & Mace, R. L. (1998). *The universal design file: Designing for people of all ages and abilities*. Center for Universal Design.

Subrahmanyam, S. (2025). Development of Research Design. In Segura, L., Vasconcelos, A., & Slomski, V. (Eds.), *Evolving Designs, Applications, Technological Advances, and the Future of Qualitative Research* (pp. 31–58). IGI Global Scientific Publishing., DOI: 10.4018/979-8-3373-4432-4.ch002

Subrahmanyam, S. (2025). Emerging Technologies and Shift in Employment in the Transportation Sector. In Olga Titovna, E., Elango, D., Kumar, P., Mehra, P., & Verma, R. (Eds.), *Integrating Sustainability and Autonomous Transportation for Enhanced Urban Mobility* (pp. 149–178). IGI Global Scientific Publishing., DOI: 10.4018/979-8-3373-0882-1.ch006

Subrahmanyam, S. (2025). Human Factors and Workforce Adaptation in Digital Transition. In Martínez-Falcó, J., Sánchez-García, E., & Marco-Lajara, B. (Eds.), *Navigating Digital Transformation Through Business Process Management* (pp. 95–126). IGI Global Scientific Publishing., DOI: 10.4018/979-8-3373-3336-6.ch005

Subrahmanyam, S. (2025). Leveraging AI and ML for Enhanced Efficiency and Innovation in Manufacturing. In Vyas, V., Bi, X., Pandey, D., Xu, Z., & Pandey, B. (Eds.), *AI's Role in Enhanced Automotive Safety* (pp. 319–344). IGI Global Scientific Publishing., DOI: 10.4018/979-8-3373-0442-7.ch021

Subrahmanyam, S. (2025). Sustainability and AI. In Younas, A., & Kassim, A. (Eds.), *Transforming Business Through Digital Sustainability Models* (pp. 363–392). IGI Global Scientific Publishing., DOI: 10.4018/979-8-3373-0608-7.ch016

Subrahmanyam, S., & Baz, B. E. (2025). AI and Diversity: Ensuring Fairness in Automated Decision-Making. In El Imrani, O. (Ed.), *Navigating Responsible Business Practices Through Ethical AI* (pp. 19–50). IGI Global Scientific Publishing., DOI: 10.4018/979-8-3373-3104-1.ch002

Supply Chain Digital. (2024, August 7). *4 challenges in logistics industry in 2024*. https://www.dhl.com/discover/en-vn/logistics-advice/logistics-insights/challenges-in-logistics-and-supply-chain-management-best-practices

Supply Chain, D. H. L. (2025). DHL's Smart Warehouse initiative: Human-robot collaboration in logistics. https://www.dhl.com/global-en/delivered/innovation/warehouse-robotics-and-automation.html

Szabó, B., Őrsi, B., & Csukonyi, C. (2024). Robots for surgeons? Surgeons for robots? Exploring the acceptance of robotic surgery in the light of attitudes and trust in robots. *BMC Psychology*, *12*(1), 45. DOI: 10.1186/s40359-024-01529-8 PMID: 38267998

TabSquare. (2021, March 23). *Jollibee at Gadong, #Brunei is now successfully live with our Android Kiosk solution integrated with Aloha POS*. Facebook. https://www.facebook.com/TabSquare/posts/jollibee-at-gadong-brunei-is-now-successfully-live-with-our-android-kiosk-soluti/4072903862799467/

Tahta Kuswanto, D. R., & Rahmat Syah, T. Y. (2021). The influence of the self-service technology service quality, esatisfaction, and e-trust on e-loyalty. *American International Journal of Business Management*, *4*(8), 89–96. https://www.aijbm.com/wp-content/uploads/2021/08/J488996.pdf

Tan, L., Liu, C., Wang, Y., Li, Y., Zhao, J., Wang, S., & Zhong, B. (2025). The effect of human-robot collaboration on frontline employees' service performance: A resource perspective. *International Journal of Hospitality Management*, *126*, 104033. Advance online publication. DOI: 10.1016/j.ijhm.2024.104033

Tawalbeh, L., Muheidat, F., Tawalbeh, M., & Quwaider, M. (2020). IoT Privacy and Security: Challenges and Solutions. *Applied Sciences (Basel, Switzerland)*, *10*(12), 12. Advance online publication. DOI: 10.3390/app10124102

Tawari, K. N. (2019, August 21). The impact of self-service technologies and other non-technology drivers on customer satisfaction in the Irish supermarket industry. NORMA@NCI Library. https://norma.ncirl.ie/4026/

Taylor, E. (2016, July 24). Supermarket self-checkouts and retail theft: The curious case of the SWIPERS. *Criminology & Criminal Justice*, *16*(5), 552–567. DOI: 10.1177/1748895816643353

Taylor, S. A. (1993, October). The Roles of Service Quality, Consumer Satisfaction, and Value in Quinn's (1992) Paradigm of Services. *Journal of Marketing Theory and Practice*, *2*(1), 14–26. DOI: 10.1080/10696679.1993.11501636

Thabane, L., Ma, J., Chu, R., Cheng, J., Ismaila, A., Rios, L. P., Robson, R., Thabane, M., Giangregorio, L., & Goldsmith, C. H. (2010). A tutorial on pilot studies: The what, why and how. *BMC Medical Research Methodology*, *10*(1), 1. Advance online publication. DOI: 10.1186/1471-2288-10-1 PMID: 20053272

Theeraworawit, M., Suriyankietkaew, S., & Hallinger, P. (2022). Sustainable Supply Chain Management in a Circular Economy: A Bibliometric Review. *Sustainability (Basel), 14*(15), 9304. DOI: 10.3390/su14159304

Thomson, R., & Alvarez, P. (2024). AI-powered automation and the transformation of organizational structures: A socio-ethical analysis. *Journal of Organizational Transformation & AI, 12*(2), 44–63.

Tian, B., Kaul, H., & Janardhanan, M. (2024). Balancing heterogeneous assembly line with multi-skilled human-robot collaboration via Adaptive cooperative co-evolutionary algorithm. *Swarm and Evolutionary Computation, 91*, 101762. Advance online publication. DOI: 10.1016/j.swevo.2024.101762

Time In Brunei. (2021, November 10). *Self-Checkout your own grocery at Hua Ho Department Store*. Facebook. https://www.facebook.com/timeinbrunei/posts/self-checkout-your-own-grocery-at-hua-ho-department-store-the-first-in-brunei-da/2028381677311122/

Tjahjono, H. K. (2011). The Configuration among Social Capital, Distributive and Procedural Justice and Its Consequences to Individual Satisfaction. *International Journal of Information and Management Sciences*, 87–103. DOI: 10.6186/IJIMS.2011.22.1.6

Tjahjono, H. K., Fachrunnisa, O., & Palupi, M. (2019). Configuration of organisational justice and social capital: Their impact on satisfaction and commitment. *International Journal of Business Excellence, 17*(3), 336–360. DOI: 10.1504/IJBEX.2019.097957

Topol, E. (2021). *Deep medicine: How artificial intelligence can make healthcare human again*. Basic Books.

Toupas, P., Tsamis, G., Giakoumis, D., Votis, K., & Tzovaras, D. (2023). From Detection to Action Recognition: An Edge-Based Pipeline for Robot Human Perception. *2023 5th International Conference on Control and Robotics, ICCR 2023*. DOI: 10.1109/ICCR60000.2023.10444804

Toyota, U. K. Magazine. (2021). *Toyota Human Support Robot: What is it and how can it be used?* https://mag.toyota.co.uk/toyota-human-support-robot/#:~:text=The%20original%20prototype%20was%20developed,practical%20adoption%20of%20the%20HSR

Treasury Board of Canada Secretariat. (2021). *Algorithmic Impact Assessment Tool*. https://www.canada.ca/en/government/system/digital-government/digital-government-innovations/responsible-use-ai/algorithmic-impact-assessment.html

Trist, E. L., & Emery, F. E. (1973). *Towards a Social Ecology: Contextual Appreciations of the Future in the Present*. Plenum Press.

Turkar, S. (2023, August 15). *At 7% cagr, self service technologies market size to hit USD 57.7 bn by 2028, says Intellectual Market Insights Research*. LinkedIn. https://www.linkedin.com/pulse/7-cagr-self-service-technologies-market-size-hit-usd-577-turkar/?trk=pulse-article_more-articles_related-content-card

Uğuz Arsu, Ş., & Sipahi Döngül, E. (2023). The impact of robots and robotic systems on businesses in the workplace: Systematic review. *Journal of Organizational Change Management*, *36*(7), 1193–1221. Advance online publication. DOI: 10.1108/JOCM-11-2022-0340

Uhl-Bien, M., Marion, R., & McKelvey, B. (2006). Complexity leadership theory: Shifting leadership from the industrial age to the knowledge era. *The Leadership Quarterly*, *18*(4), 298–318. DOI: 10.1016/j.leaqua.2007.04.002

Ul Hassan, M., Iqbal, M. S., & Habibah, U. (2020). Self-service technology service quality: Building Loyalty and intention through Technology Trust in pakistani service sector. *SAGE Open*, *10*(2), 215824402092441. DOI: 10.1177/2158244020924412

Ursavaş, M. F. (2022). Unified Theory of Acceptance and Use of Technology Model (UTAUT). *Springer Texts in Education*, 111–133. DOI: 10.1007/978-3-031-10846-4_6

UTB'S SCHOOL OF COMPUTING AND INFORMATICS SHOWCASES CUTTING-EDGE HOLOGRAPHIC EDUCATION TECHNOLOGY | Universiti Teknologi Brunei. (n.d.). https://www.utb.edu.bn/news/utb-s-school-of-computing-and-informatics-showcases-cutting-edge-holographic-education-technology/

Vakulenko, Y., Hellström, D., & Oghazi, P. (2018). Customer value in self-service kiosks: A systematic literature review. *International Journal of Retail & Distribution Management*, *46*(5), 507–527. DOI: 10.1108/IJRDM-04-2017-0084

Vakulenko, Y., Oghazi, P., & Hellström, D. (2019). Innovative Framework for self-service kiosks: Integrating customer value knowledge. *Journal of Innovation & Knowledge*, *4*(4), 262–268. DOI: 10.1016/j.jik.2019.06.001

van Dijk, J. (2020). *The Digital Divide*. Polity Press.

Vansteenkiste, M., & Ryan, R. M. (2013). On psychological growth and vulnerability: Basic psychological need satisfaction and need frustration as a unifying principle. *Journal of Psychotherapy Integration*, *23*(3), 263–280. DOI: 10.1037/a0032359

Villani, V., Pini, F., Leali, F., & Secchi, C. (2018). Survey on human–robot collaboration in industrial settings: Safety, intuitive interfaces and applications. *Mechatronics*, *55*, 248–266. DOI: 10.1016/j.mechatronics.2018.02.009

Vincent, J. (2019). Google dissolves AI ethics board after controversy. *The Verge*. https://www.theverge.com/2019/4/4/18295310

Vining, R., McDonald, N., McKenna, L., Ward, M., Doyle, B., Liang, J., Hernandez, J., & Guilfoyle, J. Arwa Shuhaiber, Geary, U., Fogarty, M., & Brennan, R. (2022). Developing a Framework for Trustworthy AI-Supported Knowledge Management in the Governance of Risk and Change. *Lecture Notes in Computer Science*, 318–333. https://doi.org/DOI: 10.1007/978-3-031-17615-9_22

Walker, F., Boelhouwer, A., Alkim, T., Verwey, W. B., & Martens, M. H. (2018). Changes in trust after driving level 2 automated cars. *Journal of Advanced Transportation*, *2018*, 1045186. Advance online publication. DOI: 10.1155/2018/1045186

Wamba, S. F., & Queiroz, M. M. (2020). Industry 4.0 and supply chain digitalization: A systematic review. *Production Planning and Control*, *33*(2–3), 91–104. DOI: 10.1080/09537287.2020.1810761

Wang, X., Zhang, X., Cheng, Y., Tian, F., Chen, K., & Ordóñez, P. (2022). Artificial Intelligence-Enabled Knowledge Management. *Routledge EBooks*, 153–168. https://doi.org/DOI: 10.4324/9781003112150-13

Wang, B., Chen, S., & Xiao, G. (2024). Advancing healthcare through mobile collaboration: A survey of intelligent nursing robots research. *Frontiers in Public Health*, *12*, 1368805. Advance online publication. DOI: 10.3389/fpubh.2024.1368805 PMID: 39659720

Wang, B., Zheng, L., Wang, Y., Fang, W., & Wang, L. (2024). Towards the industry 5.0 frontier: Review and prospect of XR in product assembly. *Journal of Manufacturing Systems*, *74*, 777–811. DOI: 10.1016/j.jmsy.2024.05.002

Wang, D., Ma, E., & Leung, X. Y. (2024). Working with robots: A job design perspective of hospitality employees' collaboration intentions with service robots. *Journal of Hospitality and Tourism Management*, *61*, 66–77. DOI: 10.1016/j.jhtm.2024.09.012

Wang, K. J., Lin, C. J., Tadesse, A. A., & Woldegiorgis, B. H. (2023). Modeling of human–robot collaboration for flexible assembly—A hidden semi-Markov-based simulation approach. *International Journal of Advanced Manufacturing Technology*, *126*(11–12), 5371–5389. Advance online publication. DOI: 10.1007/s00170-023-11404-2

Wang, R., Zhao, D., Gupte, A., & Min, B.-C. (2024). Initial Task Allocation in Multi-Human Multi-Robot Teams: An Attention-Enhanced Hierarchical Reinforcement Learning Approach. *IEEE Robotics and Automation Letters*, *9*(4), 3451–3458. Advance online publication. DOI: 10.1109/LRA.2024.3366414

Wang, X. V., Kemény, Z., Váncza, J., & Wang, L. (2021). Human–robot collaboration in manufacturing: A survey of communication, safety, and control. *Robotics and Computer-integrated Manufacturing*, *68*, 102012. DOI: 10.1016/j.rcim.2020.102012

Wang, Y., Wang, S., Wang, J., Wei, J., & Wang, C. (2018). An empirical study of consumers' intention to use ride-sharing services: Using an extended technology acceptance model. *Transportation*, *47*(1), 397–415. DOI: 10.1007/s11116-018-9893-4

Wang, Y., Yu, C., & Fesenmaier, D. R. (2020). Accessibility and its ethical implications in digital tourism. *Tourism Management Perspectives*, *33*, 100606.

WAWASAN BRUNEI 2035. (n.d.). Www.wawasanbrunei.gov.bn. https://www.wawasanbrunei.gov.bn/en/SitePages/home.aspx

West, S. M., Whittaker, M., & Crawford, K. (2019). Discriminating systems: Gender, race, and power in AI. *AI Now Institute*. https://ainowinstitute.org/discriminatingsystems.html

Westerman, G., Bonnet, D., & McAfee, A. (2014). *Leading digital: Turning technology into business transformation*. Harvard Business Press.

Wilson, H. J., & Daugherty, P. R. (2018). Collaborative intelligence: Humans and AI are joining forces. *Harvard Business Review*, *96*(4), 114–123.

Wilson, H. J., & Daugherty, P. R. (2022). *Radically human: How new technology is transforming business and shaping our future*. Harvard Business Review Press.

Wilson, S. (2020). Pandemic leadership: Lessons from New Zealand's approach to COVID-19. *Leadership*, *16*(3), 279–293. DOI: 10.1177/1742715020929151

World Economic Forum. (2023). *Future of Jobs Report 2023*. https://www.weforum.org/reports/future-of-jobs-report-2023

WSJ. (2025). Amazon is on the cusp of using more robots than humans in its warehouses. *The Robot Report*. Retrieved from https://www.livemint.com/companies/news/amazon-is-on-the-cusp-of-using-more-robots-than-humans-in-its-warehouses-11751343089546.html

Wurman, P. R., Barrett, S., & Kawamoto, S. (2022). Coordinating autonomous robots in large-scale logistics environments. *AI Magazine*, *43*(1), 17–30. DOI: 10.1609/aimag.v43i1.20020

Wurman, P. R., D'Andrea, R., & Mountz, M. (2008). Coordinating hundreds of cooperative, autonomous vehicles in warehouses. *AI Magazine*, *29*(1), 9–9.

Wu, T.-J., & Zhang, R.-X. (2024). Exploring the impacts of intention towards human-robot collaboration on frontline hotel employees' positive behavior: An integrative model. *International Journal of Hospitality Management*, *123*, 103912. DOI: 10.1016/j.ijhm.2024.103912

Xinhua. (2024, June 16). War against the scammers - Brunei cybersecurity agency takes down 1,023 phishing websites. *The Star*. Retrieved June 21, 2024, from https://www.thestar.com.my/aseanplus/aseanplus-news/2024/06/16/war-against-the-scammers---brunei-cybersecurity-agency-takes-down-1023-phishing-websites

Xu, X., Lu, Y., Vogel-Heuser, B., & Wang, L. (2021). Industry 4.0 and Industry 5.0—Inception, conception and perception. *Journal of Manufacturing Systems*, *61*, 530–535. DOI: 10.1016/j.jmsy.2021.10.006

Yang, B., Fu, X., Sidiropoulos, N. D., & Hong, M. (2017, July). Towards k-means-friendly spaces: Simultaneous deep learning and clustering. In *international conference on machine learning* (pp. 3861-3870). PMLR.

Yang, Y., Lu, Q. B., Liu, M. J., Wang, Y. X., Zhang, A. R., Jalali, N., . . . Fang, L. Q. (2020). Epidemiological and clinical features of the 2019 novel coronavirus outbreak in China. medrxiv, 2020-02. DOI: 10.1101/2020.02.10.20021675

Yang, G. Z., Cambias, J., Cleary, K., Daimler, E., Drake, J., Dupont, P. E., Hata, N., Kazanzides, P., Martel, S., Patel, R. V., Santos, V. J., & Taylor, R. H. (2017). Medical robotics—Regulatory, ethical, and legal considerations for increasing levels of autonomy. *Science Robotics*, *2*(4), eaam8638. Advance online publication. DOI: 10.1126/scirobotics.aam8638 PMID: 33157870

Yang, G., Lv, K., & Fan, S. (2023). Robots for sustainable development: Supporting SDGs through intelligent automation. *Nature Communications*, *14*, 2543. DOI: 10.1038/s41467-023-38215-y

Yang, G.-Z., Bellingham, J., Dupont, P. E., Fischer, P., Floridi, L., Full, R., & Wood, R. (2021). The grand challenges of Science Robotics. *Science Robotics*, *6*(50), eabf0821. Advance online publication. DOI: 10.1126/scirobotics.abf0821 PMID: 34043587

Yang, J., & Klassen, K. J. (2008). How financial markets reflect the benefits of self-Service Technologies. *Journal of Enterprise Information Management*, *21*(5), 448–467. DOI: 10.1108/17410390810904238

Yi, B., Lv, J., Chen, J., Wang, X., & Li, K. (2023). Digital Twin Constructed Spatial Structure for Flexible and Efficient Task Allocation of Drones in Mobile Networks. *IEEE Journal on Selected Areas in Communications*, *41*(11), 3430–3443. Advance online publication. DOI: 10.1109/JSAC.2023.3313193

Yoo, B., & Donthu, N. (2014). Developing a Scale to Measure the Perceived Quality of an Internet Shopping Site (PQISS). *Proceedings of the 2000 Academy of Marketing Science (AMS) Annual Conference*, 471–471. DOI: 10.1007/978-3-319-11885-7_129

Yousuf, A. (2017). Transaction Costs: A Conceptual Framework. *International Journal of Engineering and Management Sciences*, *2*(3), 131–139. DOI: 10.21791/IJEMS.2017.3.13.

Yu, F., Lu, C., Yin, L., & Zhang, B. (2024). A Self-Learning Memetic Algorithm for Human-Robot Collaboration Scheduling in Energy-Efficient Distributed Mixed Fuzzy Welding Shop. *IEEE Transactions on Automation Science and Engineering*. Advance online publication. DOI: 10.1109/TASE.2024.3448435

Zaveri, A. A., Faisal, N., Nazar, M., Shehmir, S., Parveen, M., Imtiaz, S., Matcharan, A., Akhter, S., & Sami, N. (2022, June 24). Easy-Mart: Self checkout System for Supermarkets. *Journal of Independent Studies and Research Computing*, *20*(1). Advance online publication. DOI: 10.31645/JISRC.22.20.1.2

Zhang, B., Dafoe, A., & Dafoe, A. (2019). Artificial intelligence: American attitudes and trends. Center for the Governance of AI, Future of Humanity Institute, University of Oxford.

Zhang, B., & Dafoe, A. (2019). *Artificial intelligence: American attitudes and trends*. Center for the Governance of AI, University of Oxford.

Zhang, C., & Chen, Y. (2020). Sensor-based fault diagnosis of industrial robots: A review. *IEEE Transactions on Industrial Informatics*, *16*(7), 4446–4457. DOI: 10.1109/TII.2019.2958743

Zhou, Q., Li, B., Han, L., & Jou, M. (2023). Talking to a bot or a wall? How chatbots vs. human agents affect anticipated communication quality. *Computers in Human Behavior*, *143*, 107674. DOI: 10.1016/j.chb.2023.107674

Zhu, G., So, K. K., & Hudson, S. (2017). Inside the sharing economy. *International Journal of Contemporary Hospitality Management*, *29*(9), 2218–2239. DOI: 10.1108/IJCHM-09-2016-0496

Zonta, T., da Costa, C. A., da Rosa Righi, R., de Lima, M. J., da Trindade, E. S., & Li, G. P. (2020). Predictive maintenance in the Industry4.0: A systematic review. *Computers & Industrial Engineering*, *150*, 106889. DOI: 10.1016/j.cie.2020.106889

About the Contributors

Muhammad Anshari is an Associate Professor in Business Information Systems at the Universiti Brunei Darussalam School of Business and Economics (UBDSBE), and currently serves as Deputy Director for the Institute of Policy Studies. He has been named on the Stanford University List of Top 2% World Researchers since 2021 for the categories of ICT, Business & Management. Together with colleagues from UBDSBE and IPS, winning Outstanding Paper awards at the Emerald Literati Awards 2023. Patent Impact: His research has also been referenced in two patents, US Patent 10,999,416, 2021, and US Patent 11,489,906, 2022. Following the completion of his PhD at FBEPS UBD, Dr. Anshari worked as a research fellow at National Taiwan University. Before that, he was a research fellow at King Saud University, Saudi Arabia. He has also completed eight research grant projects and three research industry consultancies. His research focuses on Business Information Systems, Knowledge Management Systems, Digital Business, Digital Humanities, and Digital Health, contributing primarily to SDG No. 9 on Industry, Innovation, and Infrastructure.

Patricia Ordóñez de Pablos is a professor in the Department of Business Administration in the Faculty of Business and Economics at The University of Oviedo, Spain. She completed her education in The London School of Economics, UK. Her teaching and research interests focus on the areas of strategic management, knowledge management, organizational learning, intellectual capital and information technologies, with special interest in Asia (Bhutan, China, Myanmar). She is Editor-in-Chief of the International Journal of Learning and Intellectual Capital (IJLIC) and International Journal of Asian Business and Information Management (IJABIM), respectively. She has edited books for IGI Global, Elsevier, Routledge, and Springer. In 2021, 2022 and 2023, she earned placement on Stanford University's "Ranking of the World Scientists: World's Top 2% Scientists" list. Additionally she is listed in

World Top Scientists of Research.com in the Areas of Social Science and Humanities in Spain and Engineering and Technology in Spain.

Mohammad Nabil Almunawar is currently an professor at School of Business and Economics, Universiti of Brunei Darussalam (UBDSBE), Brunei Darussalam. He was the former dean of UBDSBE. He received his bachelor degree in 1983 from Bogor Agricultural University, Indonesia, master Degree (MSc Computer Science) from the Department of Computer Science, University of Western Ontario, London, Canada in 1991 and Ph.D. from the University of New South Wales (School of Computer Science and Engineering, UNSW) in 1998. Dr Almunawar has published more than 250 papers in refereed journals, book chapters and international conferences. He has more than 30 years teaching experiences in the area computer and information systems. His overall research interests include applications of IT in Management, Electronic Business/Commerce, Health Informatics, information security, cloud computing, Big Data and Internet of Things. He is also interested in object-oriented technology, databases and multimedia retrieval.

***

Wafiq Lamit is an academic with expertise in information management and digital transformation strategies.

Muralidhar L B is a professionally qualified dental surgeon. He holds Ph.D. from Canara Bank School of Management Studies, Bangalore University after his MBA from Christ University. He has 18 years of total teaching experience spanning across dental education, healthcare management and general management and 4 years of corporate experience. He has worked as research associate in Institute for Social and Economic Change, Bangalore in a project entitled as "Management By and For Common Man: An Exploration of the Concept and Direction for Future"post Ph.D. He had secured 28th rank in S.S.L.C examination for Karnataka State in 1993. His 26 patents have been published and 2 design patents have been granted. He has 41 research publications to his credit out of which 18 are Scopus-indexed and 5 are ABDC indexed. He has co-authored 11 books. He is currently working as Assistant Professor in Department of Business Management Studies, School of Commerce, Jain Deemed to be University, Bangalore.

Amit Kumar has done MLIS from the Central University of Haryana and is pursuing his PhD in Library and Information Science from Maharshi Dayanand University, Rohtak. He is currently working as an Assistant Librarian at Central Library, Gurugram University. His areas of interest are Research ethics, Bibliometric studies, E-resource management, and Meta-analysis. His contribution is in formulation of search query and data analysis.

Raj Kumar is a respected former professor at the Institute of Management Studies and Research, Maharshi Dayanand University, Rohtak, Haryana. He has contributed significantly to the academic field through his teaching and research in management studies. Based in Rohtak, Haryana, Prof. Kumar has been instrumental in shaping the careers of many students and has built a reputation for excellence in the academic community. His work at the university has left a lasting impact on the development of management education.

Ilyas Masudin is a Professor of Logistics and Supply Chain at the Department of Industrial Engineering, University of Muhammadiyah Malang. He is currently the Dean of Engineering Faculty of the University of Muhammadiyah Malang.

Arun R is an Assistant Professor in the Department of Commerce at Jain (Deemed to be University), Bengaluru. Holding a Master of Commerce (Finance) from Mangalore University, Arun has demonstrated academic excellence through prestigious qualifications, including successful clearances of the UGC NET (Commerce) thrice and KSET once. Dedicated to fostering academic excellence, Arun is passionate about teaching and research, striving to inspire students and contribute valuable insights to the field of commerce and finance.

Shilpa R., is working as an Associate Professor & Vice Principal, Department of Management Studies, RJS Institute of Management Studies, has 16 years of experience in teaching. She has worked with the T.John Institute of Technology for 6 years and Quantum Leap for 1 year. She has completed an MBA, M Phil and PhD at Jain (Deemed to be university) specialize in the area of OB/ HRM. She has also published research papers in Scopus, ABDC, and UGC care journals and presented papers at national and international conferences. Authorized Book chapters for international journals and project guide for MBA students. Attended FDPs, conferences, and workshops and also organized programs for the students. Member of ISTE and MTC Global

Rakesh N, M.Com, UGC-NET, PGDFM, is an enthusiastic academician specializing in Finance and Business Analytics, with a passion for integrating analytical tools into financial education and research. A 5th Rank Holder from Bangalore University (2015–2017), he brings academic distinction and practical insight to the classroom. Currently pursuing a Ph.D. in Corporate Governance, his research focuses on regulatory frameworks, board effectiveness, and ethical accountability in corporate leadership. His teaching interests include Financial

Management, Investment Analysis, and Financial Analytics, and he actively incorporates tools such as Excel and Jamovi into pedagogy. Mr. Rakesh is committed to bridging the gap between theoretical knowledge and practical application through case-based learning and data-driven strategies. He has presented research papers at national forums and continuously mentors students in research initiatives, Placement Activities, Toastmasters Faculty Coordinator, demonstrating his dedication to academic excellence and industry relevance in the evolving financial ecosystem.

Pushpalatha S is a Research scholar in the Department of management at Saveetha University. Research focuses on Human Resource Management, with a particular interest in "Enhancing IT company performance through AI driven E-hrm: a study of HR and software maturity models"

N Sathyanarayana has over 15 years of teaching experience at both undergraduate and postgraduate levels. He holds a Master of Business Administration degree from Visvesvaraya Technological University (VTU) and a Master of Commerce degree from Periyar University. He has completed his Ph.D. in Banking and Finance from Dayananda Sagar University. Currently, he serves as an Assistant Professor in the School of Commerce at Jain (Deemed-to-be University), Bangalore. He has published 24 research papers in international journals, including three indexed in Scopus, and seven papers in national journals. Additionally, he has presented 14 research papers at international conferences and 11 at national conferences. He has co-authored two books at the national level and contributed nine Scopus-indexed book chapters.

Satya Subrahmanyam is a distinguished academic leader, researcher, and institutional strategist with over 27 years of experience in higher education, leadership, and research management. He holds two PhDs and a Postdoctoral qualification from the Polytechnic Institute of Leiria, Portugal. Currently serving as Professor and Dean (Research) at Holy Spirit University of Kaslik (USEK), Lebanon, he spearheads research initiatives that enhance academic excellence and institutional reputation. He played a pivotal role in the development of the Catholic University in Erbil, Kurdistan, where he established new departments, led research programs, and collaborated on international educational projects. As a founding member and Managing Partner of Vignan Institute of Technology and Management, India, he contributed to its transformation into a premier institution. With expertise in AI-driven business strategies, HR analytics, and leadership in technology adoption, he has published extensively, guided PhD research, and led training programs in corporate and academic settings.

Heru Susanto is a researcher and academic with extensive experience in the fields of information systems, cybersecurity, and mobile health technologies. He is currently affiliated with KAKS-BRIN and UTB-SB, also THU, where he focuses on the development of mobile physiological sensor systems, cloud computing for healthcare, and data analytics for long-term patient care. His research interests also include Internet of Things (IoT) in health applications, big data management, and digital transformation in modern management. Dr. Susanto has published numerous scientific papers and actively participates in international conferences as a speaker and reviewer.

Yashwant Waykar is a dedicated educator and researcher at Dr. Babasaheb Ambedkar Marathwada University, specializing in Software Engineering, Machine Learning, Artificial Intelligence, and Learning Management Systems (LMS). With over 14 years of experience and a Ph.D., Dr. Waykar has authored 40+ publications in prestigious journals indexed in SCOPUS, UGC CARE, and Web of Science. Their research has earned accolades and includes several book chapters and four patents (3 granted, 1 filed) in LMS and AI. Dr. Waykar's teaching in AI, Web Development, and Software Testing is highly regarded, fostering a dynamic learning environment. They have completed three research projects funded by UGC and Dr. B.A.M.U. and have mentored 3 M.Phil and 2 Ph.D. students, guiding them in Machine Learning and LMS research.

Hamza Yusuf, affiliated with RMIT University, Australia, is a dedicated learner and young researcher with a strong focus on Big Data applications across public and private organisations. He is proficient in cutting-edge technologies related to big data processing and has a growing interest in Decentralized Finance (DeFi), Artificial Intelligence (AI), and Cybersecurity. His research interests centre on leveraging data-driven innovation to enhance organisational performance, security, and decision-making.

Index

A

B

C

D

E

H

I

K

L

O

P

R

S

T

U

www.ingramcontent.com/pod-product-compliance
Lightning Source LLC
LaVergne TN
LVHW081535300926
849161LV00006B/1441